Lecture Notes in Computer Scie

T0237856

Commenced Publication in 1973
Founding and Former Series Editors:
Gerhard Goos, Juris Hartmanis, and Jan van Leeuwen

Olivier Bournez   Igor Potapov (Eds.)

Reachability Problems

3rd International Workshop, RP 2009
Palaiseau, France, September 23-25, 2009
Proceedings

 Springer

Volume Editors

Olivier Bournez
Ecole Polytechnique, Laboratoire d' Informatique (LIX)
91128 Palaiseau Cedex, France
E-mail: Olivier.Bournez@lix.polytechnique.fr

Igor Potapov
University of Liverpool, Department of Computer Science
Ashton Building, Liverpool L69 3BX, England
E-mail: potapov@liverpool.ac.uk

Library of Congress Control Number: 2009934050

CR Subject Classification (1998): G.2.2, E.1, I.2.8, D.2, D.3, D.1.6, D.3.2, F.3, F.4

LNCS Sublibrary: SL 1 – Theoretical Computer Science and General Issues

ISSN   0302-9743
ISBN-10   3-642-04419-0 Springer Berlin Heidelberg New York
ISBN-13   978-3-642-04419-9 Springer Berlin Heidelberg New York

springer.com

© Springer-Verlag Berlin Heidelberg 2009
Printed in Germany

Typesetting: Camera-ready by author, data conversion by Scientific Publishing Services, Chennai, India
Printed on acid-free paper   SPIN: 12756056   06/3180   5 4 3 2 1 0

Preface

The Third International Workshop on Reachability Problems, RP 2009 was held in Ecole Polytechnique, September 23-25, 2009, in Palaiseau, near to Paris, France.

Reachability Problems 2009 was hosted as an edition of the annual LIX Fall Colloquium. The LIX Fall Colloquium is the annual colloquium organized by the computer science laboratory of Ecole Polytechnique. The topics of this colloquium change every year. Previous editions include *Emerging Trends in Visual Computing (ETVC 2008)* in 2008, *Complex Industrial Systems: Modeling, Verification and Optimization* in 2007, and *Emerging Trends in Concurrency Theory* in 2006.

The Reachability Problems workshops series aims at gathering together scholars from diverse disciplines and backgrounds interested in reachability problems that appear in *algebraic structures, computational models, hybrid systems, verification*, etc. Reachability is a fundamental problem in the context of many models and abstractions which describe various computational processes. Analysis of the computational traces and predictability questions for such models can be formalized as a set of different reachability problems. In general reachability can be formulated as follows: Given a computational system with a set of allowed transformations (functions), decide whether a certain state of a system is reachable from a given initial state by a set of allowed transformations. The same questions can be asked not only about reachability of exact states of the system but also about a set of states expressed in term of some property as a parameterized reachability problem. Another set of predictability questions can be seen in terms of reachability of eligible traces of computations; unavoidability of some dynamics and a possibility to avoid undesirable dynamics using a limited control.

The purpose of the conference is to promote exploration of new approaches for the predictability of computational processes by merging mathematical, algorithmic and computational techniques. Topics of interest include (but are not limited to): reachability problems in infinite state systems, rewriting systems, dynamical and hybrid systems; reachability problems in logic and verification; reachability analysis in different computational models, counter/ timed/ cellular/ communicating automata; Petri-Nets; computational aspects of algebraic structures (semigroups, groups and rings); frontiers between decidable and undecidable reachability problems; predictability in iterative maps and new computational paradigms.

The first venue of Reachability Problems was Turku, Finland in 2007, as a satellite event of the Developments in Language Theory DLT 2007. The second edition was held in Liverpool in 2008. The proceedings of the previous RP workshops appeared as follows:

- Mika Hirvensalo, Vesa Halava, Igor Potapov, Jarkko Kari: Proceedings of the Satellite Workshops of DLT 2007. TUCS General Publication No 45, June 2007. ISBN: 978-952-12-1921-4.
- V. Halava and I. Potapov: Proceedings of the Second Workshop on Reachability Problems in Computational Models (RP 2008). Electronic Notes in Theoretical Computer Science. Volume 223, Pages 1-264 (26 December 2008).

The five keynote speakers at the 2009 conference were:

- Ahmed Bouajjani, "On the Reachability Problem for Dynamic Networks of Concurrent Pushdown Systems."
- Thomas Henzinger, "Formalisms for Specifying Markovian Population Models."
- Oded Maler, "Reachability for Continuous and Hybrid Systems."
- Alexander Shen, "Algorithmic Information Theory and Foundations of Probability."
- Moshe Vardi, "Model Checking as a Reachability Problem."

Each of the submitted papers received at least three reviews by members of the Program Committee, with the help of external reviewers. The full list of the 20 members of the Program Committee can be found on page VII. The list of external reviewers can be found on page VIII. The Program Committee is grateful for the highly appreciated and high-quality work produced by these external reviewers. Based on these reviews, the Program Committee decided to accept 15 papers, in addition to the 5 invited talks.

Reachability Problems 2009 benefited from all the infrastructure and equipment of Ecole Polytechnique, and received some direct financial support from LIX and CNRS GdR Informatique Mathématiques. We extend to all of them our deep gratitude.

We would also like to deeply thank Evelyne Rayssac for all the high-quality work she did for this edition of the annual LIX Fall Colloquium. Her help and expertise were deeply appreciated.

It is also a great pleasure to acknowledge the team of the *EasyChair system*, and the fine cooperation with *Lecture Notes in Computer Science* of Springer, which made possible the production of this volume in time for the conference.

Finally, we thank all the authors for the high quality of their contributions, and the participants for making this edition of RP 2009 a success.

July 2009   Olivier Bournez
                                                             Igor Potapov

Conference Organization

Program Chairs

Olivier Bournez
Igor Potapov

Program Committee

Parosh Abdulla	Uppsala, Sweden
Luca de Alfaro	Santa Cruz, USA
Eugene Asarin	Paris, France
Vincent Blondel	Louvain, Belgium
Bernard Boigelot	Liège, Belgium
Ahmed Bouajjani	Paris, France
Olivier Bournez	Palaiseau, France
Cristian S. Calude	Auckland, New Zealand
Javier Esparza	Munich, Germany
Laurent Fribourg	Cachan
Vesa Halava	Turku, Finland
Oscar Ibarra	Santa Barbara, USA
Franjo Ivancic	Princeton, USA
Juhani Karhumki	Turku, Finland
Alexei Lisitsa	Liverpool, UK
Maurice Margenstern	Metz, France
Igor Potapov	Liverpool, UK
Colin Stirling	Edinburgh, UK
Wolfgang Thomas	Aachen, Germany
Hsu-Chun Yen	Taipei, Taiwan, China

Local Organization

Olivier Bournez
Igor Potapov
Evelyne Rayssac

with the help offered by the LIX laboratory
for the organization of the annual LIX Fall Colloquium.

External Reviewers

Gregory Batt
Georgios Fainekos
Peter Habermehl
Matthew Hague
Lukas Holik
Wong Karianto
Jörg Olschewski
Ahmed Rezine

Arnaud Sangnier
Sriram Sankaranarayanan
Zdenek Sawa
Mihaela Sighireanu
Michaela Slaats
Jeremy Sproston
Farn Wang
Martin Zimmermann

Table of Contents

On the Reachability Problem for Dynamic Networks of Concurrent
Pushdown Systems (invited talk) 1
  Mohamed Faouzi Atig and Ahmed Bouajjani

Formalisms for Specifying Markovian Population Models
(invited talk) 3
  Thomas Henzinger, Barbara Jobstmann, and Verena Wolf

Reachability for Continuous and Hybrid Systems (invited talk) 24
  Oded Maler

Algorithmic Information Theory and Foundations of Probability
(invited talk) 26
  Alexander Shen

Model Checking as a Reachability Problem (invited talk) 35
  Moshe Y. Vardi

Automatic Verification of Directory-Based Consistency Protocols....... 36
  Parosh Aziz Abdulla, Giorgio Delzanno, and Ahmed Rezine

On Yen's Path Logic for Petri Nets............................... 51
  Mohamed Faouzi Atig and Peter Habermehl

Probabilistic Model Checking of Biological Systems with Uncertain
Kinetic Rates.. 64
  Roberto Barbuti, Francesca Levi, Paolo Milazzo, and Guido Scatena

How to Tackle Integer Weighted Automata Positivity................. 79
  Yohan Boichut, Pierre-Cyrille Héam, and Olga Kouchnarenko

A Reduction Theorem for the Verification of Round-Based Distributed
Algorithms 93
  Mouna Chaouch-Saad, Bernadette Charron-Bost, and Stephan Merz

Computable CTL^* for Discrete-Time and Continuous-Space Dynamic
Systems 107
  Pieter Collins and Ivan S. Zapreev

An Undecidable Permutation of the Natural Numbers 120
  Eero Lehtonen

Forward Analysis of Dynamic Network of Pushdown Systems Is Easier
without Order 127
  Denis Lugiez

Counting Multiplicity over Infinite Alphabets 141
  Amaldev Manuel and R. Ramanujam

The Periodic Domino Problem Is Undecidable in the Hyperbolic
Plane 154
  Maurice Margenstern

Games with Opacity Condition 166
  Bastien Maubert and Sophie Pinchinat

Abstract Counterexamples for Non-disjunctive Abstractions 176
  Kenneth L. McMillan and Lenore D. Zuck

Cross-Checking - Enhanced Over-Approximation of the Reachable
Global State Space of Component-Based Systems 189
  Mila Majster-Cederbaum and Christoph Minnameier

Games on Higher Order Multi-stack Pushdown Systems 203
  Anil Seth

Limit Set Reachability in Asynchronous Graph Dynamical Systems 217
  V.S. Anil Kumar, Matt Macauley, and Henning S. Mortveit

Author Index 233

On the Reachability Problem for Dynamic Networks
of Concurrent Pushdown Systems

Mohamed Faouzi Atig and Ahmed Bouajjani

LIAFA, CNRS & Univ. Paris Diderot (Paris 7)
{atig,abou}@liafa.jussieu.fr

We consider the problem of checking safety properties for concurrent programs. We assume that programs may have (potentially recursive) procedure calls as well as (un-bounded) dynamic creation of parallel threads. Each procedure can have a finite number of local variables, and there is a finite number of global variables that can be accessed by all parallel threads. We assume that these variables range over a finite data domain (e.g., booleans).

We consider concurrent pushdown dynamic networks as a formal model for this class of programs. In fact, sequential programs can naturally be modeled as pushdown systems, and then, concurrent programs can be modeled as networks where each process can behave as a pushdown systems (i.e., it can modify the global store and operate on the stack representing its local context), and additionally it can create new processes in the network. At each point in time, only one process is running (and can act on the global store) and all the others are idle. A scheduling policy is used along the computations to switch the contexts, i.e., to freeze the execution of some process at some point and resume the execution of some idle one. The most liberal scheduling policy is the one which may introduce context switches at any point in time and without any distinction between processes. For this policy, a computation of the program may have an infinite number of context switches, and the number of context switches in each of the potential computations of the program is in general unbounded. It is easy to see that this model is Turing powerful.

Other policies can be defined by imposing various conditions on the occurrences of context switches. These conditions can concern, e.g., the size of the stacks (for instance in asynchronous programs switches can occur only if the stack of the active thread is empty), the classes of the processes (for instance priorities between threads may be considered), the number of allowed context-switches globally, or per thread, or per class of threads, etc. We show that by considering special scheduling policies, it is possible to obtain models for significant classes of programs/applications for which the reachability problem is decidable. The presented results cover recent work published in [ABT08] and [ABQ09] showing that the considered reachability problems can be reduced to reachability/coverability problems in some classes of Petri nets.

O. Bournez and I. Potapov (Eds.): RP 2009, LNCS 5797, pp. 1–2, 2009.

References

[ABQ09]   Atig, M.F., Bouajjani, A., Qadeer, S.: Context-Bounded Analysis for Concurrent Programs with Dynamic Creation of Threads. In: 15th Intern. Conf. on Tools and Algorithms for the Construction and Analysis of Systems (TACAS 2009). LNCS, vol. 5505, pp. 107–123. Springer, Heidelberg (2009)

[ABT08]   Atig, M.F., Bouajjani, A., Touili, T.: Analyzing Asynchronous Programs with Preemption. In: IARCS Ann. Intern. Conf. on Foundations of Software Technology and Theoretical Computer Science (FSTTCS 2008), Leibniz Intern. Proc. in Informatics, vol. 2. Schloss Dagstuhl - Leibniz-Zentrum fuer Informatik, Germany (2008)

Formalisms for Specifying Markovian Population Models[*]

Thomas A. Henzinger[1,2], Barbara Jobstmann[1], and Verena Wolf[1,3]

[1] EPFL, Switzerland
[2] IST Austria (Institute of Science and Technology Austria)
[3] Saarland University, Germany

Abstract. We compare several languages for specifying Markovian population models such as queuing networks and chemical reaction networks. These languages —matrix descriptions, stochastic Petri nets, stoichiometric equations, stochastic process algebras, and guarded command models— all describe continuous-time Markov chains, but they differ according to important properties, such as compositionality, expressiveness and succinctness, executability, ease of use, and the support they provide for checking the well-formedness of a model and for analyzing a model.

1 Introduction

Markov chains are an omnipresent modeling approach in the applied sciences. Often, they describe *population processes*, that is, they operate on a multidimensional discrete state space, where each dimension of a state represents the number of individuals of a certain type. Depending on the application area, "individuals" may be customers in a queuing network, molecules in a chemically reacting volume, servers in a computer network, etc.

Here, we are particularly interested in dynamical models of biochemical reaction networks, such as signaling pathways, gene expression networks, and metabolic networks. They are an important emerging application area of continuous-time Markov chains and operate on an abstraction level where a state of the system is given by an n-dimensional vector of chemical populations, that is, the system involves n different types of molecules and the i-th coordinate represents the number of molecules of type i. Molecules collide randomly and may undergo chemical reactions, which change the state of the system. Classical modeling approaches in biochemistry are based on a system of ordinary differential equations that assume a continuous deterministic change of chemical concentrations. Over the last decade, however, various experimental results have shown that the discreteness and randomness of the chemical reactions need to be taken into account. Thus, discrete-state continuous-time Markov models have gained in importance for describing the dynamics in the cell [31,35,32,42,44,45,39].

---

[*] This research was supported in part by the Excellence Cluster on Multimodal Computing and Interaction and the Swiss National Science Foundation.

O. Bournez and I. Potapov (Eds.): RP 2009, LNCS 5797, pp. 3–23, 2009.

There are many different formalisms for the specification of Markovian population models. Most popular are matrix descriptions, stochastic Petri nets, stochastic process algebras, and languages based on guarded commands. Moreover, Markov chains for biochemical reaction networks are often specified based on rules for chemical reactions, called stoichiometric equations. While all of these formalisms describe the same kind of system —a continuous-time Markov chain— they vary considerably in their ease of use, support of analysis techniques, and other properties. In this paper we give a brief survey of specification formalisms for Markovian population models and discuss some properties of these formalisms which are of particular importance in modeling. These properties include *compositionality* (how does the formalism support the construction of complex models from simpler parts?), *expressiveness and succinctness* (which systems can be specified in the formalism and how large are the specifications?), *executability* (how easy is it to compute the possible direct successor states of a given state?), and *well-formedness* (how easy is it to check if a model has a unique solution?). We illustrate our remarks with examples of Markovian population models.

The main message of this paper is that, even if one agrees on the underlying mathematical model, the choice of language for specifying the model has significant implications on the modeling process itself, as well as on the possibilities for subsequent analysis.

2   Continuous-Time Markov Chains

Let S be a countable set. We consider a (homogeneous) continuous-time Markov chain $(X^{(t)})_{t \geq 0}$ on a probability space $(\Omega, \mathcal{F}, \Pr)$ with state space S and transition function

$$P_{ij}^{(t)} = \Pr(X^{(t)} = j \mid X^{(0)} = i), \quad i, j \in S, t \geq 0.$$

If initial probabilities $\Pr(X^{(0)} = i)$ are specified for each $i \in S$, the transient state probabilities $p_j^{(t)} := \Pr(X^{(t)} = j)$, are given by

$$p_j^{(t)} = \sum_{i \in S} p_i^{(0)} \cdot P_{ij}^{(t)}.$$

The transition functions $P_{ij}^{(t)}$ of a Markov chain are usually represented by their derivatives $q_{ij} = P_{ij}'(0)$ at $t = 0$, called *rates*. Here, we focus on Markov chains arising from population models. We therefore rule out "pathological cases" by assuming that the rates are finite and that $\sum_{j \in S} q_{ij} = 0$ for all $i \in S$. Note that this ensure that the sample paths $X^{(t)}(\omega)$ are right-continuous step functions (at least until a certain random point in time). Let Q be the matrix with components q_{ij}. Note that the diagonal elements are nonpositive and the off-diagonal elements are nonnegative. The matrix Q is called the (infinitesimal) *generator* of X since, if $\sup_{i \in S} |q_{ii}| < \infty$, the transition functions can be "generated" from Q. They are the unique solution of the Kolmogorov backward equations

$$P'(t) = Q \cdot P(t), \tag{1}$$

Fig. 1. The intensity graph of a Poisson process with rate λ

where the components of $P(t)$ are the values $P_{ij}(t)$. As a general solution, this gives

$$P(t) = \exp(Qt) = \sum_{k=0}^{\infty} (Qt)^k / k!.$$

Algorithms for the computation of the vector $p^{(t)}$ with entries $p_j^{(t)}$ are usually based on the numerical integration of the linear system of differential equations

$$p'(t) = Q \cdot p(t), \tag{2}$$

with initial condition $p^{(0)}$. Another approach is the approximation of the matrix exponential $\exp(Qt)$, which gives an approximation of $p(t) = p^{(0)} \cdot P(t) = p(0) \cdot \exp(Qt)$. In the case of an infinite or very large state space, the computation of $p(t)$ is computationally very expensive or even infeasible. Accurate approximations are, however, possible if the model is truncated appropriately.

For every $i \in S$, the limit probability $\pi_i = \lim_{t \to \infty} p_i(t)$ exists, but π_i may be zero for all states $i \in S$. Under certain conditions, however, $\sum_{i \in S} \pi_i = 1$ and the vector π with entries π_i is computed as the unique solution of the linear equation system

$$0 = \pi \cdot Q, \quad \sum_{i \in S} \pi_i = 1. \tag{3}$$

The distribution π is then called *steady-state distribution* or *stationary distribution*. Note that in this case $\pi_i > 0$ for all $i \in S$.

Each Markov chain has an associated state-transition graph, called *intensity graph*. It is a directed graph whose node set corresponds to the state space of the chain. It has an edge from state i to state j labeled by q_{ij} whenever $q_{ij} > 0$. The Markov chain is uniquely determined by its intensity graph.

Example 1. Consider a Markov chain X that has the infinitesimal generator

$$Q = \begin{pmatrix} -\lambda & \lambda & 0 & \cdots \\ 0 & -\lambda & \lambda & 0 & \cdots \\ \vdots & \ddots & \ddots & \ddots & \vdots \end{pmatrix},$$

where $\lambda > 0$. The process X is called Poisson process and is often used to model the number of arrivals of identical entities during a time interval $[0, t)$, where $\lambda = k \cdot t$ assuming an average of k arrivals per time unit. The intensity graph is shown in Fig. 1. □

3   Specifying Continuous-Time Markov Chains

In this section, we focus on the syntax of specification formalisms for large (or infinite) Markov chains with continuous time that describe *population models*, that is, models with state space $S = \mathbb{Z}_+^n = \{0, 1, \ldots\}^n$, where the i-th state variable represents the number of instances of the i-th species. Depending on the application area, "species" stands for types of system components, molecules, customers, etc. The application areas that we have in mind are chemical reaction networks, performance evaluation of computer systems, logistics, epidemics, etc.

3.1   Matrix Descriptions

A Markov chain may be specified by defining the elements of its generator matrix Q.

Example 2. Consider an epidemic process where individuals of a population are infected by a certain communicable disease. A state of the system is a pair $(x, y) \in \mathbb{Z}_+^2$, where x is the number of infected individuals and y is the number of individuals that are not infected [36]. Given positive rate constants a, b, c, d, e, the positive elements of the (infinite) generator matrix Q are given by

$$
\begin{aligned}
q_{(x,y),(x+1,y)} &= a   &&\text{for } x \geq 0 \text{ and } y \geq 0, \\
q_{(x,y),(x-1,y)} &= b \cdot x   &&\text{for } x > 0 \text{ and } y \geq 0, \\
q_{(x,y),(x,y+1)} &= c   &&\text{for } x \geq 0 \text{ and } y \geq 0, \\
q_{(x,y),(x,y-1)} &= d \cdot y   &&\text{for } x \geq 0 \text{ and } y > 0, \\
q_{(x,y),(x-1,y+1)} &= e \cdot x \cdot y &&\text{for } x > 0 \text{ and } y > 0.
\end{aligned}
$$

All remaining off-diagonal entries are 0 and for each row the element on the diagonal is the negative sum of the remaining row entries.   □

If the matrix exhibits a particular structure, it can also be described as the Kronecker product of smaller matrices that describe parts of the system. A general framework for descriptions based on the Kronecker product is provided by *stochastic automata networks* (SANs) [33,14]. A stochastic automaton is equivalent to a state-transition graph in which transitions are labeled by rates. Several automata can interact with each other and the state-transition graph of the global automaton determines the intensity graph of a Markov chain. The generator matrix of the Markov chain is then the Kronecker product of the matrices that represent the different automata and their interactions (compare also Section 4.1).

3.2   Stochastic Petri Nets

Petri nets are a pictorial language for describing systems of concurrent activities. A classical Petri net is a labeled directed bipartite graph whose node set is the disjoint union of a set P of *places* and a set T of *transitions*. The directed edges, called *arcs*, are given by a set $A \subseteq (P \times T) \cup (T \times P)$ and are labeled with a *multiplicity function* $l : A \to \mathbb{N}$. Stochastic Petri nets (SPN) [20] are an extension of classical Petri nets that associate a firing rate λ_τ with each transition $\tau \in T$.

A Petri net represents an infinite state-transition system with a set of states called *markings*. A marking is a function $m : P \to \mathbb{N}$ that maps every place of the Petri net to a nonnegative integer representing the number of *tokens* in that

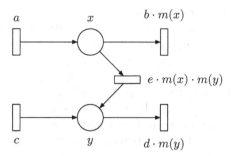

Fig. 2. Stochastic Petri net of the epidemic process in Example 2

place. Given a marking m, a transition $\tau \in T$ is *enabled in* m if all places p with an arc a leading to τ have at least $l(a)$ tokens in m, i.e., $m(p) \geq l(a)$. Note that transitions with no incoming arcs are always enabled. A transition τ is fired by removing $l(a)$ tokens from every place with an arc a leading to t and adding $l(b)$ tokens to every place with an arc b coming from t. The firing of a transition results in a new marking m' and corresponds to a transition from m to m' in the underlying transition system. In an SPN the firing rate $\lambda_\tau > 0$ of transition τ determines the random delay during which τ has to be enabled before it can fire. The underlying graph is the intensity graph of a Markov chain if the transitions are labeled with their respective firing rates.

Example 2 (cont.). Fig. 2 shows a stochastic Petri net for the epidemic process. The net has two places x and y depicted as circles, and five transitions depicted as rectangles. (We omit the multiplicity labeling of the arcs because all arcs have multiplicity 1.) The firing rate of each transition is given by the transition label. Here, we use functions that depend on the current marking m. The initial marking m_0 is the empty marking, i.e., $m_0(x) = m_0(y) = 0$. □

3.3 Stoichiometric Equations

Markov chain models for networks of biochemical reactions are usually specified by means of stochiometric equations. A stochiometric equation describes a reaction type. For instance, $A + B \to C$ means that if a molecule of type A hits a molecules of type B, they may form a complex molecule C. We call the species that are consumed by a reaction *reactants*; in the above example, A and B are reactants. Species that are produced by a reaction are called *products*.

Assume that the system involves n different chemical species $S_1, \ldots, S_n$. Consider a set $\{R_1, \ldots, R_m\}$ of chemical reactions, where the j-th reaction is given by the stoichiometric equation

$$R_j: \quad l_{j,1}S_1 + \cdots + l_{j,n}S_n \to k_{j,1}S_1 + \cdots + k_{j,n}S_n.$$

The stoichiometric coefficients $l_{j,1}, \ldots, l_{j,n}$ and $k_{j,1}, \ldots, k_{j,n}$ are nonnegative integers and describe how many molecules of each type are consumed and produced by the reaction. In the equation, we may omit a species if its coefficient is 0, and we may omit coefficients that are 1. Assume that $x = (x_1, \ldots, x_n)$ is the current state of the system, that is, we have x_i molecules of species i in the system. If a reaction of type R_j occurs, then the successor state is $x + v_j$, where the *change vector* v_j is given by $v_j = (k_{j,1} - l_{j,1}, \ldots, k_{j,n} - l_{j,n})$.

For a given state x, an instance of reaction R_j may occur whenever there are enough reactants in the system, i.e., whenever all entries of the vector $x - (l_{j,1}, \ldots, l_{j,n})$ are nonnegative. In this case, there is a transition in the underlying state-transition graph between state x and state $x + v_j$.[1] The rate of reaction R_j in state x determines the corresponding transition label in the intensity graph.

Stochastic Chemical Kinetics. If the reaction R_j is an elementary reaction, meaning that each instance corresponds to a single mechanistic step, then the transition rate $\alpha_j(x)$ between state x and $x + v_j$ is given by

$$\alpha_j(x) = c_j \cdot \prod_{i=1}^{n} \binom{x_i}{l_{j,i}}, \tag{4}$$

where $c_j > 0$ is a constant. This definition reflects the law of mass action kinetics, which states that the rate at which a chemical reaction occurs is proportional to the product of the reactant concentrations. Stochastic chemical kinetics considers populations of chemical species and replaces the product of reactant concentrations by $\prod_{i=1}^{n} \binom{x_i}{l_{j,i}}$, which is the number of distinct reactant combinations [17]. Usually, the constant c_j appears above the reaction arrow in the stoichiometric equation.

The following example shows that stoichiometric equations can be used to describe population models from other application areas.

Example 2 (cont.). We describe the epidemic process as a network of the "reactions":

$$R_1 : \emptyset \xrightarrow{a} S_x \qquad\qquad R_2 : S_x \xrightarrow{b} \emptyset \qquad\qquad R_3 : \emptyset \xrightarrow{c} S_y$$
$$R_4 : S_y \xrightarrow{d} \emptyset \qquad\qquad R_5 : S_x + S_y \xrightarrow{e} 2S_y$$

Here, the symbol $\emptyset$ means that all stoichiometric coefficients are zero. Note that if the transition rates are defined as in Eq. (4), they agree with the rates of Example 2. □

Since stoichiometric equations are classically used to describe biochemical reactions, we present an example from biology next.

Example 3. An enzyme-catalyzed substrate conversion is specified by the three reactions $R_1 : E + S \xrightarrow{c_1} ES$, $R_2 : ES \xrightarrow{c_2} E + S$, $R_3 : ES \xrightarrow{c_3} E + P$. This network involves four chemical species, namely, enzyme (E), substrate (S), complex (ES), and product (P) molecules. The change vectors are $v_1 = (-1, -1, 1, 0)$, $v_2 = (1, 1, -1, 0)$, and $v_3 = (1, 0, -1, 1)$. For $(x_1, x_2, x_3, x_4) \in \mathbb{Z}_+^4$, the rate functions are $\alpha_1(x_1, x_2, x_3, x_4) = c_1 \cdot x_1 \cdot x_2$, $\alpha_2(x_1, x_2, x_3, x_4) = c_2 \cdot x_3$, and $\alpha_3(x_1, x_2, x_3, x_4) = c_3 \cdot x_3$. □

---

[1] We assume for simplicity that each change vector v_j has at least one nonzero entry, and that all change vectors are distinct.

Systems Biology Markup Language. For software tools in systems biology, a standard language for the specification of systems is the Systems Biology Markup Language (SBML) [26]. It is an XML-based format that describes biochemical reaction networks by a list of components. Each component may describe dynamic behaviors by reactions, events, and mathematical rules, or give details about reacting species or compartments. SBML also offers several mechanisms such as unit and parameter definitions to ensure the unambiguous understanding of quantitative descriptions.

Example 3 (cont.). In Fig. 3, we show a part of an SBML description of the enzyme-catalyzed substrate conversion. (The SBML description is taken from the SBML homepage [26].) Lines 22–27 define the species ES, P, S, and E. In lines 64–84, we can see the description of the reaction $R_3 : ES \xrightarrow{c_3} E + P$. Note that SBML uses an extended version of stoichiometric equations to describe reactions. Like a stoichiometric equation, every reaction has a set of reactants and a set of products. However, the rate function is defined independently (cf. Fig. 3, lines 72–83) and need not follow Eq. (4). □

3.4 Stochastic Process Algebras

Stochastic process algebras can be used to specify continuous-time Markov chains based on a high-level description language that emphasizes the construction of complex processes from simple processes [24,18,4,34]. They provide several types of operators, such as prefix, choice, parallel composition, and recursion, in order to support different ways to combine processes. Typically, these languages are accompanied by structured operational semantics that define a state-transition graph, whose states are process terms. The graph can then be transformed into the intensity graph of a Markov chain.

Originally, stochastic process algebras were designed to *explicitly* model different molecules of the same species, that is, the model distinguishes instances of components being in the same local state. Since for population models this may lead to an enormous blow-up of the description, symmetry representations have been developed [22]. They support the specification of the local state and number of instances for a component type. In this way, as in the other languages, the separate spatial identity of each molecule is hidden on the syntactical level.

Besides the symmetry representations, the recently developed process algebra Bio-PEPA [9] can be used to specify population models. Bio-PEPA focuses on applications in systems biology, however, it can be used to model arbitrary Markovian population models.

Example 2 (cont.). In Bio-PEPA [9], the five reactions of the epidemic process are modeled as five actions (r1 to r5). The two species are models as two sequential processes (X and Y) that synchronize on action r5 in the model component. Below we show an input file for the Bio-PEPA workbench of the epidemic process.

```
1 <?xml version="1.0" encoding="UTF-8"?>
2 <sbml level="2" version="3" xmlns="http://www.sbml.org/sbml/level2/version3">
3   <model name="EnzymaticReaction">
4     <listOfUnitDefinitions>
5       <unitDefinition id="per_second">
6         <listOfUnits>
7           <unit kind="second" exponent="-1"/>
8         </listOfUnits>
..      ...
19    <listOfCompartments>
20      <compartment id="cytosol" size="1e-14"/>
21    </listOfCompartments>
22    <listOfSpecies>
23      <species compartment="cytosol" id="ES" initialAmount="0"    name="ES"/>
24      <species compartment="cytosol" id="P"  initialAmount="0"    name="P"/>
25      <species compartment="cytosol" id="S"  initialAmount="1e-20" name="S"/>
26      <species compartment="cytosol" id="E"  initialAmount="5e-21" name="E"/>
27    </listOfSpecies>
28    <listOfReactions>
..      ...
64      <reaction id="R3" reversible="false">
65          <listOfReactants>
66            <speciesReference species="ES"/>
67          </listOfReactants>
68          <listOfProducts>
69            <speciesReference species="E"/>
70            <speciesReference species="P"/>
71          </listOfProducts>
72          <kineticLaw>
73            <math xmlns="http://www.w3.org/1998/Math/MathML">
74              <apply>
75                <times/>
76                <ci>cytosol</ci>
77                <ci>c3</ci>
77                <ci>ES</ci>
78              </apply>
79            </math>
80            <listOfParameters>
81              <parameter id="c3" value="0.1" units="per_second"/>
82            </listOfParameters>
83          </kineticLaw>
84      </reaction>
85    </listOfReactions>
86  </model>
87 </sbml>
```

Fig. 3. Part of the SBML description in XML syntax of an enzymatic reaction

```
r1 = [ a ];
r2 = [ b * X ];
r3 = [ c ];
r4 = [ d * Y ];
r5 = [ e * X * Y ];
X = r1>> + r2<< + r5<<;
Y = r3>> + r4<< + r5>>;
(X <r5> Y)
```

The first five lines specify the actions and the corresponding rates. Line 6 and 7 specify in which reactions the components take part and what role they play in the reaction. E.g., r1>>, which is a shortcut for (r1,1) >> X, means X is an reactant in reaction r1 with stoichiometry coefficient 1 and r2<< means X is a product in reaction r2. The plus operator (+) is the sequential composition operator defining that the actions are sequentially interleaved. Finally, the last line specifies that the model is the parallel composition of processes X and Y that synchronize on reaction r5.   □

3.5 Guarded Commands

Similar to Petri nets, guarded-command models (GCM) describe the state transitions of the underlying process. However, unlike Petri nets, GCM are textual. Often, the set of all transitions can be partitioned into classes of transitions. Instead of listing all states, the modeler describes the possible classes of transitions that may occur. As a representative for such transition class description, we present a syntax that is inspired by Dijkstra's guarded-command language [13], which has subsequently been used by GCM such as Reactive Modules [1] and by the language for specifying PRISM models [37]. We describe transition classes by guarded commands that operate on the state variables of the system. Recall that the state variables of the system are nonnegative integers representing numbers of molecules for each species. A guarded command takes the form

[] guard |- rate -> update

where the guard is a Boolean predicate over the variables, which determines in which states the corresponding transitions are enabled. The update is a rule that describes the change of the system variables if a corresponding transition is performed. Syntactically, update is a list of statements, each assigning to a variable an expression over variables. Assume that x is a variable. If, for instance, the update rule is that x is incremented by 1, we write x:=x+1. We assume that variables that are not listed in the update rule do not change if the transition is taken. Each guarded command also assigns a rate to the corresponding transitions, which is a function in the state variables. We do not fix an expression language for the rate functions here.

Example 2 (cont.). We define a GCM for the epidemic process.

```
variables x,y
[] true          |- a       -> x:=x+1
[] (x>0)         |- b*x     -> x:=x-1
[] true          |- c       -> y:=y+1
[] (y>0)         |- d*y     -> y:=y-1
[] (x>0)&(y>0)   |- e*x*y   -> x:=x-1; y:=y+1
```
□

Note that each guarded command specifies infinitely many transitions. For examples, the guarded command `[] true |- a -> x:=x+1` specifies one transition from each state, with constant rate a, to a successor state in which the number of x molecules is incremented and the number of y molecules remains unchanged.

Example 3 (cont.). The enzyme reaction is specified by the guarded commands:

```
variables e,s,es,p
[] (e>0)&(s>0)  |- c1*e*s -> e:=e-1;  s:=s-1; es:=es+1
[] (es>0)       |- c2*es  -> es:=es-1; e:=e+1;  s:=s+1
[] (es>0)       |- c3*es  -> es:=es-1; e:=e+1;  p:=p+1
```
□

Now, we show how to derive the underlying generator matrix from a GCM. To simplify the presentation, we assume that the updates of two commands differ whenever there is a state in which both guards are true. Moreover, we do not consider commands with empty updates, because "self-loops" do not alter the dynamics of a Markov chain. Then, each guarded command determines an entry in the row of a state s in the generator matrix whenever the guard is true in s. Assume that the state space of the underlying Markov chain is $S = \mathbb{Z}_+^n$, and $G \subseteq S$ is the subset where the guard is true. Furthermore, $s = (s_1, \ldots, s_n)$ and the update is a function $u : G \to S$. Then $q_{s,u(s)} = r(s)$, where $r : G \to \mathbb{R}_{\geq 0}$ is the rate function of the command. For instance, in Example 2, $G = \{(x, y) \in \mathbb{Z}_+^2 \mid x > 0 \text{ and } y > 0\}$ is the guard set of the last command. The update function is $u(x, y) = (x - 1, y + 1)$, and the rate function is $r(x, y) = e \cdot x \cdot y$, which yields the matrix entries $q_{(x,y),(x-1,y+1)} = e \cdot x \cdot y$ for all $x > 0$ and $y > 0$.

4   Properties of Specification Languages

In this section, we discuss several properties of specification languages which are important for the construction and the analysis of a model. We focus on the languages mentioned in the previous section.

4.1   Compositionality

Compositionality facilitates the description of complex systems. A compositional language allows the modular description of a system by combining submodels that describe parts of the system. Moreover, a modular description can be advantageous for the analysis of the model, e.g., for compositional aggregation techniques [5].

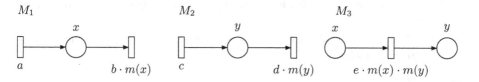

Fig. 4. Composition of stochastic Petri nets

Matrix Descriptions. We can construct the generator matrix Q of the epidemic process in Example 2 in a compositional way. We define the three matrices $A = diag([1\ 1\ 1\ldots], 1)$, $B = diag([0\ 1\ 2\ldots], 1)$, $C = diag([0\ 1\ 2\ldots], -1)$, where for $k \in \mathbb{Z}$ the notation $diag(v, k)$ refers to a matrix whose nonzero elements are the elements of the vector v that appears on the k-th diagonal of the matrix (negative values indicate that the vector appears below the main diagonal). Then the matrix

$$\hat{Q} = ((a \cdot A + b \cdot C) \oplus (c \cdot A + d \cdot C)) + e \cdot (C \otimes B)$$

agrees with Q except for the main diagonal. Thus, if $\mathbf{1}$ is the column vector with all entries equal to 1, then $Q = \hat{Q} - diag(\hat{Q} \cdot \mathbf{1}, 0)$. The matrix $\hat{Q}$ describes a network of two stochastic automata. The first automaton represents the state variable x, and $(a \cdot A + b \cdot C)$ defines its local transitions. The second automaton represents the state variable y, and $(c \cdot A + d \cdot C)$ defines its local transitions. Finally, $e \cdot (C \otimes B)$ describes the synchronous transitions of the network.

Thus, the composition of models with matrix representation requires matrix operations such as matrix sum, Kronecker product, and Kronecker sum.

Stochastic Petri Nets. In Fig. 4, we show the stochastic Petri nets of three subsystems of Example 2. Their combination yields the model shown in Fig. 2. The composition of Petri nets hinges on the identity of the places, because places with equal labels are collapsed. Thus, renaming of variables may be necessary. In the composite model, the original models are often not clearly separable.

Stoichiometric Equations. As for Petri nets, sets of stoichiometric equations may be joined, where the interfaces are specified by the names of the chemical species. For instance, the composition of the three networks M_1, M_2, M_3 of reactions specified by

M_1	M_2	M_3
$R_1 : \emptyset \xrightarrow{a} S_x$	$R_1 : \emptyset \xrightarrow{c} S_y$	$R_1 : S_x + S_y \xrightarrow{e} 2S_y$
$R_2 : S_x \xrightarrow{b} \emptyset$	$R_2 : S_y \xrightarrow{d} \emptyset$	

yields the description of the epidemic process in Example 2. Here, the composite model is constructed simply by the union of reactions.

Stochastic Process Algebras. Compositionality is one of the most important aspect of process calculi. Process algebras are equipped with various operators that can be used to combine process terms. This facilitates the description of different forms of interaction between subsystems. For instance, a detailed discussion on synchronous interaction, we refer to [23]. Note that in the Bio-PEPA example in the previous section, the epidemic process is the composition of the two process terms X and Y.

Guarded Commands. Again, we consider Example 2. The GCM for the subsystems we discussed above are

```
variables x
[] true        |- a      -> x:=x+1
[] (x>0)       |- b*x    -> x:=x-1

variables y
[] true        |- c      -> y:=y+1
[] (y>0)       |- d*y    -> y:=y-1

variables x,y
[] (x>0)*(y>0) |- e*x*y -> x:=x-1; y:=y+1
```

Similar as for stoichiometric equations, two GCM can be composed by a simple union of the guarded commands, where variables may have to be renamed.

4.2   Expressiveness and Succinctness

Two important properties of a specification language are its expressive power and its succinctness. For example, language A is as expressive as language B if every model that can be specified in B can also be specified in A. Of two equally expressive language, one may be more succinct than the other. For instance, if for some models there are descriptions in A that are exponentially smaller than all descriptions in B, then on these models, A is exponentially more succinct than B. The expressiveness and succinctness of specification languages can be compared by studying translations between languages and the cost of such translations. Other questions that fall under this topic concern the ease of extending a language to gain expressive power, and an independent characterization of which semantic objects (i.e., continuous-time Markov chains that arise from population models) can be described by expressions within a given formal syntax.

To our knowledge, no systematic and complete comparison between the various languages for describing population models has been carried out, and we make here only a few remarks.

Matrix Descriptions. The expressive power of a matrix description depends on the exact syntax of expressions for describing matrix entries and, in the case of infinite dimension, on the syntax for describing sets of entries. As first step in an analysis procedure, many other languages for specifying population models are translated into matrix descriptions. The translation from higher-level languages such as guarded commands often results in a blow-up of the size of the description.

Stochastic Petri Nets. A stochastic Petri net can be transformed into a guarded command model if we associate a variable with each place and a guarded command with each transition. Many extensions of stochastic Petri nets have been developed, such as generalized stochastic Petri nets [30], fluid stochastic Petri nets [25], etc. They can be used to describe stochastic processes that are not necessarily Markov chains. There is, however, we know of no extension of Petri nets that can specify infinitely branching Markov chains, whose intensity graph contains states with an infinite number of out-going transitions. Moreover, even though firing rates may be marking dependent, the expression syntax may not allow arbitrary rate functions.

Stoichiometric Equations. While stoichiometric equations are widely used to model networks of biochemical reactions, some models —such as the bistable toggle switch shown below— have rate functions that differ from Eq. (4). Thus, they cannot be described in the classical stoichiometric style.

Example 4. The bistable toggle switch is a prototype of a genetic switch with two competing repressor proteins and four reactions [16,43]. It involves two chemical species, A and B, and four reactions. The reactions are $\emptyset \to A$, $A \to \emptyset$, $\emptyset \to B$, and $B \to \emptyset$. Let $x = (x_1, x_2) \in \mathbb{N}_0^2$. The rate functions are $\alpha_1(x) = c_1/(c_2 + x_2^2)$, $\alpha_2(x) = c_3 \cdot x_1$, $\alpha_3(x) = c_4/(c_5 + x_1^2)$, and $\alpha_4(x) = c_6 \cdot x_2$. Here, the values c_1, c_2, c_4, and c_5 are positive constants that determine the mutual repression of A and B. The values c_3 and c_6 are positive constants that determine at which rate degradation of molecules occurs. The toggle switch can be specified using SBML syntax, as SBML provides more flexibility than stoichiometric equations.   □

A set of stoichiometric equations can always be transformed into a guarded command model or a stochastic Petri net. The number of chemical species corresponds to the number of variables (or places), and each reaction induces a guarded command (or transition).

Stochastic Process Algebras. Stochastic process algebras such as PEPA [24], TIPP [18], EMPA [4] and the stochastic pi-calculus [34] consider constant transition rates, i.e., transition rates do not depend on the state variables. This is because these languages originally were not designed for specifying population models. The extension Bio-PEPA [9] addresses this shortcoming, as was shown in the examples discussed in the previous section. Although most stochastic process algebras provide limited support for rate functions, they have recursion operators for specifying Markov chains for which no basic guarded command model can be constructed. The usefulness of such operators, however, depends on the application area.

Guarded Commands. A guarded command model can be transformed into a stochastic Petri net, where each command corresponds to a transition and each variable to a place. Note that this transformation is only possible if the guards are lower bounds on the state variables. Guarded command models are similar to stochastic Petri nets in that they allow rate functions that depend on the state variables. For instance, we can describe Example 4 using the following commands:

```
variables x1,x2
[] true  |- c1/(c2+x2^2) -> x1:=x1+1
[] x1>0  |- c3*x1         -> x1:=x1-1
[] true  |- c4/(c5+x1^2) -> x2:=x2+1
[] x2>0  |- c6*x2         -> x2:=x2-1
```

The limitations of basic guarded command models are similar to those of stochastic Petri nets; for example, infinitely branching Markov chains cannot be described by specifications consisting of finitely many guarded commands.

4.3   Executability

For the analysis of a model it is important that we can easily compute the direct transition successors of a given state from the model description. This allows us to "execute" the model, by repeatedly applying the next-state function [15].

Stochastic Petri Nets and Stochastic Process Algebras. Petri net and process algebra models provide a high-level description that is usually not directly executable. For a given marking in a stochastic Petri net, we have to inspect each place and each arc in order to determine the enabled transitions as well as their firing rates. Similarly, for a given stochastic process term, we have to consider each subterm and compute all possible transitions and their rates. Then, the composition rules determine the possible global transitions of the system and their rates. Therefore, most tools for the analysis of Markov chains with high-level specification languages construct an intermediate low-level model, such as a matrix representation [8,3,28,27].

Guarded Commands and Stoichiometric Equations. For languages based on transition classes, such as guarded commands, the construction of an intermediate low-level model is not necessary. This is one of the main strengths of guarded commands. For a given state, an on-the-fly calculation of all possible successor states and transition rates can be performed by iterating over the set of guarded commands and calling their update and rate functions [21]. Note that for stoichiometric equations we have to store for each reaction R_j the change vector v_j, the rate function α_j, and the number of necessary reactant molecules. This is essentially the representation provided by guarded commands.

4.4   Well-Formedness

In Section 3, we discussed different formalisms for the specification of a Markov chain. In most languages, however, it is possible to specify models whose underlying intensity graph (or generator matrix) do not uniquely determine a Markov chain. This is due to the fact that the Kolmogorov backward equations (see Eq. (1)) may not have a unique solution. In Section 2, we used the following sufficient but not necessary condition:

$$\sup_{i \in S} |q_{ii}| < \infty \qquad (5)$$

If the modeling formalism allows us to specify transition rates that are functions in the state variables, then this condition may not be fulfilled. Note that this can only occur if the number of reachable states is infinite. For instance, the condition in Eq. (5) is not satisfied in Example 2. In the sequel we discuss conditions that are weaker than Eq. (5) but still ensure a unique solution to Eq. (1). We focus on conditions on the matrix Q and refer to the entries of Q as q_{ij}. For the nonpositive diagonal entries, we use the abbreviation $q_i = -q_{ii}$.

A generator matrix Q is called *stable* if all entries are finite, and *conservative*, if all rows sum up to zero [2]. Except for matrix descriptions, all specification languages discussed before, guarantee by construction that the underlying generator matrices are stable and conservative. Models specified by matrix descriptions have to be checked for stability and conservation separately.

A conservative and stable generator matrix that has a unique solution to the Kolmogorov backward equations is called *regular*. The following criterion is sufficient and necessary for a generator matrix to be regular.

Theorem 1 (Reuter's Criterion). *A stable and conservative generator Q on S is regular if and only if for any real $\lambda > 0$, the system of equations*

$$\sum_{j \in S, j \neq i} q_{ij} z_j = (\lambda + q_i) z_i \text{ for all } i \in S \tag{6}$$

admits no nonnegative bounded solution other than the trivial one.

Example 5. Consider a model with the following guarded command.

```
x>0 |- 2^x -> x:=x+1
```

Recall that this model specifies a generator matrix Q with the following non-zero entries: $q_{i(i+1)} = 2^i$ and $q_{ii} = -2^i$ for all $i > 0$. Then, we obtain for Q the following equations from (6):

$$2^i z_{i+1} = (\lambda + 2^i) z_i \text{ for all } i > 0.$$

Applying simple transformations allows us to express the solutions for $i > 1$ in terms of z_1 by $z_i = \Pi_{k=1}^{i-1} (\frac{\lambda}{2^k} + 1) z_1$. We choose $\lambda = z_1 = 1$, then it reminds to show that $z_i = \Pi_{k=1}^{i-1} (\frac{1}{2^k} + 1)$ is bounded for all $i > 1$. Since $z_i = e^{\ln(z_i)}$, it suffices to show that $\ln(\Pi_{k=1}^{i-1} (\frac{1}{2^k} + 1)) = \sum_{k=1}^{i-1} \ln(\frac{1}{2^k} + 1) \leq \sum_{k=1}^{i-1} \frac{1}{2^k} \leq \frac{1}{1-1/2} - 1 = 1$ is bounded for all $i > 1$. This shows that our model has a nontrivial bounded solution and we can conclude that Q is not regular. □

Since showing that Reuter's criterion is true for a model is rather difficult, we discuss another condition that is sufficient but not necessary. The idea is to approximate the infininte unbounded generator matrix Q by a sequence of bounded submatrices.

Theorem 2 ([2] Corollary 2.16). *Let Q be a stable and conservative generator matrix over the state space S and let $S_1, S_2, \ldots$ be a sequence of subsets of S such that $S_1 \subseteq S_2 \subseteq \ldots$, $\cup_{r=1}^{\infty} S_r = S$, and $\sup_{i \in S_r} q_i < \infty$. Suppose that $z_j \geq 0$, $j \in S$ are such that*

1. $\lim_{r \to \infty} \inf_{j \notin S_r} z_j = \infty$, and
2. there $\lambda \in \mathbb{R}$ such that $\sum_{j \neq i} q_{ij} z_j \leq (\lambda + q_i) z_i$ for all $i \in S$.

Then Q is regular.

Given a suitable sequence S_r and values z_j, the sum and the number of conditions we have to check for regularity are infinite. The chosen syntax for the model description may facilitate the regularity check. For instance, in a GCM, it suffices to let the sum range over the finite set of guarded commands. We can use the guards to partition the state space into a finite number of sets such that we have to check a single condition for each set.

Example 2 (cont.). Recall the GCM of the epidemic process.

```
variables x,y
[] true          |- a       -> x:=x+1
[] (x>0)         |- b*x      -> x:=x-1
[] true          |- c       -> y:=y+1
[] (y>0)         |- d*y      -> y:=y-1
[] (x>0)&(y>0)   |- e*x*y    -> x:=x-1; y:=y+1
```

We use the four guards to partition the state space into four sets P_1, P_2, P_3, P_4, as shown in Fig. 5 by the dashed lines. Note that for suitable expression languages for the guards, we can always find this partitioning automatically. Now, for each set, we can check Theorem 2 using the corresponding guarded commands. For instance, in P_3 (right lower corner in Fig. 5), the first three guarded commands are enabled, and the second condition in Theorem 2 rewrites to $a \cdot z_{(x+1,y)} + b \cdot x \cdot z_{(x-1,y)} + c \cdot z_{(x,y+1)} \leq (\lambda + a + b \cdot x + c) \cdot z_{(x,y)}$ for all $x > 0$ and $y \geq 0$. With $z_{(x,y)} = x + y + 1$, $(x,y) \in S$ and $S_r = \{(x,y) \mid z_{(x,y)} \leq r\}$, which satisfy Condition 1 in Theorem 2, we obtain $a \cdot (x+y+2) + b \cdot x \cdot (x+y) + c \cdot (x+y+2) \leq (\lambda + a + b \cdot x + c) \cdot (x+y+1)$, which is the same as $a - b \cdot x + c \leq \lambda \cdot (x+y+1)$ and true for $(x,y) \in P_3$ if $\lambda = a + c$. $\qquad\square$

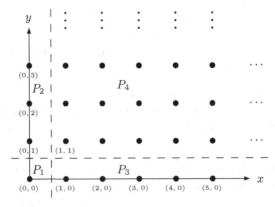

Fig. 5. State space partitioning w.r.t. the guarded command model of the epidemic process

In a similar way, it is possible to exploit the chosen syntax in order to decide whether the limit probabilities $\pi_i = \lim_{t\to\infty} p_i(t)$ of the Markov chain form a distribution. We refer to [2] for criteria that ensure the existence of a limit distribution as well as that it can be calculated according to Eq. (3).

In summary, GCM offer an efficiently checkable sufficient condition for regularity and the existence of the limit probabilities.

5 Analysis of Continuous-time Markov Chains

For the analysis of Markov chains we distinguish transient and steady-state analysis. The former refers to the computation of the vector $p(t)$ which contains the probabilities $\Pr(X(t) = x)$ for each state x reachable from a given initial distribution. Often, $p(t)$ is computed at several time instances t of interest. Steady-state analysis requires the computation of the limiting behaviour of the Markov chain, i.e., of the probability vector $\lim_{t\to\infty} p(t)$.

There are three different approaches to the analysis of continuous-time Markov chains, namely, analytical solutions, numerical solutions, and simulation. Since analytical solutions can only be obtained for Markov chains with a very simple structure, we concentrate on the latter two approaches. Numerical solutions are based on an exploration of the state space which proceeds in a breadth-first search manner, by moving the probability mass through the state space. In contrast, simulation of Markov chains is a special case of the Monte Carlo method, which relies on the repeated generation of random sample paths in the underlying state-transition graph.

5.1 Simulation

Monte Carlo simulation of Markov chains is based on the idea of generating a number of trajectories $X^{(t)}(\omega)$ using pseudo-random numbers [29]. Then, probabilities and expectations of certain random variables can be statistically estimated. The main advantage of simulation is that the memory requirements are low and therefore the analysis of systems of arbitrary size is possible. In order to achieve a high accuracy, however, a large number of trajectories have to be generated, which is very time consuming and often infeasible [12].

For the generation of trajectories, an executable model description is advantageous, as this will speed up the time needed for computing the next jump time and the successor state for given a trajectory prefix.

5.2 Numerical Solutions

Several tools exist that provide algorithms for the numerical solution of continuous-time Markov chains [28,3,27,8]. They use a matrix description which is obtained from high-level modeling formalisms such as stochastic Petri nets [3,8], guarded command languages [28,27], or stochastic process algebras [28,27]. The generator matrix of the Markov chain is stored either symbolically using multi-terminal BDDs [10] or sparse matrix packages. Moreover, if the size of the generator matrix exceeds the available memory capacity the matrix can be stored as a Kronecker product of smaller matrices.

The implemented algorithms for the computation of $p(t)$ are either based on the solution of Eq. (2) or a discretization of the Markov chain. In the former case, numerical integration methods or methods based on a Krylov subspace construction are applied [40]. They require the construction of the generator matrix Q, which is often infeasible for large Markov chains. It is, however, possible to exploit the structure of the Markov chain and approximate the solution by successively considering submatrices of Q [21].

During the discretization procedure, also called uniformization [19], a discrete-time Markov chain is constructed which has essentially the same transition graph structure as the original continuous-time Markov chain. The idea behind uniformization is that the maximal absolute value of the diagonal entries of the generator matrix can be used to "normalize" the time that the process remains in a state. Thus, an a-priori exploration of the state space is necessary to apply uniformization. For large Markov chains the memory requirements can be prohibitive, even if sparse matrix structures or symbolic representations are used. Therefore variants of the uniformization method have been developed which exploit a Kronecker representation of the matrix [7].

Most formalisms allow us to specify an infinite number of reachable states, but are limited to finite models for their numerical analysis, because the analysis requires the enumeration of all reachable states and the construction of the generator matrix. An algorithm that completely avoids the construction of any matrix and exploits a guarded command description has been proposed recently in [12]. It can be used for the approximate analysis of infinite-state systems and does not suffer from excessive memory requirements, at least for systems where the significant part of the probability mass is concentrated on a manageable subset of states. The proposed algorithm is enhanced by the executability properties of the guarded command description.

For the computation of steady-state measures, most methods also require the a-priori construction of the generator matrix from the Markov chain specification. If the state space is large, direct methods for the solution of Eq. (3) are inefficient, and iterative methods such as the Jacobi, Gauss-Seidel, or SOR method must be applied [40]. Many iterative methods have been adapted such that they exploit Kronecker representations of the generator matrix [41,6]. Other approaches are based on on-the-fly techniques [11] or reduced state spaces that are obtained by exploiting symmetries in the model structure and tailoring to the variable in question [38].

6   Conclusions

There are many different languages for describing Markov chains with continuous time. The choice of an appropriate syntax usually depends on the application area. Guarded commands provide a natural language for the description of population models. They facilitate the specification of such models, because they are compositional, succinct, and provide sufficient expressive power. Moreover, they support well-formedness checks and allow a direct execution of the model.

References

1. Alur, R., Henzinger, T.A.: Reactive modules. Formal Methods in System Design 15(1), 7–48 (1999)
2. Anderson, W.: Continuous-time Markov chains: An applications-oriented approach. Springer, Heidelberg (1991)
3. Baarir, S., Beccuti, M., Cerotti, D., De Pierro, M., Donatelli, S., Franceschinis, G.: The GreatSPN tool: recent enhancements. SIGMETRICS Perform. Eval. Rev. 36(4), 4–9 (2009)
4. Bernardo, M., Gorrieri, R.: Extended Markovian process algebra. In: Sassone, V., Montanari, U. (eds.) CONCUR 1996. LNCS, vol. 1119, pp. 315–330. Springer, Heidelberg (1996)
5. Buchholz, P.: Exact and ordinary lumpability in finite markov chains. Journal of applied probability 31(1), 59–75 (1994)
6. Buchholz, P., Dayar, T.: Block SOR preconditioned projection methods for Kronecker structured Markovian representations. SIAM Journal on Scientific Computing 26(4), 1289–1313 (2005)
7. Buchholz, P., Sanders, W.H.: Approximate computation of transient results for large Markov chains. In: Proc. of QEST 2004, pp. 126–135. IEEE Computer Society Press, Los Alamitos (2004)
8. Ciardo, G., Jones III, R.L., Miner, A.S., Siminiceanu, R.I.: Logic and stochastic modeling with SMART. Perform. Eval. 63(6), 578–608 (2006)
9. Ciocchetta, F., Hillston, J.: Bio-PEPA: a framework for modelling and analysis of biological systems. Theoretical Computer Science (to appear, 2009)
10. Clarke, E., Fujita, M., McGeer, P., Yang, J., Zhao, X.: Multi-terminal binary decision diagrams: An ecient data structure for matrix representation. In: Proc. IWLS 1993 (1993)
11. Deavours, D.D., Sanders, W.H.: "On-the-fly" solution techniques for stochastic Petri nets and extensions. In: IEEE TSE, pp. 132–141 (1997)
12. Didier, F., Henzinger, T., Mateescu, M., Wolf, V.: Approximation of event probabilities in noisy cellular processes. In: Proc. of CMSB. LNCS, Springer, Heidelberg (to appear, 2009)
13. Dijkstra, E.W.: Guarded commands, nondeterminacy and formal derivation of programs. Commun. ACM 18(8), 453–457 (1975)
14. Fernandes, P., Plateau, B., Stewart, W.J.: Numerical evaluation of stochastic automata networks. In: Proc. of MASCOTS 1995 (1995)
15. Fisher, J., Henzinger, T.A.: Executable cell biology. Nature Biotechnology 25, 1239–1249 (2007)
16. Gardner, T., Cantor, C., Collins, J.: Construction of a genetic toggle switch in Escherichia coli. Nature 403, 339–342 (2000)
17. Gillespie, D.T.: Markov Processes. Academic Press, London (1992)
18. Götz, N., Herzog, U., Rettelbach, M.: TIPP – a language for timed processes and performance evaluation. Technical Report Technical Report 4/92, IMMD VII, University of Erlangen-Nurnberg (1992)
19. Gross, D., Miller, D.: The randomization technique as a modeling tool and solution procedure for transient Markov processes. Operations Research 32(2), 926–944 (1984)

20. Haas, P.J.: Stochastic Petri Nets: Modelling, Stability, Simulation. Springer, Heidelberg (2002)
21. Henzinger, T., Mateescu, M., Wolf, V.: Sliding window abstraction for infinite Markov chains. In: Proc. CAV. LNCS, Springer, Heidelberg (to appear, 2009)
22. Hermanns, H.: An operator for symmetry representation and exploitation in stochastic process algebras. In: Proc. of PAPM 1997, pp. 55–70 (1997)
23. Hillston, J.: The nature of synchronisation. In: Proc. of PAPM 1994, pp. 51–70 (1994)
24. Hillston, J.: A Compositional Approach to Performance Modelling. Cambridge University Press, Cambridge (1996)
25. Horton, G., Kulkarni, V.G., Nicol, D.M., Trivedi, K.S.: Fluid stochastic Petri nets: Theory, applications, and solution techniques. European Journal of Operational Research 105(1), 184–201 (1998)
26. Hucka, M., Finney, A., Sauro, H.M., Bolouri, H., Doyle, J.C., Kitano, H.: The systems biology markup language (SBML): a medium for representation and exchange of biochemical network models. BIOINFORMATICS 19(4), 524–531 (2003)
27. Katoen, J.-P., Khattri, M., Zapreev, I.S.: A Markov reward model checker. In: Proc. of QEST 2005, pp. 243–244. IEEE Computer Society Press, Los Alamitos (2005)
28. Kwiatkowska, M., Norman, G., Parker, D.: PRISM: Probabilistic model checking for performance and reliability analysis. ACM SIGMETRICS Performance Evaluation Review 36(4), 40–45 (2009)
29. Law, A., Kelton, W.: Simulation Modeling and Analysis. McGraw-Hill, New York (2000)
30. Marsan, M.A., Balbo, G., Conte, G., Donatelli, S., Franceschinis, G.: Modelling with generalized stochastic petri nets. Sigm. Perform. Eval. Rev. 26(2), 2 (1998)
31. McAdams, H.H., Arkin, A.: It's a noisy business! Trends in Genetics 15(2), 65–69 (1999)
32. Paulsson, J.: Summing up the noise in gene networks. Nature 427(6973), 415–418 (2004)
33. Plateau, B.: On the stochastic structure of parallelism and synchronization models for distributed algorithms. In: Proc. of the Sigmetrics Conference on Measurement and Modeling of Computer Systems, pp. 147–154 (1985)
34. Priami, C.: Stochastic pi-calculus. The Computer Journal 38(7), 578–589 (1995)
35. Rao, C., Wolf, D., Arkin, A.: Control, exploitation and tolerance of intracellular noise. Nature 420(6912), 231–237 (2002)
36. Reuter, G.E.H.: Competition processes. In: Proc. 4th Berkeley Symp. Math. Statist. Prob., vol. 2, pp. 421–430. Univ. of California Press, Berkeley (1961)
37. Rutten, J., Kwiatkowska, M., Norman, G., Parker, D.: Mathematical Techniques for Analyzing Concurrent and Probabilistic Systems. CRM Monograph Series, vol. 23. American Mathematical Society, Providence (2004)
38. Sanders, W.H., Ers, Y., Meyer, J.: Reduced base model construction methods for stochastic activity networks. In: Proc. of PNPM 1989, vol. 11, pp. 74–84 (1989)
39. Srivastava, R., You, L., Summers, J., Yin, J.: Stochastic vs. deterministic modeling of intracellular viral kinetics. Journal of Theoretical Biology 218, 309–321 (2002)
40. Stewart, W.J.: Introduction to the Numerical Solution of Markov Chains. Princeton University Press, Princeton (1995)

41. Stewart, W.J., Atif, K., Plateau, B.: The numerical solution of stochastic automata networks. European Journal of Operational Research 86(3), 503–525 (1995)
42. Swain, P.S., Elowitz, M.B., Siggia, E.D.: Intrinsic and extrinsic contributions to stochasticity in gene expression. Proc. Natl. Acad. of Sci. 99(20), 12795–12800 (2002)
43. Tian, T., Burrage, K.: Stochastic models for regulatory networks of the genetic toggle switch. Proc. Natl. Acad. Sci. 103(22), 8372–8377 (2006)
44. Turner, T.E., Schnell, S., Burrage, K.: Stochastic approaches for modelling in vivo reactions. Computational Biology and Chemistry 28, 165–178 (2004)
45. Wilkinson, D.J.: Stochastic Modelling for Systems Biology. Chapman & Hall, Boca Raton (2006)

Reachability for Continuous and Hybrid Systems

Oded Maler

CNRS-VERIMAG, 2, av. de Vignate, 38610 Gieres, France
Oded.Maler@imag.fr

Abstract. In this talk I present some past, present and future work concerning reachability computation for continuous and hybrid systems.

The problem of *what may happen in the future* is very naturally phrased as a reachability problem in a dynamical system. Starting from an initial state or a set of states, following some dynamic rules, possibly with some dose of uncertainty, we would like to know whether something will happen, whether a particular state of affairs will occur. In the finite-state case the problem is trivially decidable by graph search algorithms (but is intractable when you have a product of many subsystems), while in the general discrete case it is undecidable as *halting* is just reachability of a particular state.

In the "early" days of hybrid systems research (early 90s), after the the fact that reachability is decidable for automata with real-valued clocks has been reformulated, there was some hope that similar things could be done for hybrid systems having a more complex (but still piecewise-trivial) dynamics, that is, automata where the continuous evolution is linear (the derivative is constant) in each discrete location.[1] But soon it was realized how easy it is to build TM, counter machine or 2PDA gadgets when you have real-valued variables. Consequently, finding the boundary of decidability became a sportive activity like building a universal TM with 2 states, 3 symbols and 1.5 heads. Practically, the best you could provably expect for even those very simple hybrid systems is "semi-decidability": running a verification tool (like the pioneering HyTech) and hoping that the iteration terminates.

My conclusion from all these investigations was that in the context of continuous systems, solving the exact "reachability problem" does not make much sense, except for mathematical amusement. Almost everything in the continuous world is already approximate and the infinite precision used to simulate TMs is unrealistic from the points of view of modeling, measurement and computation. Thus the expectations were lowered once more, from *exact* semi-decidability to approximate computation of reachable sets and then from unbounded time horizon to a finite one. Approximating tubes of trajectories of continuous dynamical systems turned out to be an interesting research domain involving graph algorithms, computational geometry in high dimension, numerical analysis and other

---

[1] Such systems are very attractive for computer scientists like us because they can be studied with elementary linear algebra and there is no need to understand those horrible differential equations.

O. Bournez and I. Potapov (Eds.): RP 2009, LNCS 5797, pp. 24–25, 2009.

branches of mathematics and computer science. It has potential applications in analog circuit verification, in the debugging of biochemical models, in the analysis of control systems and in any other domain which uses differential equations subject to some nondeterminism and in which we want more information than we can get from running arbitrary simulations.

I will start my talk with some decidability and undecidability results for piecewise-constant derivative systems. Then I will explain how reachable states of linear systems are approximated by unions of convex geometric objects such as polytopes, describe an algorithmic scheme that allows this computation to be performed much more efficiently and conclude with two recent works, one that takes us one step closer to simple numerical simulation and one which extends reachability to systems with nonlinear dynamics.

Algorithmic Information Theory
and Foundations of Probability

Alexander Shen

LIF Marseille, CNRS & Univ. Aix–Marseille, On leave from IITP, RAS, Moscow,
Supported in part by NAFIT ANR-08-EMER-008-01 grant
`alexander.shen@lif.univ-mrs.fr`

Abstract. The question how and why mathematical probability theory can be applied to the "real world" has been debated for centuries. We try to survey the role of algorithmic information theory (Kolmogorov complexity) in this debate.

1 Probability Theory Paradox

One often describes the natural sciences framework as follows: a hypothesis is used to predict something, and the prediction is then checked against the observed actual behavior of the system; if there is a contradiction, the hypothesis needs to be changed.

Can we include probability theory in this framework? A statistical hypothesis (say, the assumption of a fair coin) should be then checked against the experimental data (results of coin tossing) and rejected if some discrepancy is found. However, there is an obvious problem: The fair coin assumption says that in a series of, say, 1000 coin tossings all the 2^{1000} possible outcomes (all 2^{1000} bit strings of length 1000) have the same probability 2^{-1000}. How can we say that some of them contradict the assumption while other do not?

The same paradox can be explained in a different way. Consider a casino that wants to outsource the task of card shuffling to a special factory that produced shrink-wrapped well shuffled decks of cards. This factory would need some quality control department. It looks at the deck before shipping it to the customer, blocks some "badly shuffled" decks and approves some others as "well shuffled". But how is it possible if all $n!$ orderings of n cards have the same probability?

2 Current Best Practice

Whatever the philosophers say, statisticians have to perform their duties. Let us try to provide a description of their current "best practice" (see [7,8]).

A. *How a statistical hypothesis is applied.* First of all, we have to admit that probability theory makes no predictions but only gives recommendations: *if the probability* (computed on the basis of the statistical hypothesis) *of an event A is much smaller than the probability of an event B, then the possibility of the* *event B must be taken into consideration to a greater extent than the possibility of the event A* (assuming the consequences are equally grave). For example, if the probability of A is smaller than the probability of being killed on the street by a meteorite, we usually ignore A completely (since we have to ignore event B anyway in our everyday life).

Borel [2] describes this principle as follows: "...Fewer than a million people live in Paris. Newspapers daily inform us about the strange events or accidents that happen to some of them. Our life would be impossible if we were afraid of all adventures we read about. So one can say that from a practical viewpoint we can ignore events with probability less that one millionth... Often trying to avoid something bad we are confronted with even worse... To avoid this we must know well the probabilities of different events" (Russian ed., pp. 159–160).

B. *How a statistical hypothesis is tested.* Here we cannot say naïvely that if we observe some event that has negligible probability according to our hypothesis, we reject this hypothesis. Indeed, this would mean that any 1000-bit sequence of the outcomes would make the fair coin assumption rejected (since this specific seqeunce has negligible probability 2^{-1000}).

Here algorithmic information theory comes into play: We reject the hypothesis if we observe a *simple* event that has negligible probability according to this hypothesis. For example, if coin tossing produces thousand tails, this event is simple and has negligible probability, so we don't believe the coin is fair. Both conditions ("simple" and "negligible probability") are important: the event "the first bit is a tail" is simple but has probability $1/2$, so it does not discredit the coin. On the other hand, every sequence of outcomes has negligible probability 2^{-1000}, but if it is not simple, its appearance does not discredits the fair coin assumption.

Often both parts of this scheme are united into a statement "events with small probabilities do not happen". For example, Borel writes: "One must not be afraid to use the word "certainty" to designate a probability that is sufficiently close to 1" ([3], Russian translation, p. 7). Sometimes this statement is called "Cournot principle". But we prefer to distinguish between these two stages, because for the hypothesis testing the existence of a simple description of an event with negligible probability is important, and for application of the hypothesis it seems unimportant. (We can expect, however, that events interesting to us have simple descriptions because of their interest.)

3   Simple Events and Events Specified in Advance

Unfortunately, this scheme remains not very precise: the Kolmogorov complexity of an object x (defined as the minimal length of the program that produces x) depends on the choice of programming language; we need also to fix some way to describe the events in question. Both choices lead only to an $O(1)$ change asymptotically; however, strictly speaking, due to this uncertainty we cannot say that one event has smaller complexity than the other one. (The word "negligible" is also not very precise.) On the other hand, the scheme described, while very vague, seems to be the best approximation to the current practice.

One of the possible ways to eliminate complexity in this picture is to say that a hypothesis is discredited if we observe a very unprobable event *that was specified in advance* (before the experiment). Here we come to the following question. Imagine that you make some experiment and get a sequence of thousand bits that looks random at first. Then somebody comes and says "Look, if we consider every third bit in this sequence, the zeros and ones alternate". Will you still believe in the fair coin hypothesis? Probably not, even if you haven't thought about this event before looking at the sequence: the event is so simple that one *could* think about it. In fact, one may consider the union of all simple events that have small probability, and it still has small probability (if the bound for the complexity of a simple event is small compared to the number of coin tossing involved, which is a reasonable condition anyway). And this union can be considered as specified before the experiment (e.g., in this paper).

On the other hand, if the sequence repeats some other sequence observed earlier, we probably won't believe it is obtained by coin tossing even if this earlier sequence had high complexity. One may explain this opinion saying the the entire sequence of observations is simple since it contains repetitions; however, the first observation may be not covered by any probabilistic assumption. This could be taked into account by considering the *conditional* complexity of the event (with respect to all information available before the experiment).

The conclusion: we may remove one problematic requirement (being "simple" in a not well specified sense) and replace it by another problematic one (being specified before the observation).

4   Frequency Approach

The most natural and common explanation of the notion of probability says that probability is the limit value of frequencies observed when the number of repetitions tends to infinity. (This approach was advocated as the only possible basis for probability theory by Richard von Mises.)

However, we cannot observe infinite sequences, so the actual application of this definition should somehow deal with finite number of repetitions. And for finite number of repetitions our claim is not so strong: we do not guarantee that frequency of tails for a fair coin is exactly $1/2$; we say only that it is highly improbable that it deviates significantly from $1/2$. Since the words "highly improbably" need to be interpreted, this leads to some kind of logical circle that makes the frequency approach much less convincing; to get out of this logical circle we need some version of Cournot principle.

Technically, the frequency approach can be related to the principles explained above. Indeed, the event "the number of tails in a 1 000 000 coin tossings deviates from 500 000 more than by 100 000" has a simple description and very small probability, so we reject the fair coin assumption if such an event happens (and ignore the dangers related to this event if we accept the fair coin assumption). In this way the belief that frequency should be close to probability (if the statistical hypothesis is chosen correctly) can be treated as the consequence of the principles explained above.

5   Dynamical and Statistical Laws

We have described how the probability theory is usually applied. But the fundamental question remains: well, probability theory describes (to some extent) the behavior of a symmetric coin or dice and turns out to be practically useful in many cases. But is it a new law of nature or some consequence of the known dynamical laws of classical mechanics? Can we somehow "prove" that a symmetric dice indeed has the same probabilities for all faces (if the starting point is high enough and initial linear and rotation speeds are high enough)?

Since it is not clear what kind of "proof" we would like to have, let us put the question in a more practical way. Assume that we have a dice that is not symmetric and we know exactly the position of its center of gravity. Can we use the laws of mechanics to find the probabilities of different outcomes?

It seems that this is possible, at least in principle. The laws of mechanics determine the behavior of a dice (and therefore the outcome) if we know the initial point in the phase space (initial position and velocity) precisely. The phase space, therefore, is splitted into six parts that correspond to six outcomes. In this sense there is no uncertainty or probabilities up to now. But these six parts are well mixed since very small modifications affect the result, so if we consider a small (but not very small) part of the phase space around the initial conditions and any probability distribution on this part whose density does not change drastically, the measures of the six parts will follow the same proportion.

The last sentence can be transformed into a rigorous mathematical statement if we introduce specific assumptions about the size of the starting region in the phase space and variations of the density of the probability distribution on it. It then can be proved. Probably it is a rather difficult mathematical problem not solved yet, but at least theoretically the laws of mechanics allow us to compute the probabilities of different outcomes for a non-symmetic dice.

6   Are "Real" Sequences Complex?

The argument in the preceding section would not convince a philosophically minded person. Well, we can (in principle) compute some numbers that can be interpreted as probabilities of the outcomes for a dice, and we do not need to fix the distribution on the initial conditions, it is enough to assume that this distribution is smooth enough. But still we speak about probability distributions that are somehow externally imposed in addition to dynamical laws.

Essentially the same question can be reformulated as follows. Make 10^6 coin tosses and try to compress the resulting sequence of zeros and ones by a standard compression program, say, gzip. (Technically, you need first to convert bit sequence into a byte sequence.) Repeat this experiment (coin tossing plus gzipping) as many times as you want, and this will never give you more that 1% compression. (Such a compression is possible for less than 2^{-10000}-fraction of all sequences.) This statement deserves to be called a law of nature: it can be checked experimentally in the same way as other laws are. So the question is: does this law of nature follows from dynamical laws we know?

To see where the problem is, it is convenient to simplify the situation. Imagine for a while that we have discrete time, phase space is $[0, 1)$ and the dynamical law is

$$x \mapsto T(x) = \text{ if } 2x < 1 \text{ then } 2x \text{ else } 2x - 1.$$

So we get a sequence of states $x_0, x_1 = T(x_0), x_2 = T(x_1), \ldots$; at each step we observe where the current state is — writing 0 if x_n is in $[0, 1/2)$ and 1 if x_n is in $[1/2, 1)$.

This tranformation T has the mixing property we spoke about; however, looking at it more closely, we see that a sequence of bits obtained is just the binary representation of the initial condition. So our process just reveals the initial condition bit by bit, and any statement about the resulting bit sequence (e.g., its incompressibility) is just a statement about the initial condition.

So what? Do we need to add to the dynamical laws just one more methaphysical law saying that world was created at the random (=incompressible) state? Indeed, algorithmic transformations (including dynamical laws) cannot increase significantly the Kolmogorov complexity of the state, so if objects of high complexity exist in the (otherwise deterministic, as we assume for now) real world now, they should be there at the very beginning. (Note that it is difficult to explain the randomness observed saying that we just observe the world at random time or in a random place: the number of bits needed to encode the time and place in the world is not enough to explain an incompressible string of length, say, 10^6, if we use currently popular estimates for the size and age of the world: the logarithms of the ratios of the maximal and minimal lengths (or time intervals) that exist in nature are negligible compared to 10^6 and therefore the position in space-time cannot determine a string of this complexity.

Should we conclude then that instead of playing the dice (as Einstein could put it), God provided concentrated randomness while creating the world?

7   Randomness as Ignorance: Blum – Micali – Yao

This discussion becomes too philosophical to continue it seriously. However, there is an important mathematical result that could influence the opinion of the philosophers discussing the notions of probability and randomness. (Unfortunately, knowledge does not penetrate too fast, and I haven't yet seen this argument in traditional debates about the meaning of probability.)

This result is the existence of pseudorandom number generators (as defined by Yao, Blum and Micali; they are standard tools in computational cryptography, see, e.g., Goldreich textbook [4]). The existence is proved modulo some complexity assumtions (the existence of one-way functions) that are widely believed though not proven yet.

Let us explain what a pseudorandom number generator (in Yao – Blum – Micali) sense is. Here we use rather vague terms and oversimplify the matter, but there is a rigorious mathematics behind. So imagine a simple and fast algorithmic procedure that gets a "seed", which is a binary string of moderate size, say, 1 000 bits, and produces a very long sequence of bits out of it, say, of length 10^{10}. By necessity the output string has small complexity compared to its length (complexity is bounded by the seed size plus the length of the processing program, which we assume to be rather short). However, it may happen that the output sequences will be "indistinguishable" from truly random sequences of length 10^{10}, and in this case the transformation procedure is called pseudorandom number generator.

It sounds as a contradiction: as we have said, output sequences have small Kolmogorov complexity, and this property distinguishes them from most of the sequences of length 10^{10}. So how they can be indistinguishable? The explanation is that the difference becomes obvious only when we know the seed used for producing the sequence, but there is no way to find out this seed looking at the sequence itself. The formal statement is quite technical, but its idea is simple. Consider any simple test that looks at 10^{10}-bit string and says 'yes' or 'no' (by whatever reason; any simple and fast program could be a test). Then consider two ratios: (1) the fraction of bit strings of length 10^{10} that pass the test (among all bit strings of this length); (2) the fraction of seeds that lead to a 10^{10}-bit string that passes the test (among all seeds). The pseudorandom number generator property guarantees that these two numbers are very close.

This implies that if some test rejects most of the pseudorandom strings (produced by the generator), then it would also reject most of the strings of the same length, so there is no way to find out whether somebody gives us random or pseudorandom strings.

In a more vague language, this example shows us that randomness may be in the eye of the beholder, i.e., the randomness of an observed sequence could be the consequence of our limited computational abilities which prevent us from discovering non-randomness. (However, if somebody shows us the seed, our eyes are immediately opened and we see that the sequence has very small complexity.)

In particular, trying gzip-compression on pseudorandom sequences, we rarely would find them compressible (since gzip-compressibility is a simple test that fails for most sequences of length 10^{10}, it should also fail for most pseudorandom sequences).

So we should not exclude the possibility that the world is governed by simple dynamical laws and its initial state can be also described by several thousands of bits. In this case "true" randomness does not exist in the world, and every sequence of 10^6 coin tossing that happened or will happen in the foreseeable future produces a string that has Kolmogorov complexity much smaller than its length. However, a computationally limited observer (like ourselves) would never discover this fact.

8   Digression: Thermodynamics

The connection between statistical and dynamical laws was discussed a lot in the context of thermodynamics while discussing the second law. However, one should be very careful with exact definition and statements. For example, it is often said that the Second Law of thermodynamics cannot be derived from dynamical laws because they are time-reversible while the second law is not. On the other hand, it is often said that the second law has many equivalent formulations, and one of them claims that the perpetual motion machine of the second kind is impossible, i.e., no device can operate on a cycle to receive heat from a single reservoir and produce a net amount of work.

However, as Nikita Markaryan explained (personal communication), in this formulation the second law of thermodynamics *is* a consequence of dynamic laws. Here is a sketch of this argument. Imagine a perpetual motion machine of a second kind exists. Assume this machine is attached to a long cylinder that contains warm gaz. Fluctuations of gaz pressure provide a heat exchange between gaz and machine. On the other side machine has rotating spindle and a rope to lift some weight (due to rotation).

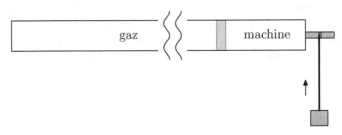

When the machine works, the gaz temperature (energy) goes down and the weight goes up. This is not enough to call the machine a perpetual motion machine of the second kind (indeed, it can contain some amount of cold substance to cool the gaz and some spring to lift the weight). So we assume that the rotation angle (and height change) can be made arbitrarily large by increasing the amount of the gaz and the length of the cylinder. We also need to specify the initial conditions of the gaz; here the natural requirement is that the machine works (as described) for most initial conditions (according to the natural probability distribibution in the gaz phase space).

Why is such a machine impossible? The phase space of the entire system can be considered as a product of two components: the phase space of the machine itself and the phase space of the gaz. The components interact, and the total energy is constant. Since the machine itself has some fixed number of components, the dimension of its component (or the number of degrees of freedom in the machine) is negligible compared to the dimension of the gaz component (resp. the number of degrees of freedom in the gaz). The phase space of the gaz is splitted into layers corresponding to different level of energy; the higher the energy is, the more volume is used, and this dependence overweights the similar dependence for the machine since the gaz has much more degrees of freedom. Since the transformation of the phase space of the entire system is measure-preserving, it is impossible that a trajectory started from a large set with high probability ends in a small set: the probability of this event does not exceed the ratio of a measures of destination and source sets in the phase space.

This argument is quite informal and ignores many important points. For example, the measure on the phase space of the entire system is not exactly a product of measures on the gaz and machine coordinates; the source set of the trajectory can have small measure if the initial state of the machine is fixed with very high precision, etc. (The latter case does not contradicts the laws of thermodynamics: if the machine use a fixed amount of cooling substance of very low temperature, the amount of work produced can be very large.) But at least these informal arguments make plausible that dynamic laws make imposiible the perpetual motion machine of the second kind (if the latter is defined properly).

9   Digression: Quantum Mechanics

Another physics topic often discussed is quantum mechanics as a source of randomness. There were many philosophical debates around quantum mechanics; however, it seems that the relation between quantum mechanical models and observations resembles the situation with probability theory and statistical mechanics; in quantum mechanics the model assigns *amplitudes* (instead of probabilities) to different outcomes (or events). The amplitudes are complex numbers and "quantum Cournot principle" says that if the (absolute value) of the amplitude of event A is smaller than for event B, then the possibility of the event B must be taken into consideration to a greater extent than the possibility of the event A (assuming the consequences are equally grave). Again this implies that we can (practically) ignore events with very small amplitudes.

The interpretation of the square of amplitude as probability can be then derived is the same way as in the case of the frequency approach. If a system is made of N independent identical systems with two outcomes 0 and 1 and the outcome 1 has amplutude z in each system, then for the entire system the amplitude of the event "the number of 1's among the outcomes deviates significantly from $N|z|^2$" is very small (it is just the classical law of large numbers in disguise).

One can then try to analyze measurement devices from the quantum mechanical viewpoint and to "prove" (using the same quantum Cournot principle) that the frequency of some outcome of measurement is close to the square of the length of the projection of the initial state to corresponding subspace outside some event of small amplitude, etc.

Acknowledgements

The material covered is not original: the topic was discussed for a long time in books and papers too numerous to mention, both the classics of the field (such as books and papers by Laplace, Borel, and Kolmogorov) and more expository writings, such as books written by Polya and Renyi. However, some remarks I haven't seen before and bear responsibility for all the errors and misunderstandings.

More information about the history of algorithmic information theory (Kolmogorov complexity, algorithmic randomness) can be found in [1].

I am indepted to many colleagues, including Vladimir Fock, Peter Gacs, Alexey Kitaev, Denis Kosygin, Leonid Levin, Leonid Levitov, Yury Makhlin, Nikita Markaryan (who explained to me the proof of the impossibility of the perpetuum mobile of the second kind), Vladimir A. Uspensky, Vladimir Vovk (who explained to me how the complexity assumption can be replaced by the requirement that the event should be specified before the experiment), Michael Vyalyi, Alexander Zvonkin, and all the participants of Kolmogorov seminar for many useful discussions.

I thank the organizers of LIX Colloquium "Reachability Problems'09" for the opportunity to present this material at the colloquium.

References

1. Bienvenu, L., Shen, A.: Algorithmic information theory and martingales, preprint, arXiv:0906.2614
2. Borel, E.: Le hazard, Alcan, Paris (1914), Russian translation: Gosizdat, Moscow – Petrograd (1923)
3. Borel, E.: Probabilité et certitude, Presse Univ. de France, Paris (1950), English translation: Probability and certainty, Walker publishers (1963), Russian translation: Fizmatgiz, Moscow (1961)
4. Goldreich, O.: Foundations of Cryptography. Basic Tools, vol. 1. Cambridge University Press, Cambridge (2007)
5. Polya, G.: Mathematics and Plausible Reasoning, vol. 1,2. Princeton University Press, Princeton (1990) (reprint)
6. Renyi, A.: Letters on probability. Wayne State University Press (1972)
7. Uspensky, V., Semenov, A., Shen, A.: Can an individual sequence of zeros and ones be random? Russian Mathematical Surveys 45(1), 121–189 (1990)
8. Shen, A.: On the logical basis of application of probability theory. In: Proc. Workshop on semiotic aspects of the formalization of intellectual activity, Telavi (Georgia, then USSR), VINITI, Moscow, pp. 144–146 (1983)

Model Checking as A Reachability Problem

Moshe Y. Vardi*

Rice University, Department of Computer Science, Houston, TX 77251-1892, U.S.A.
vardi@cs.rice.edu
http://www.cs.rice.edu/~vardi

Abstract. Model checking is a essentially a graph-searching problem. In automata-theoretic model checking we compose the design under verification with a Büchi automaton that accepts traces violating the specification. We then use graph algorithms to search the product graph for a counterexample trace. The basic theory of this approach was worked out in the 1980s, and the basic algorithms were developed during the 1990s. Both explicit and symbolic implementations, such as SPIN and and SMV, are widely used. It turns out, however, that there are still many gaps in our understanding of the algorithmic issues involved in automata-theoretic model checking. This talk covers the fundamentals of automata-theoretic model checking, reviews recent progress, and outlines areas that require further research.

References

1. Vardi, M.Y.: An automata-theoretic approach to linear temporal logic. In: Moller, F., Birtwistle, G. (eds.) Logics for Concurrency. LNCS, vol. 1043, pp. 238–266. Springer, Heidelberg (1996)
2. Vardi, M.Y.: Automata-theoretic model checking revisited. In: Cook, B., Podelski, A. (eds.) VMCAI 2007. LNCS, vol. 4349, pp. 137–150. Springer, Heidelberg (2007)

* Supported in part by NSF grants CCR-0124077, CCR-0311326, CCF-0613889, ANI-0216467, and CCF-0728882, by BSF grant 9800096, and by gift from the Intel Corporation.

Automatic Verification of Directory-Based Consistency Protocols

Parosh Aziz Abdulla[1], Giorgio Delzanno[2], and Ahmed Rezine[3]

[1] Uppsala University, Sweden
`parosh@it.uu.se`
[2] Università di Genova, Italy
`giorgio@disi.unige.it`
[3] University of Paris 7, France
`rezine.ahmed@liafa.jussieu.fr`

Abstract. We propose a symbolic verification method for directory-based consistency protocols working for an arbitrary number of controlled resources and competing processes. We use a graph-based language to specify in a uniform way both client/server interaction schemes and manipulation of directories that contain the access rights of individual clients. Graph transformations model the dynamics of a given protocol. Universally quantified conditions defined on the labels of edges incident to a given node are used to model inspection of directories, invalidation loops and integrity conditions. Our verification procedure computes an approximated backward reachability analysis by using a symbolic representation of sets of configurations. Termination is ensured by using the theory of well-quasi orderings.

1 Introduction

Several implementations of consistency and integrity protocols used in file systems, virtual memory, and shared memory multi-processors are based on client-server architectures. Clients compete to access shared resources (cache and memory lines, memory pages, open files). Each resource is controlled by a server process. In order to get access to a resource, a client needs to start a transaction with the corresponding server. Each server maintains a directory that associates to each client the access rights for the corresponding resource. In real implementations these information are stored into arrays, lists, or bitmaps and are used by the server to take decisions in response to client requests, e.g., to grant access, request invalidation, downgrade access mode or to check integrity of meta-data as in programs like fsck used to check Unix-like file-systems. Typically, a server handles a set of resources, e.g. cache lines and directory entries, whose cardinality depends on the underlying hardware/software platform. Consistency protocols however are often designed to work well independently from the number of resources to be controlled, i.e., independently from a given hardware/software configuration.

The need of reasoning about systems with an arbitrary number of resources makes verification of directory-based consistency protocols a quite challenging

O. Bournez and I. Potapov (Eds.): RP 2009, LNCS 5797, pp. 36–50, 2009.

task in general. Abstraction techniques operating on the number of resources and/or the number of clients are often applied to reduce the verification task to decidable problems for finite-state (e.g. invisible and environment abstraction in [9,13]) or Petri net-like models (e.g. counting abstraction used in [16,14,24,25]).

In this paper we propose a new approximated verification technique that operates on models in which both the number of controlled resources and of competing clients is not fixed a priori. Instead of requiring a preliminary abstraction of the model, our method makes use of powerful symbolic representations of parametric system configurations and of dynamic approximation operators applied during symbolic exploration of the state-space.

Our verification method is defined for a specification language in which system configurations are modelled by using a special type of graphs in which vertexes are partitioned into client and server nodes. Client and server nodes are labelled with a set of "states" that represent the current state of the corresponding processes. Labelled edges are used both to define client/server transactions and to describe the local information maintained by each server (e.g. a directory is represented by the set of edges incident to a given server node).

Protocol rules are specified here by rewriting rules that update the state of a node and of one of its incident edges. This very restricted form of graph rewriting naturally models asynchronous communication mechanisms. Furthermore, we admit here guards defined by means of universally quantified conditions on the set of labels of edges of a given node. This kind of guards is important to model operations like inspection of a directory or invalidation cycles without need of abstracting them by means of atomic operations like broadcast in [16,14]. In order to reason about *parameterized formulations* of consistency protocols we consider here systems in which the size of graphs (number of nodes and edges) is not bounded a priori.

The advantage of working with conditional graph rewriting is twofold. On one side it gives us enough power to formally describe each step of consistency protocols like the full-map coherence protocol [20] in a very detailed way. On the other side it allows us to define (and implement) our verification method at a very abstract level by using graph transformations.

Related Work. Parameterized verification methods based on finite-state abstractions have been applied to safety properties of consistency protocols and mutual exclusion algorithms. Among these, we mention the *invisible invariants* method [9,21] and the *environment abstraction* method [13]. Counting abstraction and Petri net-like analysis techniques are considered, e.g., in [16,14,24,25].

Differently from all the previous works, our algorithm is based on graph constraints that allow us to symbolically represent infinite-sets of configurations without need of preliminary finitization of parameters like number of clients, servers, resources, and size of directories. We apply instead dynamic approximation techniques to deal with universally quantified global conditions. We recently used a similar approach for systems with flat configurations (i.e. words) and with a single global context [7]. The new graph-based algorithm is a generalization of the approach in [7]. Indeed, the symbolic configurations we used in [7] can be viewed as graphs with a single server node and no edges, since global conditions are tested directly on the current process states.

Furthermore, the approximation we propose in this paper is more precise than the monotonic abstraction used to deal with global conditions in our previous work [3,4,5,6,1] (i.e. deletion of processes that do not satisfy the condition). Indeed, consistency property like reachability of a server in state *bad* in the case study presented in Section 4 always return false positives using monotonic abstraction (by deleting all edges that are not in Q we can always move to *bad*). In synthesis the new approach can be viewed as an attempt of introducing more precise approximated verification algorithms for parameterized systems while retaining good features of approaches like counting and monotonic abstraction in [16,3] like termination properties based on the theory of well-quasi orderings.

Concerning verification algorithms for graph rewriting systems, we are only aware of the works in [18,22]. We use here different type of graph specifications (e.g. we consider universal quantification on incoming edges) and a different notion of graph-based symbolic representation (i.e. a different entailment relation) with respect to those applied to leader election and routing protocols in [18,22].

2   A Client/Server Abstract Model

To represent configurations of client/server protocols, we define a special kind of bipartite graphs. Let Λ_s be a finite set of *server node labels*, Λ_c a finite set of *client node labels*, and Λ_e a finite set of *edge labels*. Furthermore, for $n \in \mathcal{N}$ let $\overline{n} = \{1, \ldots, n\}$. A c/s-graph is a tuple $G = (n_c, n_s, E, \lambda_c, \lambda_s, \lambda_e)$, where $\overline{n_s}$ is the set of server nodes, $\overline{n_c}$ is the set of client nodes, $E \subseteq \overline{n_s} \times \overline{n_c}$ is a set of edges connecting a server with a set of clients, and a client with at most one server (i.e. for each $j \in \overline{n_c}$ we require that there exists at most one edge incident in j in E), and $\lambda_c : \overline{n_c} \to \Lambda_c$, $\lambda_s : \overline{n_s} \to \Lambda_s$, and $\lambda_e : E \to \Lambda_e$ are labelling functions.

In the rest of the paper we use the operations on c/s-graphs defined in Fig. 1. A client/server system is a tuple $S = (I, R)$ consisting of a (possibly infinite) set I of c/s-graphs (initial configurations), and a finite set R of rules. We consider here a restricted type of graph rewriting rules to model both the interaction between clients and servers and the manipulation of directories viewed as the set of incident edges in a given server nodes.

The rules have the general form $l \Rightarrow r$ where l is a pattern that has to match (the labels and structure) of a subgraph in the current configuration in order for the rule to be fireable and r describes how the subgraph is rewritten as the effect of the application of the rule. In this paper we are interested in modelling asynchronous communication patterns. Thus, we consider the following patterns: the empty graph · (it matches with any graph); $(\!(\ell)\!)$ that denotes an isolated client node with label ℓ; $(\!(\ell)\!)$ that denotes a server node with label ℓ, $[\ell] \xleftarrow{\sigma}$ that denotes a client node with label ℓ and incident edge with label σ; $(\!(\ell)\!) \xleftarrow{\sigma}$ that denotes a server node with label ℓ and an incident edge with label σ.

Furthermore, we also admit a special type of rules in which the rewriting step can be applied to a given server node if a universally quantified condition

Given a graph $G = (n_c, n_s, E, \lambda_c, \lambda_s, \lambda_e)$, we define:

- $\mathsf{edges}(G) = E$, $\mathsf{edges}_s(i, G) = \{e \mid e = (i, j) \in E\}$ for $i \in \overline{n_s}$, and $\mathsf{edges}_c(j, G) = \{e \mid e = (i, j) \in E\}$ for $j \in \overline{n_c}$; $\mathsf{label}_e(e, G) = \lambda_e(e)$ for $e \in E$ and $\mathsf{label}_e(i, G) = \{\lambda_e(e) \mid e \in \mathsf{edges}_s(i, G)\}$ for $i \in \overline{n_s}$;
- $\mathsf{add}_e(e, \sigma, G) = (n_c, n_s, E \cup \{e\}, \lambda_c, \lambda_s, \lambda'_e)$ where $\lambda'_e(e) = \sigma$, $\lambda'_e(o) = \lambda_e(o)$ in all other cases;
- $\mathsf{update}_e(e \leftarrow \sigma, G) = (n_c, n_s, E, \lambda_c, \lambda_s, \lambda'_e)$ where $\lambda'_e(e) = \sigma$, and $\lambda'_e(o) = \lambda_e(o)$ in all other cases;
- $\mathsf{del}_e(e, G) = (n_c, n_s, E', \lambda_c, \lambda_s, \lambda'_e)$, where $E' = E \setminus \{e\}$, $\lambda'_e(o) = \lambda_e(o)$ for $o \in E'$.
- $\mathsf{nsize}_c(G) = n_c$, and $\mathsf{label}_c(i, G) = \lambda_c(i)$ for $i \in \overline{n_c}$;
- $\mathsf{add}_c(P, G) = (n_c + 1, n_s, E, \lambda'_c, \lambda_s, \lambda_e)$ where $\lambda'_c(n_c + 1) = P$ and $\lambda'_c(o) = \lambda_c(o)$ in all other cases;
- $\mathsf{update}_c(i_1 \leftarrow P_1, \ldots, i_m \leftarrow P_m, G) = (n_c, n_s, E, \lambda'_c, \lambda_s, \lambda_e)$ where $\lambda'_c(i_k) = P_k$ for $k : 1, \ldots, m$, and $\lambda'_c(o) = \lambda_c(o)$ in all other cases;
- $\mathsf{del}_c(i, G) = (n_c - 1, n_s, E', \lambda'_c, \lambda_s, \lambda'_e)$ where, given the mapping $h_i : \overline{n_c} \to \overline{n_c - 1}$ defined as $h_i(j) = j$ for $j < i$ and $h_i(j) = j - 1$ for $j > i$, $E' = \{(k, h_i(l)) \mid (k, l) \in E\}$, $\lambda'_c(k) = \lambda_c(p)$ for each $k \in \overline{n_c - 1}$ such that $k = h_i(p)$ and $p \in \overline{n_c}$, $\lambda'_e((k, l)) = \lambda_e((k, q))$ for $(k, l) \in E'$ such that $l = h_i(q)$ for $q \in \overline{n_c}$, $\lambda'_x(o) = \lambda_x(o)$ in all other cases for $x \in \{e, c\}$;
- nsize_s, label_s, add_s, update_s, and del_s are defined for server nodes in a way similar to the client node operations.

Fig. 1. Definition of basic graph operations

on the labels of the corresponding incident edges is satisfied. Specifically, we consider the rule schemes illustrated in Fig. 2, where ℓ and ℓ' are node labels of appropriate type, σ and σ' are edge labels, and $\forall Q$ is a condition with $Q \subseteq \Lambda_e$.

With the first two types of rules, we can non-deterministically add a new node to the current graph (e.g. to dynamically inject new servers and clients). With rule *start_transaction*, we non-deterministically select a server and a client (not connected by an already existing edge) add a new edge between them in the current graph (e.g. to dynamically establish a new communication). With rules of types *client/server_steps*, we update the labels of a node with label ℓ and one of its incident edges (non-deterministically chosen) with label σ (e.g. to define asynchronous communication protocols). With rule *test*, we update the node label of a server node i only if all edges incident to i have labels in the

$$\cdot \Rightarrow (\!(\ell)\!) \qquad\qquad (new\_client\_node)$$
$$\cdot \Rightarrow (\ell) \qquad\qquad (new\_server\_node)$$
$$(\!(\ell)\!) \Rightarrow [\ell'] \xleftarrow{\sigma} \qquad\qquad (start\_transaction)$$
$$(\ell) \xleftarrow{\sigma} \Rightarrow (\ell') \xleftarrow{\sigma'} \qquad\qquad (server\_step)$$
$$[\ell] \xleftarrow{\sigma} \Rightarrow [\ell'] \xleftarrow{\sigma'} \qquad\qquad (client\_step)$$
$$(\ell) \Rightarrow (\ell') : \forall Q \qquad\qquad (test)$$
$$[\ell] \xleftarrow{\sigma} \Rightarrow (\!(\ell')\!) \qquad\qquad (stop\_transaction)$$

Fig. 2. Rewriting rules with conditions on egdes

set $Q \subseteq \Lambda_e$. With rule *stop_transaction*, we non-deterministically select a client node with label ℓ and incident edge with label σ, and delete such an edge from the current graph (e.g. to terminate a conversation).

It is important to remark that the a server has not direct access to the local state of a client. Thus, it cannot check conditions on the global sets of its current clients in an atomic way. For checking global conditions a server can however check the set of it incident edges, i.e., a local snapshot of the current condition of clients. A consistency protocol should guarantee that the information on the edges (directory) is consistent with the current state of clients.

2.1   Transition Relation

Let G be a c/s-graph. The formula $\forall Q$ is satisfied in server node i if $\mathsf{label}_e(i, G) \subseteq Q$. The operational semantics is defined via a binary relation $\Rightarrow_r$ on c/s-graphs such that $G_0 \Rightarrow G_1$ if and only if one of the following conditions hold:

- r is a *new_client_node* rule and $G_1 = \mathsf{add}_c(\ell, G_0)$;
- r is a *new_server_node* rule and $G_1 = \mathsf{add}_s(\ell, G_0)$;
- r is a *server_step* rule and there exist nodes i and j in G_0 with edge $e = (i, j) \in \mathsf{edges}(G)$ such that $\mathsf{label}_s(i, G_0) = \ell$, $\mathsf{label}_e(e, G_0) = \sigma$, $G_1 = \mathsf{update}_e(e \leftarrow \sigma', \mathsf{update}_s(i \leftarrow \ell', G_0))$;
- r is an *client_step* rule and there exist nodes i and j in G_0 with edge $e = (i, j) \in \mathsf{edges}(G)$ such that $\mathsf{label}_n(j, G_0) = \ell$, $\mathsf{label}_e(e, G_0) = \sigma$, $G_1 = \mathsf{update}_e(e \leftarrow \sigma', \mathsf{update}_c(j \leftarrow \ell', G_0))$;
- r is a *start_transaction* rule, there exists in G_0 a client node j with no incident edges in E such that $\mathsf{label}_c(j, G_0) = \ell$, and $G_1 = \mathsf{add}_e((i, j), \sigma, \mathsf{update}_c(j \leftarrow \ell', G_0))$ for a server node i;
- r is a *stop_transaction* rule, there exist nodes i and j in G_0 such that $\mathsf{label}_c(j, G_0) = \ell$, $e = (i, j) \in \mathsf{edges}(G_0)$, $\mathsf{label}_e(e, G_0) = \sigma$, and $G_1 = \mathsf{del}_e(e, \mathsf{update}_c(j \leftarrow \ell', G_0))$.
- r is a *test* rule, there exist node i in G_0 such that $\mathsf{label}_s(i, G_0) = \ell$, $\mathsf{label}_e(i, G_0) \subseteq Q$, and $G_1 = \mathsf{update}_s(j \leftarrow \ell', G_0)$.

Finally, we define $\Rightarrow$ as $\bigcup_{r \in R} \Rightarrow_r$.

Example 1. As an example, consider a set of labels Λ_n partitioned in the two sets $\Lambda_c = \{idle, wait, use\}$ and $\Lambda_s = \{ready, check, ack\}$, and a set of edge labels $\Lambda_e = \{req, pend, inv, lock\}$. The following set R of rules models a client-server protocol (with any number of clients and servers) in which a server grants the use of a resource after invalidating the client that is currently using it.

$$(r_1) \ (\!|idle|\!) \Rightarrow [\!wait]\! \xleftrightarrow{req}$$
$$(r_2) \ (\!(ready)\!) \xleftrightarrow{req} \Rightarrow (\!(check)\!) \xleftrightarrow{pend}$$
$$(r_3) \ (\!(check)\!) \xleftrightarrow{lock} \Rightarrow (\!(check)\!) \xleftrightarrow{inv}$$
$$(r_4) \ (\!(check)\!) \Rightarrow (\!(ack)\!) : \forall\{pend, req\}$$
$$(r_5) \ (\!(ack)\!) \xleftrightarrow{pend} \Rightarrow (\!(ready)\!) \xleftrightarrow{lock}$$
$$(r_6) \ [\!use]\! \xleftrightarrow{inv} \Rightarrow (\!|idle|\!)$$
$$(r_7) \ [\!wait]\! \xleftrightarrow{lock} \Rightarrow [\!use]\! \xleftrightarrow{lock}$$

With rule r_1 a client non-deterministically creates a new edge connecting to a server. With rule r_2 a server processes a request by changing the edge to *pending*, and then moves to state *check*. With rule r_3 a server sends invalidation messages to the client that is currently using the resource (marked with the special edge *lock*). With rule r_4 a server moves to the acknowledge step whenever all incident edges have state different from *lock* and *inv*. With rule r_5 a server grants the pending request. With rule r_6 a clients releases the resource upon reception of an invalidation request. With rule r_7 a waiting client moves to state *use*.

Now, let us consider an initial graph G_0 with one server node with label *ready* and two client nodes with label *idle*. Then, the following sequence (of graphs) represents an evolution of the graph system (G_0, R):

$$G_0 = \langle idle \rangle, \langle idle \rangle, \langle ready \rangle \Rightarrow \langle idle \rangle, [wait] \xleftarrow{req} \langle ready \rangle \Rightarrow$$
$$[wait] \xleftarrow{req} \langle ready \rangle \xleftarrow{req} [wait] \Rightarrow [wait] \xleftarrow{pend} \langle check \rangle \xleftarrow{req} [wait] \Rightarrow$$
$$[wait] \xleftarrow{pend} \langle ack \rangle \xleftarrow{req} [wait] \Rightarrow [wait] \xleftarrow{lock} \langle ready \rangle \xleftarrow{req} [wait] \Rightarrow$$
$$[use] \xleftarrow{lock} \langle ready \rangle \xleftarrow{req} [wait] \Rightarrow [use] \xleftarrow{lock} \langle check \rangle \xleftarrow{pend} [wait] \Rightarrow$$
$$[use] \xleftarrow{inv} \langle check \rangle \xleftarrow{pend} [wait] \Rightarrow \langle inv \rangle, \langle check \rangle \xleftarrow{pend} [wait] \Rightarrow$$
$$\langle inv \rangle, \langle ack \rangle \xleftarrow{pend} [wait] \Rightarrow \langle inv \rangle, \langle ready \rangle \xleftarrow{lock} [wait] \Rightarrow$$
$$[inv], [ready] \xleftarrow{lock} [use]$$

2.2   Pattern Reachability

In this paper we are interested in studying reachability of graphs containing specific patterns (subgraphs). Patterns can be used to represent bad configurations of a graph system. In Example 1 any graph containing the pattern $[use] \xleftarrow{\sigma} \langle ready \rangle \xleftarrow{\sigma'} [use]$, for $\sigma, \sigma' \in \Lambda_e$, represents a violation to the exclusive use of a resource controlled by a server node.

To formally define the notion of pattern, we introduce an ordering $\preceq$ on c/s-graphs such that $G \preceq G'$ iff $n_c = \mathsf{nsize}_c(G) \leq m_c = \mathsf{nsize}_c(G')$, $n_s = \mathsf{nsize}_s(G) \leq m_s = \mathsf{nsize}_s(G')$, and there exist injective mappings $h_c : \overline{n_c} \to \overline{m_c}$ and $h_s : \overline{n_s} \to \overline{m_s}$ such that

- $\mathsf{label}_c(i, G) = \mathsf{label}_c(h_c(i), G')$ for $i : 1, \ldots, n_c$,
- $\mathsf{label}_s(i, G) = \mathsf{label}_s(h_s(i), G')$ for $i : 1, \ldots, n_s$,
- for each $e = (i, j) \in \mathsf{edges}(G)$, $e' = (h_s(i), h_c(j)) \in \mathsf{edges}(G')$ and $\mathsf{label}_e(e, G) = \mathsf{label}_e(e', G')$.

A set of c/s-graphs $U \subseteq C$ is *upward closed* with respect to $\preceq$ if $c \in U$ and $c \preceq c'$ implies $c' \in U$. For a c/s-graph G, we use $\widehat{G}$ to denote the upward closure of G, i.e., the set $\{G' | G \preceq G'\}$. For sets of c/s-graphs $D, D' \subseteq C$ we use $D \Rightarrow D'$ to denote that there are $G \in D$ and $G' \in D'$ with $G \Rightarrow G'$.

The *Pattern Reachability Problem* for graph systems is defined as follows:

PATTERN REACHABILITY PROBLEM (PRP)

Instance

  - A graph system $\mathcal{P} = (I, R)$.
  - A finite set C_F of c/s-graphs

Question $G_0 \Rightarrow^* \widehat{C_F}$ for $G_0 \in I$?

Typically, $\widehat{C_F}$ (which is an infinite set) is used to characterize sets of *bad* configurations which we do not want to occur during the execution of the system. In such a case, the system is safe iff $\widehat{C_F}$ is not reachable. Therefore, checking safety properties amounts to solving PRP (i.e., to the reachability of upward closed sets). In [7] we show that control state reachability for counter machines can be reduced to PRP. From this property, it follows that PRP is undecidable.

3   Approximated Verification Algorithm

In this section we propose an approximated verification algorithm based on the notion of *graph constraints*, a special symbolic representation of an infinite sets of c/s-graphs.

A *graph constraint* (gc) is a graph $\Psi = (n_c, n_s, E, \rho_c, \rho_s, \rho_e)$, with client nodes $\{1, \ldots, n_c\}$, server nodes $\{1, \ldots, n_s\}$, edges in $E \subseteq \overline{n_s} \times \overline{n_c}$, and labels defined by maps $\rho_c : \overline{n_c} \to \Lambda_c$, $\rho_s : \overline{n_s} \to (\Lambda_s \times 2^{\Lambda_e})$, and $\rho_e : E \to \Lambda_e$.

Notice that, in a graph constraint Ψ, the label of a server node is a pair (ℓ, Q) where ℓ is a node label and $Q \subseteq \Lambda_e$ is a subset of edge labels, called *padding set*. In this section we adapt the operations on c/s-graphs to graph constraints. Specifically, given $\Psi = (n_c, n_s, E, \rho_c, \rho_s, \rho_e)$, $i \in \overline{n_s}$, $j \in \overline{n_c}$, $\rho_s(i) = (\ell, Q)$, $\rho_c(j) = \ell'$, and $e \in E$, then $\mathsf{label}_s(i, \Psi) = \ell$, $\mathsf{label}_p(i, \Psi) = Q$, $\mathsf{label}_c(j, \Psi) = \ell'$, and $\mathsf{label}_e(e, \Psi) = \rho_e(e)$. The other operations are defined as for c/s-graphs.

For a graph constraint Ψ to be well-formed (wfgc), we require that $\mathsf{label}_e(i, \Psi) \subseteq \mathsf{label}_p(i, \Psi)$ for each $i \in \overline{n_s}$.

Let Ψ be a wfgc, and G be a c/s-graph. In order to define the denotation of a wfgc Ψ we introduce the relation $\preceq$ such that, given a c/s-graph G, $\Psi \preceq G$ iff $n_c = \mathsf{nsize}_c(\Psi) \leq m_c = \mathsf{nsize}_c(G)$, $n_s = \mathsf{nsize}_s(\Psi) \leq m_s = \mathsf{nsize}_s(G)$, and there exist injective mappings $h_c : \overline{n_c} \to \overline{m_c}$ and $h_s : \overline{n_s} \to \overline{m_s}$ such that

  - $\mathsf{label}_c(i, \Psi) = \mathsf{label}_c(h_c(i), G)$ for $i : 1, \ldots, n_c$,
  - $\mathsf{label}_s(i, \Psi) = \mathsf{label}_s(h_s(i), G)$ and $\mathsf{label}_e(h_s(i), G') \subseteq \mathsf{label}_p(i, \Psi) = Q$ for $i : 1, \ldots, n_s$;
  - for each $e = (i, j) \in \mathsf{edges}(\Psi)$, $e' = (h_s(i), h_c(j)) \in \mathsf{edges}(G)$ and $\mathsf{label}_e(e, \Psi) = \mathsf{label}_e(e', G)$.

The denotation of a graph constraint Ψ is then defined as $[\![\Psi]\!] = \{G \mid G \text{ is a c/s-graph}, \Psi \preceq G\}$.

Approximated Predecessor Relation The set of predecessors of a set S of c/s-graphs computed with respect to a rule r is defined as

$$pre_r(S) = \{G \mid G \Rightarrow_r S\}$$

Given a wfgc Ψ we now define a relation $\rightsquigarrow_r$ working on wfgc's that we use to overapproximate the set $[\![\Psi]\!] \cup pre_r([\![\Psi]\!])$. We consider here the union of these two sets in order to be able to discard graph constraints that denote graphs already contained in $[\![\Psi]\!]$. For brevity, we describe here the computation of predecessors for rules of the form *server-step*, *client-step*, and *test*. The complete definition is given in [26]. Specifically, for graph constraints Ψ, with $n_s = \mathsf{nsize}_s(\Psi)$ and $n_c = \mathsf{nsize}_c(\Psi)$, and Ψ', and a rule $r \in R$, the relation $\Psi \rightsquigarrow_r \Psi'$ is defined as follows:

server-step: r is the rule $(\!(\ell)\!) \xleftarrow{\;\sigma\;} \; \Rightarrow \; (\!(\ell')\!) \xleftarrow{\;\sigma'\;}$ and one of the following conditions hold

- $i \in \overline{n_s}$, $j \in \overline{n_c}$, $e = (i,j) \in \mathsf{edges}(\Psi)$, $\mathsf{label}_s(i, \Psi) = \ell'$, $\mathsf{label}_e(e) = \sigma'$, and

$$\Psi' = \mathsf{update}_e(e, \sigma, \mathsf{update}_s(i \leftarrow (\ell, Q), \Psi))$$

where $Q = \mathsf{label}_p(i) \cup \{\sigma\}$.

In this case we update the label of an existing edge (i,j) and of the node i with the labels σ and ℓ, respectively. They represent the preconditions for firing the rule. Furthermore, we augment the padding set of i with label σ. Notice that here we apply an approximation, i.e., as soon as we add σ we allow any number of occurrences of edges with label σ but we do not count them. The label of client node j is not modified.

- $i \in \overline{n_s}$, $j \in \overline{n_c}$, $\mathsf{edges}(j, \Psi) = \emptyset$ (j has no incident edges), $\mathsf{label}_s(i, \Psi) = \ell'$, $\sigma' \in \mathsf{label}_p(i, \Psi)$, and

$$\Psi' = \mathsf{add}_e((i,j), \sigma, \mathsf{update}_s(i \leftarrow (\ell, Q), \Psi))$$

where $Q = \mathsf{label}_p(i, \Psi) \cup \{\sigma\}$.

Although not explicitly present, we assume here that the edge (i,j) with label σ' is in the upward closure of Ψ (this can happen only if j is not involved in other explicit edges). We add the edge (i,j) with label σ since its presence is a precondition for the firing the rule. Furthermore, we update the label of i as in the first case.

- $i \in \overline{n_s}$, $\mathsf{label}_s(i, \Psi) = \ell'$, $\sigma' \in \mathsf{label}_p(i, \Psi)$, and

$$\Psi' = \mathsf{add}_e((i, n_c + 1), \sigma, \mathsf{add}_c(\ell'', \mathsf{update}_s(i \leftarrow (\ell, Q), \Psi)))$$

where $Q = \mathsf{label}_p(i, \Psi) \cup \{\sigma\}$, and ℓ'' is non-deterministically chosen from Λ_c. Although not explicitly present, we assume here that both the client node $n_c + 1$ (with some label taken from Λ_c) and the edge $(i, n_c + 1)$ with label σ are in the upward closure of Ψ. We add them to Ψ since their presence is a precondition for the firing of r. We update the label of i as in the other two cases. Notice that the dimension of the graph constraint is increased by one, since we insert the new node $n_c + 1$.

For this kind of rules, there are two remaining cases to consider (the edge and the server node, or the edge and both server and client nodes are not explicitly present in Ψ). However these cases give rise to graph constraints that are redundant with respect to Ψ. Thus, we can discard them without loss of precision (we recall that our aim is to symbolically represent $[\![\Psi]\!] \cup pre_r([\![\Psi]\!]))$.

client-step: r is the rule $[\ell] \xleftrightarrow{\sigma} \Rightarrow [\ell'] \xleftrightarrow{\sigma'}$ and one of the following conditions hold

- $i \in \overline{n_s}$, $j \in \overline{n_c}$, $e = (i,j) \in \mathsf{edges}(\Psi)$, $\mathsf{label}_c(j, \Psi) = \ell'$, $\mathsf{label}_e(e) = \sigma'$, and

$$\Psi' = \mathsf{update}_e(e, \sigma, \mathsf{update}_s(i \leftarrow (\mathsf{label}_s(i, \Psi), Q), \mathsf{update}_c(j \leftarrow \ell, \Psi)))$$

where $Q = \mathsf{label}_p(i, \Psi) \cup \{\sigma\}$.

In this case we update the label of an existing edge (i,j) and of the node j with the labels σ and ℓ as a precondition for the firing of the rule r. Furthermore, we add σ to the set of admitted edge labels of server node i.

- $i \in \overline{n_s}$, $j \in \overline{n_c}$, $\mathsf{edges}(j, \Psi) = \emptyset$ (j has no incident edges), $\mathsf{label}_c(j, \Psi) = \ell'$, $\sigma' \in \mathsf{label}_p(i, \Psi)$, and

$$\Psi' = \mathsf{add}_e((i,j), \sigma, \mathsf{update}_s(i \leftarrow (\mathsf{label}_s(i, \Psi), Q), \mathsf{update}_c(j \leftarrow \ell, \Psi)))$$

where $Q = p(i, \Psi) \cup \{\sigma\}$.

Although not explicitly present, we assume here that the edge (i,j) is in the upward closure of Ψ. We add the edge (i,j) with label σ since its presence is a precondition for the firing of the rule. Furthermore, we update the label of i and j as in the first case.

- $j \in \overline{n_c}$, $\mathsf{edges}(j, \Psi) = \emptyset$ (j has no incident edges), $\mathsf{label}_c(j, \Psi) = \ell'$, and

$$\Psi' = \mathsf{add}_e((n_s + 1, j), \sigma, \mathsf{add}_s((\ell'', \Lambda_e), \mathsf{update}_c(j \leftarrow \ell, \Psi)))$$

where $\ell'' \in \Lambda_s$. Although not explicitly present, we assume here that the edge $(n_s + 1, j)$, for a new server node $n_s + 1$ with a label in Λ_s, is in the upward closure of Ψ. We add the node and the edge with label σ since its presence is a precondition for firing the rule. Furthermore, we update the label of j as in the first case.

- $i \in \overline{n_s}$, $\sigma' \in \mathsf{label}_p(i, \Psi)$, and

$$\Psi' = \mathsf{add}_e((i, n_c + 1), \sigma, \mathsf{add}_c(\ell, \mathsf{update}_s(i \leftarrow (\mathsf{label}_s(i, \Psi), Q), \Psi)))$$

where $Q = \mathsf{label}_p(i, \Psi) \cup \{\sigma\}$.

Although not explicitly present, we assume here that both the node $n_c + 1$ and the edge $(i, n_c + 1)$ are in the upward closure of Ψ. We add it with label σ to the set of edges and update the label of i including σ in the set of admitted edges. Notice that there are remaining cases (client, server, and edge are not explicitly present in Ψ). However these cases give rise to a graph constraint that is redundant with respect to Ψ. Thus, we can discard it without loss of precision (we recall that our aim is to symbolically represent $[\![\Psi]\!] \cup pre_r([\![\Psi]\!])$).

Test: r is the rule $(\!(\ell)\!) \Rightarrow (\!(\ell')\!)$: $\forall Q$ and one of the following conditions hold

- $i \in \overline{n_s}$, $\mathsf{label}_s(i, \Psi) = \ell'$, $R = \mathsf{label}_p(i, \Psi) \cap Q$, $\mathsf{label}_e(e) \in R$ for each $e \in \mathsf{edges}(i, \Psi)$, and
$$\Psi' = \mathsf{update}_s(i \leftarrow (\ell, R), \Psi)$$

In this rule the padding $\mathsf{label}_p(i, \Psi)$ associated to a node i with label ℓ' plays a crucial role. We first check that the current set of labels of edges incident to i is contained into the intersection R of $\mathsf{label}_p(i, \Psi)$ and Q. If this condition is satisfied, we restrict the padding of node i to be the set R (precondition for firing the rule) and update the label of i to ℓ. This rule cannot be applied whenever there are edges in $\mathsf{edges}(i, \Psi)$ with labels not in R. If R is the empty set, then the node i must be isolated.

Given a wfgc Ψ, we define $\Psi \rightsquigarrow$ as the set $\{\Psi' \mid \Psi \overset{r}{\rightsquigarrow} \Psi', \ r \in R\}$. The following property then holds.

Lemma 1. $([\![\Psi]\!] \cup pre([\![\Psi]\!])) \subseteq ([\![\Psi]\!] \cup [\![\Psi \rightsquigarrow]\!])$.

Entailment Test We now define an entailment relation $\sqsubseteq$ used to compare denotations of graph constraints. Let Ψ and Ψ' be two wfgc such that $\mathsf{nsize}_c(\Psi) = n_c$, $\mathsf{nsize}_s(\Psi) = n_s$, $\mathsf{nsize}_c(\Psi') = m_c$, and $\mathsf{nsize}_s(\Psi') = m_s$. The relation $\Psi \sqsubseteq \Psi'$ holds iff $n_c \leq m_c$, $n_s \leq m_s$, and there exist injective mappings $h_c : \overline{n_c} \to \overline{m_c}$ $h_s : \overline{n_s} \to \overline{m_s}$ such that

- $\mathsf{label}_s(i, \Psi) = \mathsf{label}_s(h_s(i), \Psi')$ for $i \in \overline{n_s}$,
- $\mathsf{label}_c(j, \Psi) = \mathsf{label}_c(h_c(j), \Psi')$ for $j \in \overline{n_c}$,
- $\mathsf{label}_p(h_s(i), \Psi') \subseteq \mathsf{label}_p(i, \Psi)$ for $i \in \overline{n_s}$,
- for each $e = (i, j) \in E$, $e' = (h_s(i), h_c(j)) \in E'$ and $\mathsf{label}_e(e, \Psi) = \mathsf{label}_e(e', \Psi')$.

The following property then holds.

Lemma 2. *Given Ψ and Ψ', $\Psi \sqsubseteq \Psi'$ implies $[\![\Psi']\!] \subseteq [\![\Psi]\!]$.*

We naturally extend the entailment relation to finite sets of constraints as follows. Given two sets of graph constraints Φ, Φ', $\Phi \sqsubseteq \Phi'$ iff for each $\Psi' \in \Phi'$ there exists $\Psi \in \Phi$ such that $\Psi \sqsubseteq \Psi'$.

3.1 Backward Reachability

We use the relation $\rightsquigarrow$ to define a symbolic backward reachability algorithm for approximating solutions to PRP. We start with a finite set Φ_F of graph constraints denoting an infinite set of bad graph configurations. We generate a sequence $\Phi_0, \Phi_1, \Phi_2, \cdots$ of finite sets of constraints such that $\Phi_0 = \Phi_F$, and $\Phi_{j+1} = \Phi_j \cup (\Phi_j \rightsquigarrow)$. Since $[\![\Phi_0]\!] \subseteq [\![\Phi_1]\!] \subseteq [\![\Phi_2]\!] \subseteq \cdots$, the procedure terminates when we reach a point j where $\Phi_j \sqsubseteq \Phi_{j+1}$. Notice that the termination condition implies that $[\![\Phi_j]\!] = (\bigcup_{0 \leq i \leq j} [\![\Phi_i]\!])$. By Lemma 1, Φ_j denotes an over-approximation of the set of all predecessors of $[\![\Phi_F]\!]$. This means that if $(I \cap [\![\Phi_j]\!]) = \emptyset$, then there exists no $G \in [\![\Phi_F]\!]$ with $G_0 \Rightarrow^* G$ for $G_0 \in I$. Thus, the procedure can be used as a semi-test for checking PRP.

According to the general results in [2], the termination of our (approximated) symbolic backward reachability procedure can be ensured by proving that the entailment relation of graph constraints is a well-quasi ordering (wqo). The latter property follows from the fact that a c/s-graph with n_s server nodes and n_c client nodes can be given an alternative representation as a bag of tuples of a special form. A wfgc can be represented as a *bag* (multiset) containing the (multiset) of isolated client nodes in G together with tuples of the form (s_i, Q_i, M_i) for $i \in \{1, \ldots, n_s\}$, where

- $s_i \in \Lambda_s$ is the label of the server node i,
- $Q_i \in 2^{\Lambda_c}$ is the padding associated to i,
- if i has client nodes $j_1, \ldots, j_{k_i}$ connected to it M_i is a bag $\{p_1, \ldots, p_{k_i}\}$ such that $p_l = (\sigma_l, c_l)$, where σ_l is the label of the edge incident to node j_l and c_l is the label of node j_l.

Given bags m_1 and m_2 associated resp. to wfgc's G_1 and G_2, $m_1 \leq m_2$ holds if: each isolated client node in m_1 can be injected into an isolated client node in m_2; each tuple (s, Q, M) in m_1 can be injected into a tuple (s', Q', M') in m_2 such that $s = s'$, $Q' \subseteq Q$ and M is contained into M' (multiset containment). From closure properties of wqo's under bag and tuple composition operators, we have that $\leq$ is a wqo. Furthermore, we have that $m_1 \leq m_2$ implies $G_1 \sqsubseteq G_2$. Thus, the entailment relation of graph constraints is a well-quasi ordering (wqo).

4   A Case Study

We have implemented a prototype version, SYMGRAPH [26], of our approximated verification algorithm and tested on a model of the *full-map cache coherence protocol* described in [20]. This protocol is defined for a multiprocessor with shared memory and local caches in which the memory controller maintains a directory for each memory line with information about its use, i.e., the line is shared between different caches or used in exclusive mode by a given cache. The directory is used to optimize the invalidation and downgrade phase required when a processor sends a new request for exclusive or shared use. Memory controllers associate a special flag *mode_ex* to each line to remember when the line is in exclusive use (i.e. without need to inspect the full-map).

For reason of space, we only give the key ideas behind our model of this protocol. The initial graph configurations consist of any number of isolated client nodes with label *inv* (cache controller for a given line in state *invalid*) and server nodes in state *idle* (memory controller for a given line in state *idle*).

During its life cycle the same cache line can be associated to different memory lines. However, at any given instant a cache line is either invalid or contains a copy of a given memory block. A memory line however can be copied into several cache lines. A cache controller in invalid state sends a request for exclusive or shared access using one of the two following rules

$$\langle\langle inv \rangle\rangle \Rightarrow [wait] \xleftrightarrow{req\_ex} \qquad \langle\langle inv \rangle\rangle \Rightarrow [wait] \xleftrightarrow{req\_sh}$$

A cache controller in *wait* state moves to *exclusive* (*shared*) state upon reception of message *ex* (*sh*) along the edge that connects it to the memory controller as specified by the rules

$$[wait] \xleftrightarrow{ex} \Rightarrow [exclusive] \xrightarrow{ex} \qquad [wait] \xleftrightarrow{sh} \Rightarrow [shared] \xrightarrow{sh}$$

A cache controller moves to invalid state upon reception of a *req_inv* message as specified by the rules

$$[shared] \xleftrightarrow{req\_inv} \Rightarrow \langle\langle inv \rangle\rangle \qquad [exclusive] \xleftrightarrow{req\_inv} \Rightarrow \langle\langle inv \rangle\rangle$$

A cache controller in exclusive state moves to shared state upon reception of a *req_dg* message as specified by the rule

$$[exclusive] \xleftrightarrow{req\_dg} \Rightarrow [shared] \xrightarrow{sh}$$

A memory controller that receives a *req_ex* message from a channel (edge) updates the label on the corresponding egde to *pend* and then moves to state *inv_loop* as specified by the rule

$$\langle\langle idle \rangle\rangle \xleftrightarrow{req\_ex} \Rightarrow \langle\langle inv\_loop \rangle\rangle \xleftrightarrow{pend}$$

While in the *inv_loop* state, the memory controller sends an invalidation request *req_inv* to all caches connected to it with edges marked *sh* or *ex* as specified by rules

$$\langle\langle inv\_loop \rangle\rangle \xleftrightarrow{ex} \Rightarrow \langle\langle inv\_loop \rangle\rangle \xleftrightarrow{req\_inv} \qquad \langle\langle inv\_loop \rangle\rangle \xleftrightarrow{sh} \Rightarrow \langle\langle inv\_loop \rangle\rangle \xleftrightarrow{req\_inv}$$

The memory controller moves to state *ack_inv* after testing that all requests have been processed by the cache controllers connected to it as specified by the rule

$$\langle\langle inv\_loop \rangle\rangle \Rightarrow \langle\langle ack\_inv \rangle\rangle : \forall\{pend, req\_sh, req\_ex\}$$

In state *ack_inv* the memory controller grants the access to the waiting cache controller connected to it with an edge labelled *pend* by using the rule

$$\langle\langle ack\_inv \rangle\rangle \xleftrightarrow{pend} \Rightarrow \langle\langle idle_{ex} \rangle\rangle \xrightarrow{ex}$$

The state *idle_{ex}* is used here to remember that a cache is using the line in exclusive state (the *mode_ex* flag in [20] is set to 1).

Request for shared access are treated in a similar way. However a downgrade request req_dg instead of an invalidation message is sent to all caches connected with an edge to the memory controller. The invalidation/downgrade loop can be avoided when the request is processed in the special state $idle_{ex}$. The complete client/server model for this protocol is given in [26].

For this case study we consider the following pattern reachability problems (PRP) that represent violation to mutual exclusion and consistency properties. For proving mutual exclusion, we consider a number of PRPs defined by taking as target set of configurations the denotations of a graph with a memory node m and two cache nodes c, c' both linked to m (to model the fact that the cache lines stored in c, c' correspond to that controlled by m) and such that c, c' and the corresponding incident edges have a conflicting state. Formally, we consider graph constraints defined as follows $G = \{1, 2, \{e = (1, 1), e' = (1, 2)\}, \rho_c, \rho_s, \rho_e\}$ where $\rho_s(1) = (\ell, \Lambda_e)$, $\ell \in \{idle, idle_{ex}\}$, $\rho_c(1) = ex$, $\rho_e(e) = ex$, and either $(\rho_c(2) = ex$ and $\rho_e(e') = ex)$ or $(\rho_c(2) = sh$ and $\rho_e(e') = sh)$.

We can also formulate other types of consistency properties as PRP. For instance, to check that $idle_{ex}$ corresponds to a memory (line) state in which one cache controller has exclusive access we can first add the following rule:

$$(idle_{ex}) \Rightarrow (bad) : \forall Q$$

where bad is a new memory label and Q is an appropriate set of edge labels (see [26]). The graph $G = \{1, 0, \emptyset, \rho_s, \emptyset, \emptyset\}$ with $\rho_s(1) = (bad, \Lambda_e)$ represents the set of violations to the consistency of the $mode\_ex$ flag with respect to the current state of the fullmap. Our prototype implementation of the symbolic backward procedure with graph constraints verifies the above mentioned properties automatically [26].

5   Conclusions and Related Work

We have presented a new algorithm for parameterized verification of directory-based consistency protocols based on a graph representation (graph constraints) of infinite collections of configurations. The algorithm computes an overapproximation of the set of backward reachable configurations denoted by an initial set of graph constraints. We apply the new algorithm to different versions of a non-trivial case-study discussed in [20]. We plan to investigate how to extend this approach to deal with parameterized systems in which some of the nodes play both the role of server and client in different instances of a given communication protocol.

References

1. Abdulla, P.A., Bouajjani, A., Cederberg, J., Haziz, F., Rezine, A.: Monotonic abstraction for programs with dynamic memory heaps. In: Gupta, A., Malik, S. (eds.) CAV 2008. LNCS, vol. 5123, pp. 341–354. Springer, Heidelberg (2008)

2. Abdulla, P.A., Čerāns, K., Jonsson, B., Tsay, Y.-K.: General decidability theorems for infinite-state systems. LICS 1996, 313–321 (1996)
3. Abdulla, P.A., Ben Henda, N., Delzanno, G., Rezine, A.: Regular model checking without transducers. In: Grumberg, O., Huth, M. (eds.) TACAS 2007. LNCS, vol. 4424, pp. 721–736. Springer, Heidelberg (2007)
4. Abdulla, P.A., Ben Henda, N., Delzanno, G., Rezine, A.: Handling parameterized systems with non-atomic global conditions. In: Logozzo, F., Peled, D.A., Zuck, L.D. (eds.) VMCAI 2008. LNCS, vol. 4905, pp. 22–36. Springer, Heidelberg (2008)
5. Abdulla, P.A., Delzanno, G., Rezine, A.: Parameterized verification of infinite-state processes with global conditions. In: Damm, W., Hermanns, H. (eds.) CAV 2007. LNCS, vol. 4590, pp. 145–157. Springer, Heidelberg (2007)
6. Abdulla, P.A., Delzanno, G., Haziza, F., Rezine, A.: Parameterized tree systems. In: Suzuki, K., Higashino, T., Yasumoto, K., El-Fakih, K. (eds.) FORTE 2008. LNCS, vol. 5048, pp. 69–83. Springer, Heidelberg (2008)
7. Abdulla, P.A., Delzanno, G., Rezine, A.: Approximated Context-sensitive Analysis for Parameterized Verification FORTE 2009 (2009)
8. Abdulla, P.A., Jonsson, B., Nilsson, M., d'Orso, J.: Regular model checking made simple and efficient. In: Brim, L., Jančar, P., Křetínský, M., Kucera, A. (eds.) CONCUR 2002. LNCS, vol. 2421, pp. 116–130. Springer, Heidelberg (2002)
9. Arons, T., Pnueli, A., Ruah, S., Xu, J., Zuck, L.: Parameterized verification with automatically computed inductive assertions. In: Berry, G., Comon, H., Finkel, A. (eds.) CAV 2001. LNCS, vol. 2102, pp. 221–234. Springer, Heidelberg (2001)
10. Boigelot, B., Legay, A., Wolper, P.: Iterating transducers in the large. In: Hunt Jr., W.A., Somenzi, F. (eds.) CAV 2003. LNCS, vol. 2725, pp. 223–235. Springer, Heidelberg (2003)
11. Bouajjani, A., Habermehl, P., Vojnar, T.: Abstract regular model checking. In: Alur, R., Peled, D.A. (eds.) CAV 2004. LNCS, vol. 3114, pp. 372–386. Springer, Heidelberg (2004)
12. Bouajjani, A., Muscholl, A., Touili, T.: Permutation Rewriting and Algorithmic Verification. Inf. and Comp. 205(2), 199–224 (2007)
13. Clarke, E., Talupur, M., Veith, H.: Environment abstraction for parameterized verification. In: Emerson, E.A., Namjoshi, K.S. (eds.) VMCAI 2006. LNCS, vol. 3855, pp. 126–141. Springer, Heidelberg (2005)
14. Delzanno, G.: Constraint-Based Verification of Parameterized Cache Coherence Protocols. FMSD 23(3), 257–301 (2003)
15. Emmi, M., Jhala, R., Kohler, E., Majumdar, R.: Verifying reference counted objects. In: TACAS 2009 (to appear, 2009)
16. Esparza, J., Finkel, A., Mayr, R.: On the Verification of Broadcast Protocols. LICS (1999)
17. Finkel, A., Schnoebelen, P.: Well-structured transition systems everywhere! TCS 256(1-2), 63–92 (2001)
18. Joshi, S., König, B.: Applying the graph minor theorem to the verification of graph transformation systems. In: Gupta, A., Malik, S. (eds.) CAV 2008. LNCS, vol. 5123, pp. 214–226. Springer, Heidelberg (2008)
19. Kesten, Y., Maler, O., Marcus, M., Pnueli, A., Shahar, E.: Symbolic model checking with rich assertional languages. TCS 256, 93–112 (2001)
20. Pong, F., Dubois, M.: Correctness of a Directory-Based Cache Coherence Protocol: Early Experience. In: SPDP 1993, pp. 37–44 (1993)

21. Pnueli, A., Ruah, S., Zuck, L.: Automatic deductive verification with invisible invariants. In: Margaria, T., Yi, W. (eds.) TACAS 2001. LNCS, vol. 2031, pp. 82–97. Springer, Heidelberg (2001)
22. Saksena, M., Wibling, O., Jonsson, B.: Graph Grammar Modeling and Verification of Ad Hoc Routing Protocols. In: Ramakrishnan, C.R., Rehof, J. (eds.) TACAS 2008. LNCS, vol. 4963, pp. 18–32. Springer, Heidelberg (2008)
23. Vardi, M.Y., Wolper, P.: An automata-theoretic approach to automatic program verification. LICS 1986, 332–344 (1986)
24. Yavuz-Kahveci, T., Bultan, T.: A symbolic manipulator for automated verification of reactive systems with heterogeneous data types. STTT 5(1), 15–33 (2003)
25. Yavuz-Kahveci, T., Bultan, T.: Verification of parameterized hierarchical state machines using action language verifier. In: MEMOCODE 2005, pp. 79–88 (2005)
26. Symgraph: http://www.disi.unige.it/person/DelzannoG/Symgraph/

On Yen's Path Logic for Petri Nets

Mohamed Faouzi Atig and Peter Habermehl

LIAFA, CNRS & Univ. of Paris 7, Case 7014, 75205 Paris 13, France
{atig,haberm}@liafa.jussieu.fr

Abstract. In [13], Yen defines a class of formulas for paths in Petri nets and claims that its satisfiability problem is EXPSPACE-complete. In this paper, we show that in fact the satisfiability problem for this class of formulas is as hard as the reachability problem for Petri nets. Moreover, we salvage almost all of Yens results by defining a fragment of this class of formulas for which the satisfiability problem is EXPSPACE-complete by adapting his proof.

1 Introduction

Petri nets (or equivalently, vector addition systems) are one of the most popular mathematical model for the representation and analysis of parallel processes [2]. The reachability problem for Petri nets is one of the key problems in the area of automatic verification since many other problems (e.g. the liveness problem) were shown to be recursively equivalent to the reachability problem (see [4,6]). It is well known that the reachability problem for Petri nets is decidable [11,10,7,8]. However, the precise complexity of the reachability problem for Petri nets remains open (all known algorithms require non-primitive recursive space). The best known lower bound is exponential space given by Lipton in [9].

On the other hand, to obtain a uniform approach for deciding and studying the complexity of many Petri nets problems, Yen has defined in [13] a class of formulas for paths in Petri nets, each of them is of the form:

$$\exists \mu_1,\ldots,\mu_n \exists \sigma_1,\ldots,\sigma_n \left(\left(\mu_0 \xrightarrow{\sigma_1} \mu_1 \xrightarrow{\sigma_2} \cdots \mu_{n-1} \xrightarrow{\sigma_n} \mu_n \right) \wedge \phi(\mu_1,\ldots,\mu_n,\sigma_1,\ldots,\sigma_n) \right)$$

where ϕ belongs to a certain set of predicates (constraining the markings and transitions sequence occurring in the formula) and μ_0 is the initial marking of the given Petri net. The above formula means that any marking μ_i can be reached from μ_{i-1} ($1 \le i \le n$) in the Petri net through the firing sequence of transitions σ_i and such that the predicate $\phi(\mu_1,\ldots,\mu_n,\sigma_1,\ldots,\sigma_n)$ holds. In [13], Yen claims that the satisfiability problem for such class of formulas (i.e., the problem of, given a Petri net and a formula, determining whether there exists a path in the Petri net satisfying the given formula) is complete for exponential space. This class of formulas is a useful and an interesting one since it is powerful enough to express many Petri nets properties. In particular, Petri nets problems such as *boundedness*, *coverability*, *fair-nontermination*, and *regularity detection* are reducible to the satisfiability problem for this class of formulas [13]. Moreover, Yen's result has been cited and used in several papers [1,15,14,5,3].

O. Bournez and I. Potapov (Eds.): RP 2009, LNCS 5797, pp. 51–63, 2009.

In this paper, we prove that the reachability problem for Petri nets is in fact as hard as the satisfiability problem for this class of formulas. However, we can salvage almost all of Yen's results by defining an interesting and useful fragment of this class of formulas of paths in Petri nets for which the satisfiability problem is EXPSPACE-complete. In proving the upper bound for this fragment, we correct an error in the proof given in [13]. Essentially, the fragment requires the marking μ_n to be bigger than μ_1 allowing the path satisfying the formula to be repeated.

The *regularity detection* problem can not be expressed using our fragment and therefore to the best of our knowledge it's complexity (given as EXPSPACE in [13]) remains unclear.

2 Preliminaries

Let $\mathbb{Z}$ (resp. $\mathbb{N}$) denote the set of (resp. nonnegative) integers, and $\mathbb{Z}^k$ (resp. $\mathbb{N}^k$) the set of vectors of k (resp. nonnegative) integers.

Let Σ be a finite alphabet. We denote by Σ^* (resp. Σ^+) the set of all finite (resp. non empty) words over Σ and by ε the empty word. We use $|\Sigma|$ to denote the number of symbols in Σ. We denote by $\mathbb{N}^\Sigma$ (resp. $\mathbb{Z}^\Sigma$) the set of all mappings from Σ to $\mathbb{N}$ (resp. to $\mathbb{Z}$) and by $\mathbf{0}$ the mapping that maps every symbol in Σ to 0. (Notice that $\mathbb{N}^\Sigma \subseteq \mathbb{Z}^\Sigma$.)

Let Σ and Σ' be two finite alphabets such that $\Sigma \subseteq \Sigma'$. Given a mapping μ in $\mathbb{Z}^{\Sigma'}$, we write $\mu|_\Sigma$ to denote the mapping that maps every $a \in \Sigma$ to $\mu(a)$.

Let Σ be a finite alphabet and μ_1 and μ_2 two mappings from Σ to $\mathbb{Z}$, we denote by $\mu_1 \odot \mu_2$ the inner product of μ_1 and μ_2 (i.e., $\mu_1 \odot \mu_2 = \sum_{a \in \Sigma} \mu_1(a)\mu_2(a)$).

The *Parikh image* $\sharp : \Sigma^* \mapsto \mathbb{N}^\Sigma$ maps a word w to a mapping $\sharp(w)$ from Σ to $\mathbb{N}$ such that $\sharp(w)(a)$ is the number of occurrences of a in w.

A Petri net $\mathcal{N} = (P, T, F, \mu_0)$ consists of a finite set P of places, a finite set T of transitions disjoint from P, a weight function $F : (P \times T) \cup (T \times P) \mapsto \mathbb{N}$, and an initial marking $\mu_0 \in \mathbb{N}^P$. A marking is a map from P to $\mathbb{N}$. For a marking μ of $\mathcal{N}$ and a place $p \in P$, we say that, in μ, the place p contains $\mu(p)$ tokens. For markings μ, μ', we write $\mu + \mu'$ for the marking obtained by point wise addition of place contents. We write $\mu \le \mu'$ if $\mu(p) \le \mu'(p)$ for all $p \in P$, and we write $\mu < \mu'$ if $\mu \le \mu'$ and $\mu(p') \ne \mu'(p')$ for some place $p' \in P$. The marking $\mathbf{0}$ maps every $p \in P$ to 0.

A transition $t \in T$ is enabled at a marking μ if and only if $F(p, t) \le \mu(p)$ for all $p \in P$. If a transition t is enabled at a marking μ, then t may be fired yielding to a new marking μ' defined as follows: $\mu'(p) = \mu(p) - F(p, t) + F(t, p)$ for all $p \in P$. We then write $\mu \xrightarrow{t} \mu'$ to denote that the marking μ' is reached from μ by firing the transition t. A sequence of transitions $\sigma = t_1 \cdots t_n$ is a firing sequence from μ_0 if and only if $\mu_0 \xrightarrow{t_1} \mu_1 \xrightarrow{t_2} \cdots \xrightarrow{t_n} \mu_n$ for some sequence of markings $\mu_1, \ldots, \mu_n$. Furthermore, we call $\mu_0 \xrightarrow{\sigma} \mu_n$ a *computation* of $\mathcal{N}$.

A marking μ is said to be reachable in $\mathcal{N}$ if and only if $\mu = \mu_0$ or there is some $\sigma \in T^+$ such that $\mu_0 \xrightarrow{\sigma} \mu$. The reachability problem for a Petri net $\mathcal{N}$ is, for a given marking μ, to determine whether μ is reachable in $\mathcal{N}$.

We define the size $s(\mathcal{N})$ of a Petri Net $\mathcal{N}$ as in [13], i.e. numbers are encoded in binary and the size of a Petri Net is then $\lceil logk \rceil + \lceil logr \rceil$ (where k is the number of places and r is the number of transitions) + the sum of the sizes of the elements of F +

the size of μ_0. The firing of a transition may result in removing (or adding) $2^{s(\mathcal{N})}$ tokens from (to) a place.

Finally, we recall that the reachability problem for Petri nets is decidable.

Theorem 1 ([11,9]). *The reachability problem for Petri nets is* EXPSPACE-*hard.*

3 Yen's Path Logic for Petri Nets

In this section we define the class of path formulas for Petri nets considered by Yen in [13]. We essentially follow his definitions. Let $\mathcal{N} = (P, T, F, \mu_0)$ be a Petri net. Each path formula consists of the following elements:

1. *Variables:* There are two types of variables, namely, marking variables $\mu_1, \mu_2, \ldots$ and variables for transition sequences $\sigma_1, \sigma_2, \ldots$, where each μ_i denotes a marking of $\mathcal{N}$ and each σ_i denotes a finite sequence of transition rules.
2. *Terms:* Terms are defined recursively as follows:
   - For every mapping $\mathbf{c} \in \mathbb{N}^P$, $\mathbf{c}$ is a term.
   - For all $j > i$, $\mu_j - \mu_i$ is a term, where μ_i and μ_j are marking variables.
   - $\mathcal{T}_1 + \mathcal{T}_2$ and $\mathcal{T}_1 - \mathcal{T}_2$ are terms if $\mathcal{T}_1$ and $\mathcal{T}_2$ are terms. (Consequently, every mapping $\mathbf{c} \in \mathbb{Z}^P$ is also a term.)
3. *Atomic predicates:* There are two types of atomic predicates, namely, transition predicates and marking predicates.
   (a) Transition predicates:
      - $\mathbf{z} \odot \sharp(\sigma_i) \geq c$ and $\mathbf{z} \odot \sharp(\sigma_i) > c$ are predicates, where $i > 1$, $c \in \mathbb{N}$ is a constant, and $\mathbf{z}$ is a mapping from T to $\mathbb{Z}$.
      - $\sharp(\sigma_1)(t) \geq c$ and $\sharp(\sigma_1)(t) \leq c$ are predicates, where $c \in \mathbb{N}$ is a constant and $t \in T$ is a transition rule of $\mathcal{N}$.
   (b) Marking predicates:
      - $\mu(p) \geq z$ and $\mu(p) > z$ are predicates, where μ is a marking variable, $p \in P$ is a place of $\mathcal{N}$, and $z \in \mathbb{Z}$ is an integer.
      - $\mathcal{T}_1(p_1) = \mathcal{T}_2(p_2)$, $\mathcal{T}_1(p_1) < \mathcal{T}_2(p_2)$, and $\mathcal{T}_1(p_1) > \mathcal{T}_1(p_2)$ are predicates, where $\mathcal{T}_1$ and $\mathcal{T}_2$ are terms and $p_1, p_2 \in P$ are two places of $\mathcal{N}$.

A *predicate* is either a marking predicate, a transition predicate, or of the form $\bigvee_{1 \leq i \leq k} \bigwedge_{1 \leq j \leq m_i} \varphi_i^j$ (i.e., in the disjunctive normal form[1]) where each φ_i^j is a marking or transition predicate. A *Path formula* f is a formula of the form:

$$\exists \mu_1, \ldots, \mu_n \exists \sigma_1, \ldots, \sigma_n \left((\mu_0 \xrightarrow{\sigma_1} \mu_1 \xrightarrow{\sigma_2} \cdots \mu_{n-1} \xrightarrow{\sigma_n} \mu_n) \wedge \phi(\mu_1, \ldots, \mu_n, \sigma_1, \ldots, \sigma_n) \right)$$

where ϕ is a predicate.

Given a Petri net $\mathcal{N}$ and a path formula f, we use $\mathcal{N} \models f$ to denote that f is true in $\mathcal{N}$. The *satisfiability problem* for such a path formula f asks if there exists an execution of $\mathcal{N}$ of the form $\mu_0 \xrightarrow{\sigma_1} \mu_1 \xrightarrow{\sigma_2} \cdots \mu_{n-1} \xrightarrow{\sigma_n} \mu_n$ such that $\phi(\mu_1, \ldots, \mu_n, \sigma_1, \ldots, \sigma_n)$ holds. In this case, we say $\mathcal{N}$ satisfies the path formula f (i.e., $\mathcal{N} \models f$).

The following result can be shown following [13].

---

[1] In [13], a predicate can be any positive boolean combination of predicates. In fact, we can show that our results (in particular Theorem 3) still hold even if we consider this general case.

Lemma 1. *Given a Petri net $\mathcal{N} = (P, T, F, \mu_0)$ and a formula f, we can construct in polynomial time, a Petri net $\mathcal{N}' = (P', T', F', \mu'_0)$ and a formula f' containing no transition predicates such that $\mathcal{N} \models f$ if and only if $\mathcal{N}' \models f'$.*

Therefore, it is sufficient to consider formulas containing only marking predicates in order to decide satisfiability.

4 From the Reachability Problem to the Satisfiability Problem

In the following, we prove that the reachability problem for Petri nets is polynomially reducible to the satisfiability problem for path formulas.

Theorem 2. *Given a Petri net $\mathcal{N} = (P, T, F, \mu_0)$ and a marking $\mu \in \mathbb{N}^P$, we can construct, in polynomial time, a Petri net $\mathcal{N}' = (P', T', F', \mu'_0)$ and a path formula f such that the marking μ is reachable by $\mathcal{N}$ if and only if $\mathcal{N}' \models f$.*

The rest of this section is devoted to the proof of Theorem 2. We first construct a Petri net $\mathcal{N}' = (P', T', F', \mu'_0)$ with $P \subseteq P'$ such that a marking $\mu \in \mathbb{N}^P$ is reachable in $\mathcal{N}$ if and only if there is a marking $\mu' \in \mathbb{N}^{P'}$ such that $\mu'|_P = \mu$ and μ' is reachable in $\mathcal{N}'$. Then, we construct a path formula f for the Petri net $\mathcal{N}'$ such that $\mathcal{N}'$ satisfies the formula f if and only if there is a marking $\mu' \in \mathbb{N}^{P'}$ such that $\mu'|_P = \mu$ and μ' is reachable in $\mathcal{N}'$. This implies that the marking μ is reachable in $\mathcal{N}$ if and only if $\mathcal{N}'$ satisfies the formula f.

4.1 Constructing the Petri Net $\mathcal{N}'$

The Petri net $\mathcal{N}' = (P', T', F', \mu'_0)$ is built up from $\mathcal{N}$ in a way described in Fig. 1. Formally, $\mathcal{N}'$ contains all transitions and places of $\mathcal{N}$. In addition, three new places q_0, q_1, q_2 and two new transitions r_1 and r_2 are added to $\mathcal{N}'$. Initially, $\mathcal{N}'$ has just one token in the place q_0 and $\mu_0(p)$ tokens in each place $p \in P$ (i.e., $\mu'_0(q_0) = 1, \mu'_0(q_1) = \mu'_0(q_2) = 0$, and $\mu'_0|_P = \mu_0$). The transition r_1 (resp. r_2) consumes exactly one token from the place q_0 (resp. q_1) and produces only one token in the place q_1 (resp. q_2), i.e., $F'(q_0, r_1) = F'(r_1, q_1) = 1$ (resp. $F'(q_1, r_2) = F'(r_2, q_2) = 1$) and 0 otherwise. A transition $t \in T$ of $\mathcal{N}'$ consumes exactly one token from the place q_2 and $F(p, t)$ tokens from each place $p \in P$, and produces one token in the place q_2 and $F(t, p)$ token in each place $p \in P$. Formally, we have that for every $i \in \{0, 1\}, F'(q_i, t) = F'(t, q_i) = 0, F'(q_2, t) = F(t, q_2) = 1$ and for every $p \in P, F'(p, t) = F(p, t)$ and $F'(t, p) = F(t, p)$.

Then, the relation between $\mathcal{N}$ and $\mathcal{N}'$ is giving by the following lemma.

Lemma 2. *Let $\mu \in \mathbb{N}^P$ be a marking and $\sigma \in T^+$ be a sequence of transitions of $\mathcal{N}$. $\mu_0 \xrightarrow{\sigma} \mu$ is a computation of $\mathcal{N}$ if and only if $\mu'_0 \xrightarrow{r_1} \mu'_1 \xrightarrow{r_2\sigma} \mu'$ is a computation of $\mathcal{N}'$ where:*

- *$\mu'_1(q_0) = \mu'_1(q_2) = 0, \mu'_1(q_1) = 1,$ and $\mu'_1|_P = \mu_0$.*
- *$\mu'(q_0) = \mu'(q_1) = 0, \mu'(q_2) = 1,$ and $\mu'|_P = \mu$.*

As an immediate consequence of Lemma 2, we get the following result.

Corollary 1. *A marking $\mu \in \mathbb{N}^P$ is reachable by $\mathcal{N}$ if and only if there is a sequence of transitions $\sigma \in T^*$ such that $\mu'_0 \xrightarrow{r_1} \mu'_1 \xrightarrow{r_2\sigma} \mu'$ is a computation of $\mathcal{N}'$ with $\mu'(q_0) = \mu'(q_1) = 0, \mu'(q_2) = 1,$ and $\mu'|_P = \mu$.*

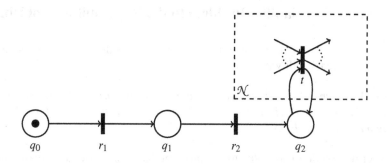

Fig. 1. The Petri net $\mathcal{N}'$

4.2 Constructing the Path Formula f for the Petri Net $\mathcal{N}'$

In the following, we construct a path formula f such that $\mathcal{N}'$ satisfies f if and only if the marking μ is reachable by $\mathcal{N}$. The path formula f is of the following form:

$$\exists \mu_1, \mu_2 \exists \sigma_1, \sigma_2 \left((\mu_0' \xrightarrow{\sigma_1} \mu_1 \xrightarrow{\sigma_2} \mu_2) \wedge \phi_1(\mu_1) \wedge \phi_2(\mu_1, \mu_2) \right)$$

where ϕ_1 and ϕ_2 are two predicates.

The predicate $\phi_1(\mu_1) = \mu_1(q_1) \geq 1$ says that only the transition rule r_1 is fired during the sequence of transitions σ_1 (i.e., $\sigma_1 = r_1$). This implies that the marking μ_1 is defined as follows: $\mu_1(q_0) = \mu_1(q_2) = 0$, $\mu_1(q_1) = 1$, and $\mu_1|_P = \mu_0$.

$$\phi_2(\mu_1, \mu_2) = (\mu_2(q_2) \geq 1) \wedge \bigwedge_{p \in P} \left(\mu_2(p) - \mu_1(p) = \mu(p) - \mu_0(p) \right)$$

Fig. 2. The predicate $\phi_2(\mu_1, \mu_2)$

The predicate ϕ_2 (given by Fig. 2) says that for each place $p \in P$, the difference between the number of tokens added to p and the number of tokens taken from p, during firing the sequence of transitions σ_2, is equal to $\mu(p) - \mu_0(p)$. This implies that $\mu_2(q_0) = \mu_2(q_1) = 0$, $\mu_2(q_2) = 1$, and $\mu_2|_P = \mu$.

Lemma 3. *The Petri net $\mathcal{N}'$ satisfies the path formula f if and only if $\mu_0' \xrightarrow{r_1} \mu_1 \xrightarrow{r_2\sigma} \mu_2$ is a computation of $\mathcal{N}'$ where $\sigma \in T^*$, and μ_1 and μ_2 are two markings defined as follows:*

- $\mu_1(q_0) = \mu_1(q_2) = 0$, $\mu_1(q_1) = 1$, *and* $\mu_1|_P = \mu_0$.
- $\mu_2(q_0) = \mu_2(q_1) = 0$, $\mu_2(q_2) = 1$, *and* $\mu_2|_P = \mu$.

As an immediate consequence of Lemma 3 and Corollary 1, we have that:

Corollary 2. *The marking μ is reachable by $\mathcal{N}$ if and only if $\mathcal{N}' \models f$.*

Hence, the reachability problem for Petri nets is polynomially reducible to the satisfiability problem for the class of path formulas.

Remark 1. It is also possible to reduce the reachability problem for Petri nets to the satisfiability problem for a path formula that contains only transition predicates.

5   From the Satisfiability Problem to the Reachability Problem

In this section, we show that the satisfiability problem for path formulas is polynomially reducible to the reachability problem for Petri nets.

Theorem 3. *Given a Petri net $\mathcal{N} = (P, T, F, \mu_0)$ and a path formula f, we can construct, in polynomial time, a Petri net $\mathcal{N}' = (P', T', F', \mu_0')$ such that $\mathcal{N} \models f$ iff the empty marking $\mathbf{0}$ is reachable by $\mathcal{N}'$.*

The rest of this section is devoted to the proof of Theorem 3. Let us suppose that the path formula f is of the form:

$$\exists \mu_1, \ldots, \mu_n \exists \sigma_1, \ldots, \sigma_n \left((\mu_0 \xrightarrow{\sigma_1} \mu_1 \xrightarrow{\sigma_2} \mu_2 \cdots \mu_{n-1} \xrightarrow{\sigma_n} \mu_n) \wedge \phi(\mu_1, \ldots, \mu_n) \right)$$

We assume, without loss of generality, that ϕ contains only marking predicates (see Lemma 1). Furthermore, because $\phi_1 \vee \phi_2$ is satisfiable if and only if ϕ_1 is satisfiable or ϕ_2 is satisfiable, we can assume that ϕ is of the form $\phi = \varphi_1 \wedge \cdots \wedge \varphi_m$ where for every $i \in \{1, \ldots, m\}$, φ_i is a marking predicate of the form[2]:

$$y_0^i + \sum_{j=1}^{n} (\mathbf{y_j^i} \odot \mu_j) \leq z_0^i + \sum_{j=1}^{n} (\mathbf{z_j^i} \odot \mu_j)$$

where $\mathbf{y_j^i}$ and $\mathbf{z_j^i}$ are two mappings from P to $\mathbb{N}$ and y_0^i and z_0^i are two nonnegative integers.

For every $i \in \{1, \ldots, m\}$, let ρ_i^- and ρ_i^+ be two mappings from $(\mathbb{N}^P)^n$ to $\mathbb{N}$ such that: for every given sequence of markings $\mu_1, \ldots, \mu_n$ of $\mathcal{N}$, we have that $\rho_i^-(\mu_1, \ldots, \mu_n) = y_0^i + \sum_{j=1}^{n} (\mathbf{y_j^i} \odot \mu_j)$ and $\rho_i^+(\mu_1, \ldots, \mu_n) = z_0^i + \sum_{j=1}^{n} (\mathbf{z_j^i} \odot \mu_j)$.

In the following, we compute a Petri net $\mathcal{N}' = (P', T', F', \mu_0')$ such that $\mathcal{N} \models f$ if and only if the empty marking $\mathbf{0}$ is reachable by $\mathcal{N}'$. A computation of $\mathcal{N}'$ can be divided in two phases: First, $\mathcal{N}'$ guesses a sequence of markings $\mu_1, \ldots, \mu_n$ of $\mathcal{N}$ such that: $\mu_0 \xrightarrow{\sigma_1} \mu_1 \xrightarrow{\sigma_2} \mu_2 \cdots \mu_{n-1} \xrightarrow{\sigma_n} \mu_n$ is a computation of $\mathcal{N}$ for some $\sigma_1, \ldots, \sigma_n \in T^+$. Then, in the second phase, $\mathcal{N}'$ checks for every $i \in \{1, \ldots, m\}$, if $\rho_i^-(\mu_1, \ldots, \mu_n) \leq \rho_i^+(\mu_1, \ldots, \mu_n)$ (i.e., the predicate $\phi(\mu_1, \ldots, \mu_n)$ is true).

The Petri net $\mathcal{N}'$ contains all places of $\mathcal{N}$. In addition, the new places $q_1, \ldots, q_n$ and $\bar{q}$ are added to $\mathcal{N}'$ such that the total number of token in all these places is always less or equal to one. The sequence of places $q_1, \ldots, q_n$ is used during the first phase to guess the sequence of markings $\mu_1, \ldots, \mu_n$ of $\mathcal{N}$, while, the place $\bar{q}$ is used during the second phase to check if the predicate $\phi(\mu_1, \ldots, \mu_n)$ is true for the guessed sequence of markings. Moreover, for every $i \in \{1, \ldots, m\}$, the Petri net $\mathcal{N}'$ has two places s_i^- and s_i^+ to keep track (in some increasing way with respect to the sequence of guessed markings) of the value of ρ_i^- and ρ_i^+, respectively, such that a marking $\mu \in \mathbb{N}^{P'}$ is reachable by $\mathcal{N}'$, if and only if one of the two following cases holds:

---

[2] According to [13] (Lemma 3.4, page 130) any marking predicate can be represented as a predicate of this form. Moreover, it is easy to see that the set of predicates of this form is slightly more general than the set of marking predicates defined in section 3.

- *During the first phase:* If $\mu(q_j) = 1$ for some $j \in \{1,\ldots,n\}$ (only one token in the place q_j and, consequently, the places $q_1,\ldots,q_{j-1},q_{j+1},\ldots,q_n$, and $\bar{q}$ are empty), then there is a sequence of markings $\mu_1,\ldots,\mu_{j-1} \in \mathbb{N}^P$ of $\mathcal{N}$ such that:

  1. $\mu_0 \xrightarrow{\sigma_1} \mu_1 \xrightarrow{\sigma_2} \mu_2 \cdots \mu_{j-1} \xrightarrow{\sigma_j} \mu|_P$ is a computation of $\mathcal{N}$, and
  2. for every $i \in \{1,\ldots,m\}$, the number of tokens in the places s_i^- and s_i^+ is $\rho_i^-(\mu_1,\ldots,\mu_{j-1},\mu|_P,\ldots,\mu|_P)$ and $\rho_i^+(\mu_1,\ldots,\mu_{j-1},\mu|_P,\ldots,\mu|_P)$, respectively.

- *During the second phase:* If $\mu(\bar{q}) = 1$ (only one token in the place $\bar{q}$ and, consequently, the places $q_1,\ldots,q_n$ are empty), then there is a sequence of markings $\mu_1,\ldots,\mu_n \in \mathbb{N}^P$ of $\mathcal{N}$ such that:

  1. $\mu_0 \xrightarrow{\sigma_1} \mu_1 \xrightarrow{\sigma_2} \mu_2 \cdots \mu_{n-1} \xrightarrow{\sigma_n} \mu_n$ is a computation of $\mathcal{N}$,
  2. for every place $p \in P$, the number of tokens in the place p is less or equal to $\mu_n(p)$, and
  3. for every $i \in \{1,\ldots,m\}$, there is a nonnegative number c_i such that the number of tokens in s_i^- (resp. s_i^+) is equal to $\rho_i^-(\mu_1,\ldots,\mu_n) - c_i$ (resp. less or equal to $\rho_i^+(\mu_1,\ldots,\mu_n) - c_i$).

Initially, the Petri net $\mathcal{N}'$ has $\mu_0(p)$ tokens in each place $p \in P$, one token in the place q_1, 0 token in the set of places $q_2,\ldots,q_n,\bar{q}$, and for every $i \in \{1,\ldots,m\}$, the places s_i^- and s_i^+ have y_0^i and z_0^i tokens, respectively.

The set of transitions of $\mathcal{N}'$ is defined in such a way that the above invariant is always preserved. Formally, the set of transitions of $\mathcal{N}'$ is defined as the smallest set satisfying the following conditions:

- **The simulation of the first phase:**

  - **Simulation of a computation of $\mathcal{N}$ from μ_{j-1} to μ_j:** For every natural number $j \in \{1,\ldots,n\}$ and for every transition $t \in T$, $\mathcal{N}'$ has a transition t_j such that:

    1. $F'(q_j,t_j) = F'(t_j,q_j) = 1$, $F'(\bar{q},t_j) = F'(t_j,\bar{q}) = 0$, and $F'(q_l,t_j) = F'(t_j,q_l) = 0$ for all $l \in \{1,\ldots,n\}$ and $l \neq j$. This means that in order to fire the transition t_j, the place q_j must contain one token.
    2. For every place $p \in P$, $F'(p,t_j) = F(p,t)$ and $F'(t_j,p) = F(t,p)$. This means that the transition t_j of $\mathcal{N}'$ has the same effect over the set of places P as the transition t of $\mathcal{N}$.
    3. For every $i \in \{1,\ldots,m\}$, $F'(s_i^-,t_j) = \sum_{p \in P} F(p,t) \sum_{k \geq j} (\mathbf{y_k^i}(p))$ and $F'(t_j,s_i^-) = \sum_{p \in P} F(t,p) \sum_{k \geq j} (\mathbf{y_k^i}(p))$. Hence, the invariant between the place s_i^- and the mapping ρ_i^- is preserved.
    4. For every $i \in \{1,\ldots,m\}$, $F'(s_i^+,t_j) = \sum_{p \in P} F(p,t) \sum_{k \geq j} (\mathbf{z_k^i}(p))$ and $F'(t_j,s_i^+) = \sum_{p \in P} F(t,p) \sum_{k \geq j} (\mathbf{z_k^i}(p))$. Hence, the invariant between the place s_i^+ and the mapping ρ_i^+ is preserved.

  - **Guessing the marking μ_j:** For every natural number $j \in \{1,\ldots,n-1\}$ and for every transition $t \in T$, $\mathcal{N}'$ has a transition t_j^{j+1} such that:

1. $F'(q_j, t_j^{j+1}) = F'(t_j^{j+1}, q_{j+1}) = 1$, $F'(\bar{q}, t_j^{j+1}) = F'(t_j^{j+1}, \bar{q}) = 0$, and
   $F'(q_l, t_j^{j+1}) = F'(t_j^{j+1}, q_{l'}) = 0$ for any $l, l' \in \{1, \ldots, n\}, l \neq j$ and $l' \neq j+1$.
   This corresponds to moving the token from the place q_j to the place q_{j+1}.
2. For every place $p \in P$, $F'(p, t_j^{j+1}) = F(p,t)$ and $F'(t_j^{j+1}, p) = F(t,p)$. This
   means that the transition t_j^{j+1} of $\mathcal{N}'$ has the same effect over the set of
   places P as the transition t of $\mathcal{N}$.
3. For every $i \in \{1, \ldots, m\}$, $F'(s_i^-, t_j^{j+1}) = \sum_{p \in P} F(p,t) \sum_{k \geq j} (\mathbf{y_k^i}(p))$ and
   $F'(t_j^{j+1}, s_i^-) = \sum_{p \in P} F(t,p) \sum_{k \geq j} (\mathbf{y_k^i}(p))$. Hence, the invariant between the
   place s_i^- and the mapping ρ_i^- is preserved.
4. For every $i \in \{1, \ldots, m\}$, $F'(s_i^+, t_j^{j+1}) = \sum_{p \in P} F(p,t) \sum_{k \geq j} (\mathbf{z_k^i}(p))$ and
   $F'(t_j^{j+1}, s_i^+) = \sum_{p \in P} F(t,p) \sum_{k \geq j} (\mathbf{z_k^i}(p))$. Hence, the invariant between the
   place s_i^+ and the mapping ρ_i^+ is preserved.

Notice that firing the transition rule t_j^{j+1} in $\mathcal{N}'$ simulates the firing of the transition rule t in $\mathcal{N}$ over the set of places P. This guarantees that the guessed sequence of transitions σ_j contains at least one transition.

- **Guessing the marking μ_n:** For every transition $t \in T$, $\mathcal{N}'$ has a transition t_n^{n+1} such that:
  1. $F'(q_n, t_n^{n+1}) = F'(t_n^{n+1}, \bar{q}) = 1$, $F'(t_n^{n+1}, q_n) = 0$, and $F'(q_l, t_n^{n+1}) = F'(t_n^{n+1}, q_l) = 0$ for all $1 \leq l < n$. This corresponds to moving the token from the place q_n to the place $\bar{q}$.
  2. For every place $p \in P$, $F'(p, t_n^{n+1}) = F(p,t)$ and $F'(t_n^{n+1}, p) = F(t,p)$. This means that the transition t_n^{n+1} of $\mathcal{N}'$ has the same effect over the set of places P as the transition t of $\mathcal{N}$.
  3. For every $i \in \{1, \ldots, m\}$, $F'(s_i^-, t_n^{n+1}) = \sum_{p \in P} F(p,t) \sum_{k \geq j} (\mathbf{y_k^i}(p))$ and $F'(t_n^{n+1}, s_i^-) = \sum_{p \in P} F(t,p) \sum_{k \geq j} (\mathbf{y_k^i}(p))$. Hence, the invariant between the place s_i^- and the mapping ρ_i^- is preserved.
  4. For every $i \in \{1, \ldots, m\}$, $F'(s_i^+, t_n^{n+1}) = \sum_{p \in P} F(p,t) \sum_{k \geq j} (\mathbf{z_k^i}(p))$ and $F'(t_n^{n+1}, s_i^+) = \sum_{p \in P} F(t,p) \sum_{k \geq j} (\mathbf{z_k^i}(p))$. Hence, the invariant between the place s_i^+ and the mapping ρ_i^+ is preserved.

- **Simulation of the second phase:**
  - **Decreasing the number of tokens in each place of P:** For every $p \in P$, $\mathcal{N}'$ has a transition t_p such that $F'(\bar{q}, t_p) = F'(t_p, \bar{q}) = F'(p, t_p) = 1$ and 0 otherwise.
  - **Decreasing the number of tokens in each place s_i^+:** For every $i \in \{1, \ldots, m\}$, $\mathcal{N}'$ has a transition t_i^+ such that $F'(\bar{q}, t_i^+) = F'(t_i^+, \bar{q}) = 1$, $F'(s_i^+, t_i^+) = 1$, and 0 otherwise.
  - **Decreasing the number of tokens in each place s_i^-:** For every $i \in \{1, \ldots, m\}$, $\mathcal{N}'$ has a special transition t_i^- such that $F'(\bar{q}, t_i^-) = F'(t_i^-, \bar{q}) = 1$, $F'(s_i^-, t_i^-) = F'(s_i^+, \bar{t}_i^-) = 1$, and 0 otherwise. Notice that, while decrementing the number of tokens in s_i^-, we decrease also the number of tokens in s_i^+ by one.

- **The end of the second phase:** $\mathcal{N}'$ has a transition t_{end} such that $F'(\bar{q}, t_{\text{end}}) = 1$ and 0 otherwise.

Then, Theorem 3 is an immediate consequence of the following lemma:

Lemma 4. *The marking* **0** *is reachable in* $\mathcal{N}'$ *if and only if* $\mathcal{N}$ *satisfies* f.

Hence, the satisfiability problem for the class of path formulas is polynomially reducible to the reachability problem for Petri nets. As an immediate consequence of Theorem 2 and 3, we get the following result:

Corollary 3. *The satisfiability problem for the class of path formulas is as hard as the reachability problem for Petri nets.*

6   An EXPSPACE-Complete Fragment

In this section we consider a fragment of Yen's path logic for which we can show that its satisfiability problem is EXPSPACE-complete. The proof follows very closely Yen's proof [13] which is a generalization of Rackoff's proof [12] for the complexity of the boundedness problem. The basic idea is to show that if a path satisfying a formula exists, then there is a short one. This is done by induction on the number of places of the Petri Net. However we have to modify one crucial lemma whose proof in the paper of Yen [13] contains an error. To correct the lemma, Yen's logic has to be restricted. The restriction makes sure that if there is a path showing that a formula is satisfiable, then there is also a path starting at each intermediate marking of the path which satisfies the formula. This is achieved by requiring the last designated marking of the path to be bigger than the first designated marking. Formally,

Definition 1. *A path formula* f *of the form*

$$\exists \mu_1, \ldots, \mu_n \exists \sigma_1, \ldots, \sigma_n \left(\left(\mu_0 \xrightarrow{\sigma_1} \mu_1 \xrightarrow{\sigma_2} \cdots \mu_{n-1} \xrightarrow{\sigma_n} \mu_n \right) \wedge \phi(\mu_1, \ldots, \mu_n, \sigma_1, \ldots, \sigma_n) \right)$$

is called increasing if $\phi(\mu_1, \ldots, \mu_n, \sigma_1, \ldots, \sigma_n)$ *does not contain transition predicates and implies* $\mu_n \geq \mu_1$.

Notice that for $n = 1$, $\mu_n \geq \mu_1$ is always true and that an increasing path formula can also be written as $\exists \mu_1, \ldots, \mu_n \exists \sigma_1, \ldots, \sigma_n \left(\left(\mu_0 \xrightarrow{\sigma_1} \mu_1 \xrightarrow{\sigma_2} \cdots \mu_{n-1} \xrightarrow{\sigma_n} \mu_n \right) \wedge \phi(\mu_1, \ldots, \mu_n) \right)$.

For the rest of the section we consider increasing path formulas. We can suppose furthermore that the formulas ϕ are conjunctions of marking predicates, since disjunctions can be considered separately. We first give some additional definitions. Given a predicate ϕ and a set of positive integers D we define $\phi^{[D]}$ to be the predicate resulting from removing all marking predicates of the form $\mu_i(p) \geq c$ and $\mu_i(p) > c$ from ϕ for all $i \notin D$. Let (P, T, F, μ_0) be a Petri Net with k places. We suppose an ordering on P and T and can then suppose that markings are vectors of $\mathbb{N}^k$.

The *transition vector* of a transition t, denoted by $\hat{t}$ is a k-dimensional vector with $\hat{t}(i) = F(t, p_i) - F(p_i, t)$ for all i with $1 \leq i \leq k$. The set of transition vectors, denoted by $\hat{T}$ is $\{\hat{t} \mid t \in T\}$. A *generalized marking* is a mapping from P to $\mathbb{Z}$ (i.e. a vector of $\mathbb{Z}^k$).

A *generalized firing sequence* is any sequence of transitions of T. A finite sequence of vectors $w_1, \ldots, w_m \in \mathbb{Z}^k$ is said to be a *path* (of length $m-1$) if $w_1 = \mu_0$ and $w_{i+1} - w_i \in \hat{T}$ for all i with $1 \le i < m$. A path $w_1, \ldots, w_m$ corresponds to at least one generalized firing sequence $t_1, \ldots, t_{m-1}$ such that $w_{i+1} - w_i = \hat{t}_i$ for all i with $1 \le i < m$. Let $w \in \mathbb{Z}^k$. The vector w is *i bounded* if $w(j) \ge 0$ for $1 \le j \le i$. If $r \in \mathbb{N}^+$ is such that $0 \le w(j) < r$ for $1 \le j \le i$, then w is called *i-r bounded*. A path $p = w_1, \ldots, w_m \in \mathbb{Z}^k$ is called *i bounded (i-r bounded)* if each w_j in p is i bounded (i-r bounded). Given a predicate $\phi(\mu_1, \ldots, \mu_n)$, an i bounded (i-r bounded) path $w_1, \ldots, w_m$ is called an *i bounded ϕ path* if $\exists 1 \le j_1 \le j_2 \le \ldots \le j_n = m$ such that $\phi^{[\{1, \ldots, i\}]}(w_{j_1}, w_{j_2}, \ldots, w_{j_n})$ is true. Let $m'(i, \mu, \phi)$ be either the length of the shortest i bounded ϕ path whose initial generalized marking is μ, or 0 if it does not exist. Let $g(i, \phi) = max\{m'(i, \mu, \phi) \mid \mu \in \mathbb{Z}^k\}$. We have $g(i, \phi) \in \mathbb{N}$ (see [13]).

The following two lemmas are from [13].

Lemma 5. *If there is an i-r bounded ϕ-path in the Petri Net (P, T, F, μ_0), then there is an i-r bounded ϕ-path of length $\le r^{(s(\mathcal{N}))^c}$, for some constant independent of r and $s(\mathcal{N})$.*

We derive $g(i, \phi)$ recursively.

Lemma 6. $g(0, \phi) \le 2^{(s(\mathcal{N}))^c}$, *for some constant c independent of $s(\mathcal{N})$.*

Lemma 7. $g(i+1, \phi) \le (2^{(s(\mathcal{N}))}(g(i, \phi) + 1))^{(s(\mathcal{N}))^c}$ *for all $i < k$, where c is a constant independent of $s(\mathcal{N})$.*

Proof:

- Case 1. If there is an $(i+1)$-$2^{(s(\mathcal{N}))}(g(i, \phi) + 1)$ bounded ϕ path, then using Lemma 5, there exists a short one with length $\le (2^{(s(\mathcal{N}))}(g(i, \phi) + 1))^{(s(\mathcal{N}))^c}$.

- Case 2. Otherwise, let $v_1, \ldots, v_{m_0}, v_{m_0+1}, \ldots, v_m$ be an $(i+1)$ bounded ϕ path such that v_{m_0} is the first vector not $(i+1) - 2^{(s(\mathcal{N}))}(g(i, \phi) + 1)$ bounded. Without loss of generality, we assume that $v_{m_0}(i+1) > 2^{(s(\mathcal{N}))}(g(i, \phi) + 1)$. Furthermore we assume that no two of $v_1, \ldots, v_{m_0}$ can agree on the first $i+1$ positions, otherwise the path could be made shorter. Therefore $m_0 \le (2^{(s(\mathcal{N}))}(g(i, \phi) + 1))^{i+1}$. Now we show that if we take as initial marking v_{m_0}, there is an i bounded ϕ path in the Petri Net[3]. There are two cases depending on ϕ.

  1. ϕ is of the form $\phi(\mu_1)$. In this case, since $v_1, \ldots, v_m$ is an $i+1$ bounded ϕ path and ϕ is just a predicate on the marking μ_1, $v_{m_0}, v_{m_0+1}, \ldots, v_m$ is clearly an i bounded ϕ path.

  2. ϕ is of the form $\phi(\mu_1, \ldots, \mu_n)$ and it implies $\mu_n \ge \mu_1$. Since $v_1, \ldots, v_{m_0}$, $v_{m_0+1}, \ldots, v_m$ is an $(i+1)$ bounded ϕ path it is an i bounded ϕ path as well. Therefore $\exists 1 \le j_1 \le j_2 \le \ldots \le j_n = m$ such that $\phi^{[\{1, \ldots, i\}]}(v_{j_1}, v_{j_2}, \ldots, v_{j_n})$ is true. Furthermore $v_{j_n} \ge v_{j_1}$. Let $s' = t'_1, \ldots, t'_o$ be a sequence of transitions corresponding to the path $v_{j_1}, v_{j_1+1}, \ldots, v_{j_n}$. Let $s = t_{m_0}, \ldots, t_{m-1}$ be a sequence

---

[3] At this point, there is a mistake in the proof of [13] (Lemma 3.7, page 130), as it assumes that this path always exists and takes the shortest one. However *no i bounded ϕ path* might exist starting from v_{m_0}.

of transitions corresponding to the path $v_{m_0}, v_{m_0+1}, \ldots, v_m$. Then ss' is a sequence of transitions corresponding to a path $v_{m_0}, \ldots, v_m, v'_{j_1+1}, \ldots, v'_{j_n}$ where $v'_i = v_i + v_{j_n} - v_{j_1}$ for all i such that $j_1 + 1 \le i \le j_n$. Clearly the path is an i bounded ϕ path starting from v_{m_0} (since $\phi^{[\{1,\ldots,i\}]}(v_m, v'_{j_2}, \ldots, v'_{j_n})$ is true, because all predicates stay true when adding to all markings the same positive vector).

Now, we can take the shortest i bounded ϕ path p in (P, T, F, v_{m_0}). It's length is $\le g(i, \phi)$. As $v_{m_0}(i+1) > 2^{(s(\mathcal{N}))}(g(i, \phi) + 1)$ and each place of each transition vector in the Petri Net is at most $2^{(s(\mathcal{N}))}$ in absolute value, p is also $i+1$ bounded and the $(i+1)$ position will never fall below $2^{(s(\mathcal{N}))}$ in p (so that marking predicates of the form $\mu_i(p') \ge c$ and $\mu_i(p') > c$ will still hold in p). Therefore $v_1, \ldots, v_{m_0-1}, p$ is an $(i+1)$ bounded ϕ path of length $(2^{(s(\mathcal{N}))}(g(i, \phi) + 1))^{i+1} + g(i, \phi) < (2^{(s(\mathcal{N}))}(g(i, \phi) + 1))^{(s(\mathcal{N}))^c}$.

$\square$

The following theorem now follows easily [13] from the bound on g.

Theorem 4. *The satisfiability problem for increasing path formulas can be decided in* $O(2^{d*s(\mathcal{N})*log(s(\mathcal{N}))})$ *space, for some constant d independent of $s(\mathcal{N})$.*

Since unboundedness can be expressed in the logic and boundedness is EXPSPACE-hard [9] we have the following:

Theorem 5. *The satisfiability problem for increasing path formulas is* EXPSPACE-*complete.*

6.1 Some Applications

In the following we consider the applications given in [13] and discuss if they are in the increasing fragment. The following six problems are all in the fragment and therefore in EXPSPACE. They have already been shown to be in EXPSPACE before Yen's paper.

1. *Boundedness problem.* Unboundedness of a Petri Net can be formulated as $\exists \mu_1, \mu_2 \exists \sigma_1, \sigma_2 ((\mu_0 \xrightarrow{\sigma_1} \mu_1 \xrightarrow{\sigma_2} \mu_2) \wedge (\mu_2 > \mu_1))$ which is clearly an increasing path formula.

2. *Coverability.* It can be formulated as $\exists \mu_1, \exists \sigma_1 ((\mu_0 \xrightarrow{\sigma_1} \mu_1) \wedge (\mu_1 \ge v))$ which is an increasing path formula.

3. *(Strict) Self-Coverability Problem.* It can be solved by considering formulas of the form $\exists \mu_1, \mu_2 \exists \sigma_1, \sigma_2 ((\mu_0 \xrightarrow{\sigma_1} \mu_1 \xrightarrow{\sigma_2} \mu_2) \wedge ((\bigwedge_{s \in I} \mu_2(s) \ge \mu_1(s)) \wedge (\bigwedge_{s' \notin I} \mu_2(s') = \mu_1(s'))))$ where I is a set of places (For strict self-coverability, replace $\ge$ by $>$). The formulas are clearly increasing path formulas.

4. *u-Self-Coverability Problem.* This can be solved by considering formulas of the form $\exists \mu_1, \mu_2 \exists \sigma_1, \sigma_2 ((\mu_0 \xrightarrow{\sigma_1} \mu_1 \xrightarrow{\sigma_2} \mu_2) \wedge (\mu_2 - \mu_1 = u))$ where $u \in \mathbb{N}^k$. These formulas are increasing.

5. *Final-State Self-Coverability Problem.* This can be solved by considering formulas of the form $\exists \mu_1, \mu_2, \mu_3 \exists \sigma_1, \sigma_2, \sigma_3 ((\mu_0 \xrightarrow{\sigma_1} \mu_1 \xrightarrow{\sigma_2} \mu_2 \xrightarrow{\sigma_3} \mu_3) \wedge (\mu_3 \ge \mu_1) \wedge (\bigvee_{s \in F} \mu_2(s) > 0))$ for some set F of places. This formula is increasing.

6. *Fair Nontermination Problems.* All the formulas considered for these problems are of the form $\exists\mu_1,\mu_2\exists\sigma_1,\sigma_2\big((\mu_0\xrightarrow{\sigma_1}\mu_1\xrightarrow{\sigma_2}\mu_2)\wedge(\mu_2\geq\mu_1)\wedge\varphi(\sigma_1,\sigma_2)\big)$ where $\varphi(\sigma_1,\sigma_2)$ is a formula containing only transition predicates. By carefully inspecting this transition predicates one can easily see that eliminating them with Lemma 1 yields increasing formulas.

The following problems were claimed to be in EXPSPACE in [13].

1. *Regularity Detection Problem.* Nonregularity of a Petri Net is equivalent to the satisfiability of the following path formula:

$$\exists\mu_1,\mu_2,\mu_3,\mu_4\exists\sigma_1,\sigma_2,\sigma_3,\sigma_4\big((\mu_0\xrightarrow{\sigma_1}\mu_1\xrightarrow{\sigma_2}\mu_2\xrightarrow{\sigma_3}\mu_3\xrightarrow{\sigma_4}\mu_4)\wedge\varphi(\mu_1,\mu_2,\mu_3,\mu_4)\big)$$

where $\varphi(\mu_1,\mu_2,\mu_3,\mu_4)$ is $(\mu_2\geq\mu_1)\wedge(\bigvee_{i=1}^{k}\mu_2(i)>\mu_1(i))\wedge(\bigwedge_{i=1}^{k}(\mu_1(i)<\mu_2(i))\vee(\mu_3(i)\leq\mu_4(i)))\wedge(\bigvee_{i=1}^{k}\mu_3(i)>\mu_4(i)))$. Unfortunately, this formula is not increasing and we can not apply our complexity result. To the best of our knowledge the complexity of regularity is therefore still unknown.

2. *(Potential) Determinism Detection Problem.* Nondeterminism of a Petri Net can be expressed using the formula $\exists\mu_1,\exists\sigma_1\big((\mu_0\xrightarrow{\sigma_1}\mu_1)\wedge((\bigvee_{t,t',t\neq t'}(\mu_1\geq v_t)\wedge(\mu_1\geq v_{t'})))\big)$ where the v_t are the minimal vectors for which t is enabled. Clearly, the formula is increasing. Non potential determinism can then be expressed as $\exists\mu_1,\mu_2,\exists\sigma_1,\sigma_2\big((\mu_0\xrightarrow{\sigma_1}\mu_1\xrightarrow{\sigma_2}\mu_2)\wedge((\bigvee_{t,t',t\neq t'}(\mu_1\geq v_t)\wedge(\mu_1\geq v_{t'}))\wedge(\mu_2\geq\mu_1)))$. This formula is increasing and therefore the problem is in EXPSPACE.

3. *Frozen Token Detection Problem.* To decide if a Petri Net has a frozen token it is sufficient to check the formula $\exists\mu_1,\mu_2,\exists\sigma_1,\sigma_2\big((\mu_0\xrightarrow{\sigma_1}\mu_1\xrightarrow{\sigma_2}\mu_2)\wedge(\mu_1(p)>0)\wedge(\mu_2\geq\mu_1)\wedge(\sigma_2\neq\Lambda)\big)$ where p is a designated place and $\sigma_2\neq\Lambda$ denotes $\bigvee_{t\in T}\sharp_{\sigma_2}(t)>0$. Eliminating with Lemma 1 the transition predicates yields an increasing path formula and therefore the problem is in EXPSPACE.

4. *(Strong) Promptness Detection.* A Petri Net is not (strongly) prompt if and only if $\exists\mu_1,\mu_2,\exists\sigma_1,\sigma_2\big((\mu_0\xrightarrow{\sigma_1}\mu_1\xrightarrow{\sigma_2}\mu_2)\wedge((\bigwedge_{t\in T_1}\sharp_{\sigma_2}(t)\leq0)\wedge(\mu_2\geq\mu_1)\wedge(\sigma_2\neq\Lambda))\big)$ is true. Again eliminating with Lemma 1 the transition predicates yields an increasing path formula and therefore the problem is in EXPSPACE.

5. *y-Synchronization Problem.* Given a map y from the transitions T to $\mathbb{Z}$, a Petri net is not y-synchronized iff $\exists\mu_1,\mu_2,\exists\sigma_1,\sigma_2\big((\mu_0\xrightarrow{\sigma_1}\mu_1\xrightarrow{\sigma_2}\mu_2)\wedge((\bigwedge_{t\in T_1}\sharp_{\sigma_1}(t)\leq0)\wedge(((\sum_{t\in T}y(t)\sharp(\sigma_2)(t)>0)\vee(\sum_{t\in T}y(t)\sharp(\sigma_2)(t)<0))\wedge(\mu_2\geq\mu_1)))\big)$ is true. While eliminating with Lemma 1 the transition predicates we notice that the newly added places are always increasing. Thus this yields an increasing path formula and therefore the problem is in EXPSPACE.

7   Conclusion

In this paper, we have shown that the satisfiability problem for the class of path formulas considered by Yen [13] is as hard as the reachability problem for Petri nets. However for an important fragment we have shown that its satisfiability problem is EXPSPACE-complete. By doing this, we have corrected the proof given in [13]. Furthermore we show that almost all applications considered by Yen can be solved using our fragment. However, the exact complexity of the regularity detection problem remains open. It would be interesting to obtain a bigger fragment which is in EXPSPACE allowing to show the EXPSPACE complexity of the regularity detection problem.

Acknowledgement. The authors would like to thank Ahmed Bouajjani and Javier Esparza for very helpful discussions on this topic as well as Stéphane Demri who independently found the error in Yen's proof.

References

1. Esparza, J.: On the decidability of model checking for several μ-calculi and petri nets. In: Tison, S. (ed.) CAAP 1994. LNCS, vol. 787, pp. 115–129. Springer, Heidelberg (1994)
2. Esparza, J., Nielsen, M.: Decidability issues for petri nets - a survey. Bulletin of the EATCS 52, 244–262 (1994)
3. Ganty, P., Majumdar, R., Rybalchenko, A.: Verifying liveness for asynchronous programs. In: POPL, pp. 102–113. ACM, New York (2009)
4. Hack, M.H.T.: Decidability Questions for Petri Nets. PhD thesis, M.I.T (1976)
5. Haddad, S., Poitrenaud, D.: Checking linear temporal formulas on sequential recursive petri nets. In: TIME, pp. 198–205 (2001)
6. Keller, R.M.: A fundamental tehoerem of asynchronous parallel computation. In: Tse-Yun, F. (ed.) Parallel Processing. LNCS, vol. 24, pp. 102–112. Springer, Heidelberg (1975)
7. Kosaraju, S.R.: Decidability of reachability in vector addition systems (preliminary version). In: STOC, pp. 267–281. ACM, New York (1982)
8. Lambert, J.L.: A structure to decide reachability in petri nets. Theor. Comput. Sci. 99(1), 79–104 (1992)
9. Lipton, R.: The reachability problem requires exponential time. Technical Report TR 66 (1976)
10. Mayr, E.W.: An algorithm for the general petri net reachability problem. In: STOC, pp. 238–246. ACM, New York (1981)
11. Mayr, E.W.: Persistence of vector replacement systems is decidable. Acta Inf. 15, 309–318 (1981)
12. Rackoff, C.: The covering and boundedness problems for vector addition systems. Theor. Comput. Sci. 6, 223–231 (1978)
13. Yen, H.-C.: A unified approach for deciding the existence of certain petri net paths. Inf. Comput. 96(1), 119–137 (1992)
14. Yen, H.-C.: A note on fine covers and iterable factors of vas languages. Inf. Process. Lett. 56(5), 237–243 (1995)
15. Yen, H.-C.: On the regularity of petri net languages. Inf. Comput. 124(2), 168–181 (1996)

Probabilistic Model Checking of Biological Systems with Uncertain Kinetic Rates

Roberto Barbuti[1], Francesca Levi[1], Paolo Milazzo[1], and Guido Scatena[2]

[1] Dip. di Informatica, Univ. di Pisa, Largo B. Pontecorvo 3, 56127 - Pisa, Italy
{barbuti,levifran,milazzo}@di.unipi.it
[2] IMT Lucca Inst. for Advanced Studies, Piazza San Ponziano 6, 55100 - Lucca, Italy
g.scatena@imtlucca.it

Abstract. We present an abstraction of the probabilistic semantics of Multiset Rewriting to formally express systems of reactions with uncertain kinetic rates. This allows biological systems modelling when the exact rates are not known, but are supposed to lie in some intervals. On these (abstract) models we perform probabilistic model checking obtaining lower and upper bounds for the probabilities of reaching states satisfying given properties. These bounds are under- and over-approximations, respectively, of the probabilities one would obtain by verifying the models with exact kinetic rates belonging to the intervals.

Keywords: probabilistic model checking, systems biology, uncertain kinetic rates, abstract interpretation, interval Markov chains.

1   Introduction

When modelling biological systems, the rates of the reactions involved in the evolution of the systems are often not precisely known. Thus, it is necessary to model such systems with some level of approximation. However, approximations should be significant, namely they should preserve, although not precisely, the overall behaviour of the systems.

In this paper we present a formalisation of biological systems based on *Multiset Rewriting* (MSR) [1], and we investigate the use of abstract interpretation [2] on its probabilistic semantics. In particular, we use an *Interval Markov Chain* (IMC) [3,4] to abstract the *Discrete Time Markov Chain* (DTMC) probabilistic semantics of a set of MSR models. The abstraction is able to model the semantics of a biological system for which the exact kinetic rates are not precisely known, but are supposed to lie in some intervals.

We start defining MSR as the formalism used to construct *concrete models*, namely models with exact kinetic rates (Section 2). We give a *Labelled Transition System* (LTS) semantics to MSR and show how to derive, in standard way, a probabilistic semantics from it, in terms of a DTMC. On the DTMC it is possible to perform probabilistic model checking.

In order to deal with uncertainty we define *abstract models* in which the kinetic rates are given as intervals (Section 3). We give an abstract LTS semantics and

O. Bournez and I. Potapov (Eds.): RP 2009, LNCS 5797, pp. 64–78, 2009.

$$C \subset \mathcal{P}(\mathcal{M}) \xrightarrow{\;LTS\;} \mathcal{LTS} \xrightarrow{\;\mathcal{H}\;} \mathcal{MC}$$

$$\downarrow \alpha \qquad\qquad\quad \downarrow \alpha_{\mathcal{LTS}} \qquad\quad \downarrow \alpha_{\mathcal{MC}}$$

$$\mathcal{M}^{\circ} \xrightarrow{\;LTS^{\circ}\;} \mathcal{LTS}^{\circ} \xrightarrow{\;\mathcal{H}^{\circ}\;} \mathcal{MC}^{\circ}$$

Fig. 1. Schematics of the the defined theory; with ∘ we indicate abstract structures, with α abstraction functions

a method to derive an abstract probabilistic semantics from it, in terms of IMC. On the IMC it is possible to perform probabilistic model checking (that gives *lower* and *upper bounds* for the probability of reaching states satisfying given properties) by following the approach of [5].

We relate the concrete probabilistic semantics with the abstract one by means of concepts of abstract interpretation (see Figure 1). We prove the soundness of the abstract semantics with respect to the concrete one. This implies that the lower and upper bounds obtained by model checking an abstract model are valid for all the models with exact kinetic rates belonging to the specified intervals.

In Section 4, we apply probabilistic model checking to verify reachability properties in an abstract model of tumor growth [6]. We review related work in Section 5 and we conclude with a summary and further research ideas with Section 6.

2 Probabilistic Model Checking of Biological Systems

To model biological systems we adopt *Multiset Rewriting* (MSR) where the rewriting rules are enriched with non negative real kinetic constants. Multisets are states of computation and transitions between states are performed by applying rewriting rules with a probability proportional to their kinetic constants.

Let Σ be a finite set of *species names* with cardinality n. A *multiset* is a function $s : \Sigma \to \mathbb{N}$ and $\mathcal{S}(\Sigma)$ is the *universe of multisets over* Σ. We assume multiset sum $\oplus$ and difference $\ominus$, to be defined as follows: given $s', s'' \in \mathcal{S}(\Sigma)$ we have $s' \oplus s''(x) = s'(x) + s''(x)$ and $s' \ominus s''(x) = max(s'(x) - s''(x), 0)$. In what follows we shall often assume Σ to be given.

A multiset represents the configuration of a biological system model, whereas the description of the possible events is given by rewriting rules. A *rewriting rule* is a pair (l, r) where l and r, called *reactants* and *products*, are multisets. Each rule is associated with a *kinetic constant* that is, roughly, an indication of the likelihood of the represented event.

Definition 1 (Concrete Model). *A concrete model M is a triple $(\mathcal{R}, \mathcal{K}, s_0)$:*

- $\mathcal{R} = \{R_1, \ldots, R_m\}$, *with $R_i \in \mathcal{S}(\Sigma) \times \mathcal{S}(\Sigma)$, is a set of* rewriting rules;
- $\mathcal{K} = \{k_1, \ldots, k_m\}$, *with $k_i \in \mathbb{R}_{\geqslant 0}$, is a set of* kinetic constants;
- $s_0 \in \mathcal{S}(\Sigma)$ *is the* starting state.

We denote the universe of concrete models as $\mathcal{M}$. When the model $M = (\mathcal{R}, \mathcal{K}, s_0)$ is clear, for $i \in [1, m]$, we use l_i and r_i to denote the multisets of rule R_i and $\mathcal{K}[i]$ for k_i. We use the notation $(l \xrightarrow{k} r)$ for $(l, r) \in \mathcal{R}$ and $k \in \mathcal{K}$. Finally we use $R(M), K(M), S_0(M)$ to denote $\mathcal{R}, \mathcal{K}, s_0$ respectively.

Two concrete models $M_i, i \in \{1, 2\}$, are *isomorphic* $(M_1 \sim M_2)$ if and only if $R(M_1) = R(M_2) \wedge S_0(M_1) = S_0(M_2)$.

2.1 Labelled Transition System Semantics

To describe the semantics of a concrete model we adopt a *Labelled Transition System* (LTS) with a *transition* relation of the form $s' \xrightarrow{\eta, \beta} s''$ where η is the number of the applied rule, and $\beta \in \mathbb{R}_{\geqslant 0}$ is the *transition rate*.

The application of a rule R_η to a state s' is modelled by the inference rule

$$\frac{(l_\eta \xrightarrow{k_\eta} r_\eta) \quad l \subseteq s' \quad \beta = rate(l_\eta, s', k_\eta) \quad s'' = ((s' \ominus l_\eta) \oplus r_\eta)}{s' \xrightarrow{\eta, \beta} s''} \quad (1)$$

where $rate(l_\eta, s', k_\eta) = kin(l_\eta, s') \times k_\eta$ and $kin(l_\eta, s') = \prod_{x \in \Sigma} \binom{s'(x)}{l_\eta(x)}$.

To compute $kin(l_\eta, s')$ we take into account the number of possible distinct applications of the rule R_η to the state s'. Actually, this requires to compute the number of distinct combinations of the reactants l_η into the multiset s'. To compute $rate(l_\eta, s', k_\eta)$ we multiply the value of $kin(l_\eta, s')$ by the kinetic constant associated with R_η, namely k_η.

Given a concrete model $M = (\mathcal{R}, \mathcal{K}, s_0) \in \mathcal{M}$, we define the function $LTS : \mathcal{M} \mapsto \mathcal{LTS}$, such that $LTS(M) = (S, \rightarrow, s_0)$ is the LTS, obtained as usual by transitive closure of (1) starting from s_0. In the following, we use $\mathcal{LTS}$ to denote the universe of LTSs. Moreover, we use $Next(s)$ for the set of transitions from the state s; in addition, we use $TS(s', s'') = \{s' \xrightarrow{\eta, \beta} s''$ for some $\eta, \beta\}$ for describing the set of transitions from s' to s''. Given a transition $t = s' \xrightarrow{\eta, \beta} s''$, we also use $rate(t) = \beta$. Note that, $\forall s \in S, R_\eta \in \mathcal{R}$, there is at most one transition $s \xrightarrow{\eta, \beta} s' \in Next(s)$ corresponding to R_η.

2.2 Derivation of Probabilistic Semantics

We present the probabilistic semantics of a concrete model by means of a translation from LTS into *Discrete Time Markov Chain* (DTMC).

Given a countable set S we denote with $Distr(S) = \{\rho \mid \rho : S \rightarrow [0, 1] \wedge \sum_{s \in S} \rho(s) = 1\}$ the set of *probability distributions* and with $PDistr(S) = \{\rho \mid \rho : S \rightarrow [0, 1]\}$ the set of *probability pseudo–distributions*. Given a finite set S, a function $P : S \times S \mapsto \mathbb{R}$ and $s \in S$, we denote with $P(s, S) = \sum_{s' \in S} P(s, s')$.

Definition 2 (Discrete Time Markov Chain). *A* DTMC *is a tuple* (S, P, s_0), *where: S is the* set of states, $s_0 \in S$ is the *starting state and* $P : S \mapsto Distr(S)$ is *the* transition probability function.

In the following, we restrict our attention to *finitely branching* DTMC, meaning that for each $s \in S$, the set $\{s' \mid P(s)(s'') > 0\}$ is finite. Since our models have m–sized set of rules from each state we have at most m exit transitions. Moreover, we use $\mathcal{MC}$ to denote the universe of (finitely branching) DTMCs.

To derive a DTMC from the LTS, we have to calculate, for each multiset s' and s'', the probability of moving from s' to s'', by exploiting transition rates. Thus, we introduce two functions $R : S \times S \mapsto \mathbb{R}_{\geqslant 0}$ and $E : S \mapsto \mathbb{R}_{\geqslant 0}$, such that $R(s', s'') = \sum_{t \in TS(s', s'')} rate(t)$ and $E(s') = \sum_{s'' \in S} R(s', s'')$. Intuitively $R(s', s'')$ gives the rate of the transition from s' to s'', while $E(s')$ computes the exit rate of the state. The probability of moving from s' to s'' is derived from $R(s', s'')$ and $E(s')$ in standard way.

Definition 3 (Probabilistic Translation Function). *We define* $\mathcal{H} : \mathcal{LTS} \to \mathcal{MC}$ *as* $\mathcal{H}((S, \to, s_0)) = (S, P, s_0)$, *where* $P : S \to Distr(S)$ *is the* probability transition function, *s.t.* , $\forall s', s'' \neq s' \in S :$ *if* $E(s') = 0$, *then* $P(s')(s'') = 0$, *and* $P(s')(s') = 1$; $P(s')(s'') = R(s', s'')/E(s')$ *otherwise.*

2.3 Probabilistic Model Checking

In the context of probabilistic model checking [7,8] we focus our attention on a fragment of the Probabilistic CTL (PCTL) [9] able to express reachability properties. Formally, we have to evaluate the probability of a set of paths.

Let (S, P, s_0) be a DTMC. A *path* π is a non–empty (finite or infinite) ordered succession of states $s_0, s_1, \ldots$ of S. We denote the ith state of the path π by $\pi[i]$, starting from 1, and the length of π by $|\pi|$, where $|\pi| = \infty$ if π is infinite. The set of paths over S is denoted by $Paths(S)$ and its subset of finite paths is denoted as $FPaths(S)$. The *cylinder* corresponding to a path π is the set of all paths prefixed by π. Formally, for $\pi \in Paths(s)$, $C(\pi) = \{\pi \pi' \mid \pi' \in Paths(S)\}$ and $C(s)$ denotes the set of paths starting from the state s.

Definition 4 (Probability of Paths). *Let* (S, P, s_0) *be a DTMC. Let* $\Pi = \bigcup_{\pi \in FPaths(s)} C(\pi)$ *be the set of all cylinder,* $\mathcal{B}$ *be the smallest* σ–*algebra containing* Π, *and* $s \in S$ *a state. The tuple* $(Paths(S), \mathcal{B}, P_s)$ *is a probability space, where* P_s *is the unique measure satisfying, for all path* $s_0 \ldots s_n$,

$$P_s(C(s_0 \ldots s_n)) = \begin{cases} 1 & \text{if } s_0 = s \wedge n = 0 \\ P(s_0, s_1) \times \ldots \times P(s_{n-1}, s_n) & \text{if } s_0 = s \wedge n > 0 \\ 0 & \text{otherwise.} \end{cases}$$

Our reachability properties are parametric w.r.t. a set AP of propositional symbols (ranged over by $\{A, B, \ldots\}$). A symbol $A \in AP$ denotes a set of conditions on multisets that are evaluated by a corresponding notion of satisfaction $\vDash : \mathcal{S}(\Sigma) \times AP \mapsto \{true, false\}$. As usual, given $s \in \mathcal{S}(\Sigma)$ and $A \in AP$, $s \vDash A$ says that s satisfies A.

Definition 5 (Reachability Probability). *Let* $mc = (S, P, s_0)$ *be a DTMC. The probability of reaching a state satisfying* $A \in AP$, *starting from* $s \in S$, *is* $Reach_{A,mc}(s) = P_s(\{\pi \in C(s) \mid \pi[i] \vDash A \text{ for some } i \geq 0\})$.

We use $Reach(A)$ to denote $Reach_{A,mc}(s_0)$ where $mc = \mathcal{H}(LTS(M))$, for a model M clear from the context.

Example 1. We consider a chemical reactions system where molecules X and Y may bind to form complex XY and molecule X may be degraded by molecule W. With $\Sigma = \{X, Y, W, XY\}$, the system is modelled by $s_0 = \{(X, 2), (Y, 2), (W, 10)\}$, $\mathcal{R} = \{(R_1 = \{X, Y\} \xrightarrow{k_1} \{XY\}), (R_2 = \{X, W\} \xrightarrow{k_2} \{W\})\}$ and $\mathcal{K} = \{k_1 = 3, k_2 = 1\}$. Notice that we assume that the complexation is three times faster than the degradation. Figure 1 shows the derived $LTS(M)$ and $\mathcal{H}(LTS(M))$ where

$$S = \{ \quad s_0 = \{(X, 2), (Y, 2), (W, 10), (XY, 0)\} \quad s_1 = \{(X, 1), (Y, 1), (W, 10), (XY, 1)\}$$
$$s_2 = \{(X, 1), (Y, 2), (W, 10), (XY, 0)\} \quad s_3 = \{(X, 0), (Y, 0), (W, 10), (XY, 2)\}$$
$$s_4 = \{(X, 0), (Y, 1), (W, 10), (XY, 1)\} \quad s_5 = \{(X, 0), (Y, 2), (W, 10), (XY, 0)\} \ \}.$$

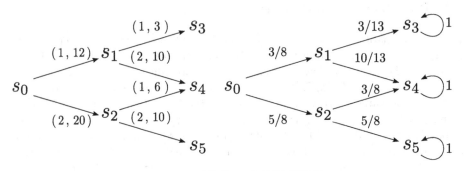

Fig. 2. $LTS(M)$, and $\mathcal{H}(LTS(M))$

The probability of obtaining at least two complexes XY is given by the probability to reach s_3. Therefore we obtain $3/8 \times 3/13 = 9/104$. This shows that, even if the rate of the complexation is (three times) greater that the one of the degradation, the concentration of reagent W makes the degradation more likely to happen than the binding of reagent X and Y.

3   Abstract Modelling and Model Checking

In order to approximate the information related to the kinetic rates of the reaction rules we adopt the *domain of intervals of (non negative) reals* $\mathbb{I}$ (the real valued version of intervals of integers [2,10,11]).

Definition 6 (Intervals). $\mathbb{I} = \{ [m, n] \mid m \in \mathbb{R}_{\geq 0}, n \in \mathbb{R}_{\geq 0} \cup \{\infty\} \wedge m \leq n \}$.

Over intervals of reals $\mathbb{I}$ we use the operations and the order defined as follows.

$$\forall \mathcal{I}, \mathcal{J} \in \mathbb{I}, \mathcal{I} = [a,\, b],\, \mathcal{J} = [c,\, d] \; :$$

$$\mathcal{I} \times^{\mathbb{I}} \mathcal{J} = [\, a \times c,\, b \times d\,], \qquad\qquad \mathcal{I} \cup^{\mathbb{I}} \mathcal{J} = [\, \min(a,\, c),\, \max(b,\, d)\,],$$

$$\mathcal{I} +^{\mathbb{I}} \mathcal{J} = [\, a + c,\, b + d\,], \qquad\qquad \mathcal{I} \sqsubseteq_{\mathbb{I}} \mathcal{J} \; \text{iff} \; (\mathcal{I} \cup_{\mathbb{I}} \mathcal{J} = \mathcal{J})\,.$$

We consider both $\cup_{\mathbb{I}}$ and $\sqsubseteq_{\mathbb{I}}$ extended component-wise to m–sized vectors of intervals, and we use the same symbols. For $x \in \mathbb{R}_{\geqslant 0}$ we use $x^{\bullet} = [x, x] \in \mathbb{I}$ for its best abstraction as interval, and we consider $^{\bullet}$ extended to vector of reals.

In abstract models each *reaction rule* does not have associated a precise kinetic constant ($\in \mathbb{R}$) but instead an interval of reals ($\in \mathbb{I}$).

Definition 7 (Abstract Model). *An* abstract model M *is a triple* $(\mathcal{R}, \mathcal{K}^{\circ}, s_0)$ *with* $\mathcal{R}$ *and* s_0 *as in the concrete case, while* $\mathcal{K}^{\circ} = \{k_1^{\circ}, \ldots, k_m^{\circ}\}$, $k_i^{\circ} \in \mathbb{I}$, *is a set of interval values.*

We denote the universe of abstract models as $\mathcal{M}^{\circ}$. We assume the notation used for concrete models extended in the oblivious way to concrete models. The order $\sqsubseteq_{\mathbb{I}}$ over intervals introduces a corresponding order $\sqsubseteq_{\mathcal{M}^{\circ}}$ over abstract models.

Definition 8 (Order on Abstract Models). *Given* $M_i^{\circ}, i \in \{1, 2\}$:
$M_1^{\circ} \sqsubseteq_{\mathcal{M}^{\circ}} M_2^{\circ}$ *iff* $M_1^{\circ} \sim M_2^{\circ} \wedge K(M_1^{\circ}) \sqsubseteq_{\mathbb{I}} K(M_2^{\circ})$.

3.1 Abstraction and Concretization

To formalise the relation between concrete and abstract models we introduce the concepts of *abstraction function* and *concretization function* [2].

Let $\mathcal{C} = \{X \in \mathcal{P}(\mathcal{M}) \,|\, \forall M_i, M_j \in X\,, M_i \sim M_j\}$ the *domain of isomorphic concrete models*. Given $X \in \mathcal{C}$ we denote with $R(X)$ and $S_0(X)$ the shared rule set and the shared starting state respectively.

Definition 9 (Order on Set of Isomorphic Concrete Models)
Given two set of isomorphic concrete models $X_i \in \mathcal{C}$, $i \in \{1, 2\}$:
$X_1 \sqsubseteq_{\mathcal{C}} X_2$ *iff* $\overline{K_1} \sqsubseteq_{\mathbb{I}} \overline{K_2}$ *where* $\overline{K_i} = \cup^{\mathbb{I}}_{M \in X_i} (K(M))^{\bullet}$.

Definition 10 (Abstraction and Concretization Functions)
We define $\alpha : \mathcal{C} \mapsto \mathcal{M}^{\circ}$ *and* $\gamma : \mathcal{M}^{\circ} \mapsto \mathcal{C}$ *s.t.* $\forall X \in \mathcal{C}, \forall M^{\circ} \in \mathcal{M}^{\circ}$:

- $\alpha(X) = (\,R(X),\, \overline{\mathcal{K}^{\circ}},\, S_0(X)\,)$ *where* $\overline{\mathcal{K}^{\circ}} \equiv \cup^{\mathbb{I}}_{M \in X} (K(M))^{\bullet}$;
- $\gamma(M^{\circ}) = \{M \,|\, \alpha(M) \sqsubseteq_{\mathcal{M}^{\circ}} M^{\circ}\}$.

Theorem 1
The pair (α, γ) *is a Galois connection between* $(\mathcal{C}, \sqsubseteq_{\mathcal{C}})$ *and* $(\mathcal{M}^{\circ}, \sqsubseteq_{\mathcal{M}^{\circ}})$.

This formalisation shows that an abstract model M° represents a (infinite) set of concrete models with the same set of rules (same multiset of reactants and products) with kinetic rates in the specified interval.

3.2   Abstract LTS Semantics

We introduce the LTS semantics associated with abstract models, adopting an abstract transition relation $s' \xrightarrow{\eta,\beta^\circ}_\circ s''$, where η is as in the concrete case, while $\beta^\circ \in \mathbb{I}$. The application of a rule R_η to a state s' is modelled by the rule

$$\frac{(l_\eta \xrightarrow{k_\eta^\circ} r_\eta) \quad l_\eta \subseteq s' \quad \beta^\circ = rate^\circ(l_\eta, s', k_\eta^\circ,) \quad s'' = ((s' \ominus l_\eta) \oplus r_\eta)}{s' \xrightarrow{\eta,\beta^\circ}_\circ s''} \tag{2}$$

where $rate^\circ(l_\eta, s', k_\eta^\circ) = kin(l_\eta, s') \times^\mathbb{I} k_\eta^\circ$.

We define the function $LTS^\circ : M^\circ \mapsto LTS^\circ$ such that $LTS^\circ((\mathcal{R}, \mathcal{K}^\circ, s_0)) = (S, \to_\circ, s_0)$ is obtained by transitive closure of (2) starting from s_0. As in the concrete case the outgoing transitions from a state have distinct labels. In the following we use $\mathcal{LTS}^\circ$ to denote the universe of abstract LTSs and we assume the notation defined for LTSs adapted in the obvious way to the abstract case.

To relate an LTS to its abstract counterpart we introduce the concept of *best abstraction* of an LTS. The *most precise* abstract LTS can be obviously obtained by replacing the rate β of each transition with $\beta^\bullet = [\beta, \beta]$.

Definition 11 (Best Abstraction of LTS). *We define* $\alpha_{\mathcal{LTS}} : \mathcal{LTS} \mapsto \mathcal{LTS}^\circ$ *s.t.* $\alpha_{\mathcal{LTS}}((S, \to, s_0)) = ((S, \to_\alpha, s_0))$ *with* $\to_\alpha = \{s' \xrightarrow{\eta,\beta^\bullet}_\circ s'' | s' \xrightarrow{\eta,\beta} s'' \in \to\}$.

Notice that α_{LTS} does not effectively introduce any approximation. For expressing the correctness of an abstract LTS with respect to a concrete one, we need an approximation order $\sqsubseteq_{\mathcal{LTS}^\circ}$. In this way, we can say that $lts^\circ \in \mathcal{LTS}^\circ$ is a *sound approximation* of $lts \in \mathcal{LTS}$ provided that $\alpha_{\mathcal{LTS}^\circ}(lts) \sqsubseteq_{\mathcal{LTS}^\circ} lts^\circ$.

Definition 12 (Abstract LTS Order). *Let* $lts_i^\circ = (S_i, \to_\circ^i, s_{0,i}), i \in \{1, 2\}$, *two abstract LTS. For* $s_1 \in S_1$ *and* $s_2 \in S_2$, $s_1 \preceq_{\mathcal{LTS}^\circ} s_2$ *(s_2 simulates s_1) iff*

1. $s_1 = s_2$;
2. $\forall t_1^\circ = (s_1 \xrightarrow{\eta,\beta_1^\circ}_\circ s') \in \to_\circ^1, \exists t_2^\circ = (s_2 \xrightarrow{\eta,\beta_2^\circ}_\circ s') \in \to_\circ^2$ *such that* $\beta_1^\circ \sqsubseteq_\mathbb{I} \beta_2^\circ$.

We say that $lts_1^\circ \sqsubseteq_{\mathcal{LTS}^\circ} lts_2^\circ$ *iff* $s_{0,1} \preceq_{\mathcal{LTS}^\circ} s_{0,2}$.

The definition of order for abstract LTS is based on a notion of simulation between states. Intuitively, a state s' simulates another state s'' if they represents the same system configuration and if each outcoming transition from s' is matched by a transition from s'' to the same arrival configuration, with a coarser transition rate interval.

The following theorem states that the abstract LTS of an abstract model is a correct approximation of the LTS, of all the corresponding concrete models.

Theorem 2. $\forall M^\circ \in \mathcal{M}^\circ, \forall M \in \gamma(M^\circ) : \alpha_{\mathcal{LTS}}(LTS(M)) \sqsubseteq_{\mathcal{LTS}^\circ} LTS^\circ(M^\circ)$.

3.3 Abstract Probabilistic Semantics

We use the *Interval Discrete-Time Markov Chain* [3,4] to define the probabilistic semantics of an abstract model.

Definition 13 (IMC). *An IMC is a tuple* (S, P^-, P^+, s), *where:* $S \subseteq \mathcal{S}(\Sigma)$ *and* $s \in S$ *are a countable set of states and the initial state;* $P^-, P^+ : S \to PDistr(S)$ *are the* lower and upper *probability transition function s.t.* $\forall s', s'' \in S$, $P^-(s')(s'') \leq P^+(s')(s'')$ *and* $P^-(s, S) \leq 1 \leq P^+(s, S)$.

Here, $P(s')(s'')$ and $P^+(s')(s'')$ define intervals of probabilities, that represent *lower and upper bounds for the transition probabilities* of moving from s' to s''. In the following we use $\mathcal{MC}^\circ$ to denote the universe of IMCs.

On a IMC, for any state s, there is a choice for an *admissible distribution* yielding the probabilities to reach successor states. A distribution is admissible for an IMC $mc^\circ = (S, P^-, P^+, s_0)$ and a state $s \in S$, iff, $\forall s' \in S$: $P^-(s)(s') \leq \rho(s') \leq P^+(s)(s')$. We use $ADistr_{mc^\circ}(s)$ for the admissible distributions for s and mc°.

The notion of path for IMC is analogous to that presented for DTMC and we use therefore the same notation.

Definition 14 (Scheduler). *Let* $mc^\circ = (S, P^-, P^+, s_0)$ *be an IMC. A scheduler is a function* $\$: FPaths(S) \mapsto ADistr_{mc^\circ}(\pi_{last})$ *for each path* $\pi \in FPaths(S)$. *We use* $Adv(mc^\circ)$ *for the set of schedulers on* mc°.

Given a scheduler $\$ \in Adv(mc^\circ)$ the probability space over paths can be defined analogously as for DTMC (see Definition 4). Thus, $P_s^\$$ stands for the probability on an IMC starting from the state s w.r.t. the scheduler $\$$.

To relate a DTMC to its abstract counterpart IMC we introduce the concept of *best abstraction* of a DTMC. As for LTS, the derived probabilities are exact.

Definition 15 (Best Abstraction of DTMC)
Let $\alpha_{MC} : \mathcal{MC} \mapsto \mathcal{MC}^\circ$ *s.t.* $\alpha_{MC}((S, P, s_0)) = ((S, P_\alpha^-, P_\alpha^+, s_0))$ *where,* $\forall s', s'' \in S$, $P_\alpha^-(s', s'') = P_\alpha^+(s', s'') = P(s', s'')$.

In the style of [12,13], we introduce an approximation order $\sqsubseteq_{\mathcal{MC}^\circ}$.

Definition 16 (Order on IMC). *Let* $mc_i^\circ = (S_i, P_i^-, P_i^+ s_{0,i}), i \in \{1, 2\}$, *two IMC. Given two states* $s_i \in S_i, i \in \{1, 2\}$, $s_1 \preccurlyeq_{\mathcal{MC}^\circ} s_2$ (s_2 *simulates* s_1) *iff* (i) $s_1 = s_2$ *and* (ii) $ADistr_{mc_1^\circ}(s_1) \subseteq ADistr_{mc_2^\circ}(s_2)$. *We say that* $mc_1^\circ \preccurlyeq_{\mathcal{MC}^\circ} mc_2^\circ$ *iff* $s_{0,1} \preccurlyeq_{\mathcal{MC}^\circ} s_{0,2}$.

3.4 Derivation of Abstract Markov Chain Semantics

We define the abstract probabilistic translation function $\mathcal{H}^\circ : \mathcal{LTS}^\circ \to \mathcal{MC}^\circ$. The abstract LTS reports on transitions the number of the rule which is applied and the interval representing a possible range for its rate. From this kind of information, both lower and upper bounds for the probabilities of moving from a state to another can be calculated. Following the guidelines of the derivation of the DTMC from the concrete LTS, we introduce $R^\circ : S \times S \mapsto \mathbb{I}$, and $E^\circ : S \mapsto \mathbb{I}$ s.t. $R^\circ(s', s'') = \sum_{t \in TS(s', s'')}^{\mathbb{I}} rate^\circ(t)$ and $E^\circ(s') = \sum_{s'' \in S}^{\mathbb{I}} R^\circ(s', s'')$.

Intuitively, $R^\circ(s')(s'')$ reports the interval of rates corresponding to the move from s' to s'', while $E^\circ(s')$ is the abstract exit rate. Both lower and upper bounds of the probability of moving from s' to s'' can be determined by $R^\circ(s')(s'')$ and by $E^\circ(s')$. For these purposes we need to consider the *worst case* and *best case* scenario, respectively. That is, the transition to be maximised (minimised) takes as rate value its upper (lower) bound and all the others take their lower (upper) bound. This reasoning has to be properly combined with the special cases when $max(E^\circ(s')) = 0$ (the state s' is stable) or $min(E^\circ(s')) = 0$ (the state s' is stable for some values of kinetic constant of some rules).

Definition 17 (Abstract Probabilistic Translation Function). *We define* $\mathcal{H}^\circ : \mathcal{LTS}^\circ \to \mathcal{MC}^\circ$ *such that* $\mathcal{H}^\circ((S, \to^\circ, s_0)) = (S, P^-, P^+, s_0)$, *where* $P^-, P^+ : S \to PDistr(S)$ *are computed, for each* $s' \in S$, *as follows:*

- *if* $max(E^\circ(s')) = 0$, *then* $P^+(s')(s'') = P^-(s')(s'') = 0$, *for each* $s' \neq s''$ *and* $P^+(s')(s') = P^-(s')(s') = 1$;
- *if* $max(E^\circ(s')) > 0$, *then*
  (a) *if* $min(E^\circ(s')) = 0$, *then* $P^+(s')(s') = 1$ *and* $P^-(s')(s') = 0$
  (b) *for each* s'', *if* $min(R^\circ(s', s'')) = 0$, *then* $P^-(s')(s'') = 0$ *else*
     $P^-(s')(s'') = min(R^\circ(s', s''))/max(E^\circ(s')) - max(R^\circ(s', s'')) + min(R^\circ(s', s''))$
  (c) *for each* s'', *if* $max(R^\circ(s', s'')) = 0$, *then* $P^+(s')(s'') = 0$ *else*
     $P^+(s')(s'') = max(R^\circ(s', s''))/min(E^\circ(s')) - min(R^\circ(s', s'')) + max(R^\circ(s', s''))$.

The following theorems states the soundness of $\mathcal{H}^\circ$ w.r.t. the approximation order $\sqsubseteq_{\mathcal{MC}^\circ}$, and that $\alpha_{\mathcal{MC}} \circ \mathcal{H} = \mathcal{H}^\circ \circ \alpha_{\mathcal{LTS}}$.

Theorem 3. *Let* $lts_i^\circ = (S_i, \to_\circ{}^i, s_{0,i}), i \in \{1, 2\}$, *two abstract LTS. If* $lts_1^\circ \sqsubseteq_{\mathcal{LTS}^\circ} lts_2^\circ$ *then* $\mathcal{H}^\circ(lts_1^\circ) \sqsubseteq_{\mathcal{MC}^\circ} \mathcal{H}^\circ(lts_2^\circ)$.

Theorem 4. *Let* $M \in \mathcal{M}$, $\alpha_{\mathcal{MC}}(\mathcal{H}(LTS(M))) = \mathcal{H}^\circ(\alpha_{\mathcal{LTS}}(LTS(M)))$.

3.5 Probabilistic Model Checking of Interval Markov Chains

By realizing probabilistic model checking on an abstract model we compute *lower* and *upper bounds* for the concrete reachability probability of all the abstracted models. On IMCs the computation of reachability probabilities considers the minimum and maximum probabilities w.r.t. all the schedulers, giving *under* and *over* approximations (for details see [5]).

Definition 18 (Reachability Probability). *Let* $mc^\circ = (S, P^-, P^+, s_0)$ *be an IMC. The* lower *and* upper bound *of the probability of reaching a state satisfying a propositional symbol* $A \in AP$, *starting from* $s \in S$, *are defined as follows:*
$Reach^-_{A,mc^\circ}(s) = \inf_{\$ \in Adv(mc^\circ)} P_s^\$(\{\pi \in C(s) \mid \pi[i] \vDash A \text{ for some } i \geq 0\})$;
$Reach^+_{A,mc^\circ}(s) = \sup_{\$ \in Adv(mc^\circ)} P_s^\$(\{\pi \in C(s) \mid \pi[i] \vDash A \text{ for some } i \geq 0\})$.

Theorem 5. *Let $mc_i^\circ = (S_i, P_i^-, P_i^+, s_{0,i}), i \in \{1,2\}$, two IMC and $s_i \in S_i$, $i \in \{1,2\}$, two states. If $s_1 \preceq_{\mathcal{MC}^\circ} s_2$ then $\forall A \in AP$:*

$$Reach^-_{A,mc_2^\circ}(s_2) \leq Reach^-_{A,mc_1^\circ}(s_1) \leq Reach^+_{A,mc_1^\circ}(s_1) \leq Reach^+_{A,mc_2^\circ}(s_2).$$

The Theorems 3, 4 and 5 show that the IMC, derived from the abstract LTS, gives *conservative bounds* for probabilistic reachability properties.

Example 2. We consider the model of Example 1 but, in this case, we assume that we are not sure about the kinetic rate of each rule, but we only estimate the interval in which they lie in. For instance, we consider $M^\circ = (\mathcal{R}, \mathcal{K}^\circ, s_0)$ where $\mathcal{R}$ and s_0 are the same of Example 1, while $\mathcal{K}^\circ = \{k_1^\circ = [1,5], k_2^\circ = [1,5]\}$.

Figure 2 shows the derived $LTS^\circ(M^\circ)$ and $\mathcal{H}^\circ(LTS^\circ(M^\circ))$, where the state space S is the same of Example 1. By computing the probability of obtaining at least two complexes XY, we obtain $[4/104, 1/2] \times^{\mathbb{I}} [1/51, 1/3] = [1/1326, 1/6]$. This result shows that, even if the rates of the reactions are not precise, we have the same behaviour of Example 1. The concentration of reagent W makes the degradation more likely to happen than the binding of reagent X and Y.

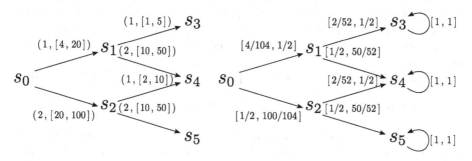

Fig. 3. $LTS^\circ(M^\circ)$ and $\mathcal{H}^\circ(LTS^\circ(M^\circ))$

4   Case Study

We briefly present the application of the proposed approach to a model of tumor growth proposed by Villasana and Radunskaya and studied with Delay Differential Equations (DDEs) in [6].

Tumor growth is based on cell divisions (or *mitosis*). The cell cycle is the process between two mitosis, and it consists of four phases: the G_1 phase (a resting phase or gap period) called pre-synthetic phase, the S phase where the replication of DNA occurs, the G_2 gap period, called the post-synthetic phase, and the mitosis phase M in which the cells segregate the duplicated sets of chromosomes between daughter cells. The three phases G_1, S, and G_2 constitute the pre-mitotic phase, also called *interphase*.

The simplest model proposed in [6] considers two populations of tumor cells: the population of tumor cells during cell cycle interphase, and the population of

tumor cells during mitosis. Such a model can be expressed as the following set
of reactions:

$$\mathcal{R} = \{R_1 : T_I \xrightarrow{a_1} T_M, \quad R_2 : T_M \xrightarrow{a_4} 2T_I, \quad R_3 : T_I \xrightarrow{d_2}, \quad R_4 : T_M \xrightarrow{d_3} \}$$

where T_I and T_M are tumor cells in interphase and in mitosis, respectively.
Reaction R_1 represents the passage of a tumor cell from the interphase to the
mitosis phase, reaction R_2 represents the mitosis, and reactions R_3 and R_4
represent tumor cell death.

Let d be the rate at which mitotic cells disappear, namely $d = d_3 + a_4$.
Figure 4 shows the results of the analytical study of the DDEs model, by setting
the parameters a_4 and d_2 to 0.5 and 0.3, respectively, and by varying a_1 and
d. There are two regions. The region in which the tumor grows is R-I, while
in R-II both kinds of tumor cells disappear. A concrete probabilistic model of
tumor growth could be trivially obtained from reactions $\mathcal{R}$. We have constructed
three abstract models of tumor growth M_1°, M_2° and M_3° by replacing rates in
the reactions with intervals. Actually, in all the three models we have replaced
a_1 with $[0.8, 0.9]$, a_4 with $0.5^\bullet$, d_2 with $0.3^\bullet$. As regards d_3, we have replaced
it with $[0.05, 0.1], [1, 1.4]$ and $[0.005, 2]$ in M_1°, M_2° and M_3°, respectively. This
corresponds to consider a region in R-I, a region in R-II and a region across the
line separating R-I and R-II (see Figure 4). Moreover, we have considered an
initial population consisting of 10 tumor cells in interphase and 10 tumor cells
in mitosis.

Formally, $M_i^\circ = (\mathcal{R}, \mathcal{K}_i^\circ, s_0)$, with $i \in \{1, 2, 3\}$, where $s_0 = \{(T_I, 10), (T_M, 10)\}$,

$$\mathcal{K}_i^\circ = [[0.8, 0.9]; 0, 5^\bullet; 0, 3^\bullet; d_3^i], \quad \text{where } d_3^1 = [0.05, 0.1], d_3^2 = [1, 1.4], d_3^3 = [0.005, 2].$$

In order to perform model checking on the abstract model we have developed
a translator [14] of abstract MSR models into equivalent MDP models (by
following the extreme distribution approach of [5]) that invokes PRISM [15]

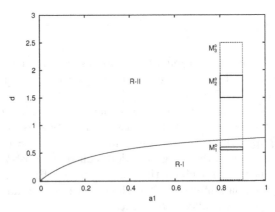

Fig. 4. The regions which describe the different behaviours of the DDEs model by
varying parameters a_1 and d

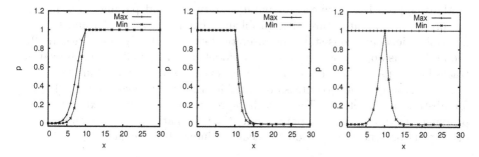

Fig. 5. Model checking results of $Reach(T_M = x)$ in, from left to right, $M_1^\circ, M_2^\circ, M_3^\circ$

for the verification of the properties of interest on the MDP model. Moreover, to obtain a finite MDP, we have limited the number of states of the model to 10^4 by applying standard abstraction techniques.

In Figure 5 we show the results of model checking of property $Reach(T_M = x)$ in M_1°, M_2° and M_3° by varying x. In M_1° both the minimum and the maximum probabilities tend to zero for small values of x while they are both equal to 1 for values greater or equal to 10 (the initial value of T_M). In M_2° it holds the opposite. In M_3° we have that both probabilities are equal to 1 when x is 10, but they tend to the interval $[0, 1]$, namely to complete uncertainty, both for greater and smaller values of x.

The obtained results agree with the analytic ones. In fact, the results on M_1° suggest tumor growth, those on M_2° suggest tumor decay while those on M_3° leave uncertainty.

Our approach is more precise with respect to analytic studies as it looks at all possible behaviours of the modeled system, rather than a single average behaviour. Moreover, a discrete probabilistic semantics is considered, instead of a continuous deterministic one.

5   Related Work

The abstraction of DTMC probabilistic semantics in terms of IMC is presented in [5,16,17,18,19]. In the context of formal studies of biological systems, in [20,21] abstract interpretation techniques are used to coarse-grain a system model and to perform static analyses, respectively.

Most of the above techniques differ substantially from our application. In particular, their goal is to address the state-explosion problem, e.g. to obtain a smaller abstract model by collapsing sets of concrete states into abstract states. An abstract model is thus derived from a single concrete model. Instead, we use abstraction to formalize uncertainty, representing with an abstract model an infinite set of concrete models.

Different kinds of abstraction of probabilistic semantics are proposed in [22,23], where abstract interpretation is applied to probabilistic programs and concurrent constraint programs, respectively.

Many approaches to uncertain parameter tuning and parameters synthesis are present in the literature. The problem is examined in [24,25] and [26] for deterministic, continuous-state and discrete-state respectively, semantics. In [27] the problem is approached by statistical model checking, while both, simulation traces based, parameter tuning and model revision are considered in [28].

The closest approach to ours is presented in [12], where a systematic technique for abstracting a set of DTMC, each representing a concrete experiment, is proposed. Their abstraction approximates the information about the multiplicity of reagents present in a solution by means of intervals of integers.

6   Conclusion

In the paper we consider models of biological systems defined by Multiset Rewriting where rewriting rules, corresponding to reactions, are enriched by real valued kinetic constraints. Our framework allows probabilistic systems with uncertain kinetics to be exhaustively model checked without any artificial assumption, obtaining conservative probabilistic bounds as result.

The computational complexity of the proposed approach is exponential in the number of uncertain parameters. The cause of this is the translation of IMC to MDP that requires for each state the computation of all the extreme distributions that grow exponentially with the number of uncertain parameters.

We plan to investigate the application of parametric DTMC [29,30] to perform parameters tuning, and the extension of our approach to CTMC, by using the theory presented in [31] where uniform CTMC [32] is used.

References

1. Cervesato, I., Durgin, N.A., Lincoln, P., Mitchell, J.C., Scedrov, A.: A meta-notation for protocol analysis. In: CSFW, pp. 55–69 (1999)
2. Cousot, P., Cousot, R.: Abstract interpretation: a unified lattice model for static analysis of programs by construction or approximation of fixpoints. In: POPL, pp. 238–252. ACM Press, New York (1977)
3. Jonsson, B., Larsen, K.G.: Specification and refinement of probabilistic processes. In: LICS, pp. 266–277. IEEE, Los Alamitos (1991)
4. Kozine, I., Utkin, L.V.: Interval-valued finite markov chains. Reliable Computing 8(2), 97–113 (2002)
5. Fecher, H., Leucker, M., Wolf, V.: *Don't Know* in probabilistic systems. In: Valmari, A. (ed.) SPIN 2006. LNCS, vol. 3925, pp. 71–88. Springer, Heidelberg (2006)
6. Villasana, M., Radunskaya, A.: A delay differential equation model for tumor growth. J. of Math. Biol. 47, 270–294 (2003)
7. Kwiatkowska, M.Z.: Model checking for probability and time: from theory to practice. In: LICS, pp. 351–360. IEEE, Los Alamitos (2003)
8. Kwiatkowska, M.Z., Norman, G., Parker, D.: Prism: Probabilistic symbolic model checker. In: Field, T., Harrison, P.G., Bradley, J., Harder, U. (eds.) TOOLS 2002. LNCS, vol. 2324, pp. 200–204. Springer, Heidelberg (2002)
9. Hansson, H., Jonsson, B.: A logic for reasoning about time and reliability. Formal Asp. Comput. 6(5), 512–535 (1994)

10. Kearfott, R.B.: Interval computations: Introduction, uses, and resources. In: Euromath Bulletin, vol. 2, pp. 95–112. European Mathematical Trust (1996)
11. Weichselberger, K.: The theory of interval-probability as a unifying concept for uncertainty. In: Cooman, G.D., Cozman, F.G., Moral, S., Walley, P. (eds.) ISIPTA, pp. 387–396 (1999)
12. Coletta, A., Gori, R., Levi, F.: Approximating probabilistic behaviors of biological systems using abstract interpretation. ENTCS, vol. 229, pp. 165–182. Elsevier, Amsterdam (2009)
13. D'Argenio, P.R., Jeannet, B., Jensen, H.E., Larsen, K.G.: Reachability analysis of probabilistic systems by successive refinements. In: de Luca, L., Gilmore, S. (eds.) PROBMIV 2001, PAPM-PROBMIV 2001, and PAPM 2001. LNCS, vol. 2165, pp. 39–56. Springer, Heidelberg (2001)
14. AMSR2PRISM, http://www.di.unipi.it/~milazzo/biosims/
15. PRISM model checker, http://www.prismmodelchecker.org
16. Huth, M.: On finite-state approximants for probabilistic computation tree logic. Theor. Comput. Sci. 246, 113–134 (2005)
17. Sen, K., Viswanathan, M., Agha, G.: Model-checking markov chains in the presence of uncertainties. In: Hermanns, H., Palsberg, J. (eds.) TACAS 2006. LNCS, vol. 3920, pp. 394–410. Springer, Heidelberg (2006)
18. Škulj, D.: Finite Discrete Time Markov Chains with Interval Probabilities. In: Lawry, J., Miranda, E., Bugarín, A., Li, S., Gil, M.A., Grzegorzewski, P., Hryniewicz, O. (eds.) SMPS, pp. 299–306. Springer, Heidelberg (2006)
19. Blanc, J.P.C., Hertog, D.D.: On Markov Chains with Uncertain Data. CentER Discussion Paper Series 50, Tilburg Univ., Center for Economic Research (2008)
20. Danos, V., Feret, J., Fontana, W., Krivine, J.: Abstract interpretation of cellular signalling networks. In: Logozzo, F., Peled, D.A., Zuck, L.D. (eds.) VMCAI 2008. LNCS, vol. 4905, pp. 83–97. Springer, Heidelberg (2008)
21. Fages, F., Soliman, S.: Formal Cell Biology in Biocham. In: Bernardo, M., Degano, P., Zavattaro, G. (eds.) SFM 2008. LNCS, vol. 5016, pp. 54–80. Springer, Heidelberg (2008)
22. Monniaux, D.: Abstract interpretation of programs as Markov decision processes. In: Sci. Comput. Program, vol. 58, pp. 179–205. Springer, Heidelberg (2005)
23. Di Pierro, A., Wiklicky, H.: Concurrent constraint programming: towards probabilistic abstract interpretation. In: PPDP, pp. 127–138. ACM, New York (2000)
24. Wilkinson, S.J., Benson, N., Kell, D.B.: Proximate parameter tuning for biochemical networks with uncertain kinetic parameters. In: Mol. bioSys., vol. 4, pp. 74–97. RSC Publishing (2008)
25. Batt, G., Belta, C., Weiss, R.: Model Checking Genetic Regulatory Networks with Parameter Uncertainty. In: Bemporad, A., Bicchi, A., Buttazzo, G. (eds.) HSCC 2007. LNCS, vol. 4416, pp. 61–75. Springer, Heidelberg (2007)
26. Manca, V.: The Metabolic Algorithm for P Systems: Principles and Applications. Theor. Comput. Sci. 404, 142–155 (2008)
27. Donaldson, R., Gilbert, D.: A Model Checking Approach to the Parameter Estimation of Biochemical Pathways. In: Heiner, M., Uhrmacher, A.M. (eds.) CMSB 2008. LNCS (LNBI), vol. 5307, pp. 269–287. Springer, Heidelberg (2008)
28. Fages, F., Soliman, S.: Model Revision from Temporal Logic Properties in Systems Biology. In: De Raedt, L., Frasconi, P., Kersting, K., Muggleton, S.H. (eds.) Probabilistic Inductive Logic Programming. LNCS, vol. 4911, pp. 287–304. Springer, Heidelberg (2008)

29. Lanotte, R., Maggiolo-Schettini, A., Troina, A.: Parametric probabilistic transition systems for system design and analysis. Form. Asp. Comput. 19, 93–109 (2007)
30. Han, T., Katoen, J.P., Mereacre, A.: Approximate Parameter Synthesis for Probabilistic Time-Bounded Reachability. In: RTSS, pp. 173–182. IEEE, Los Alamitos (2008)
31. Katoen, J.P., Klink, D., Leucker, M., Wolf, V.: Three-Valued Abstraction for Continuous-Time Markov Chains. In: Damm, W., Hermanns, H. (eds.) CAV 2007. LNCS, vol. 4590, pp. 311–324. Springer, Heidelberg (2007)
32. Baier, C., Hermanns, H., Katoen, J.P., Haverkort, B.R.: Efficient computation of time-bounded reachability probabilities in uniform continuous-time Markov decision processes. Theor. Comput. Sci. 345, 2–26 (2005)

How to Tackle Integer Weighted Automata Positivity

Yohan Boichut, Pierre-Cyrille Héam, and Olga Kouchnarenko

[1] LIFO/University of Orléans
[2] LSV INRIA/CNRS/ENS Cachan
[3] INRIA/CASSIS and LIFC/University of Franche-Comté

Abstract. This paper is dedicated to candidate abstractions to capture relevant aspects of the integer weighted automata. The expected effect of applying these abstractions is studied to build the deterministic reachability graphs allowing us to semi-decide the positivity problem on these automata. Moreover, the papers reports on the implementations and experimental results, and discusses other encodings.

1 Introduction

Weighted automata is a formalism widely used in computer science for applications in images compression [27,28], speech-to-text processing [36,13] or discrete event systems [19]. These large application areas make them intensively studied from the theoretical point of view [31,38,25,30,15,29]. The expressive power of these automata is high enough so that many natural questions are not decidable. Among them the problem to know whether for a given max/+-automaton $\mathcal{A}$, every word has a positive cost, called the positivity problem, was shown to be undecidable [31]. This problem is of special interest because systems/components comparisons modelled by max/+-automata can be based on or reduced to it.

The question we are interested in is whether the automatic verification of certain properties taking costs into account is possible on max/+-automata. As the semantics of max/+-automata model is described by an infinite structure, there is a need of finite abstractions of this semantics to perform analysis fully automatically. Here the problem of handling costs becomes apparent. Obviously, this kind of finite abstractions does not exist for max/+-automata, at least not for the cost-based verification problem investigated. Given a max/+-automaton, our research focuses on methods for semi-deciding whether in the infinite structure *there are a word and a reachable configuration containing some final state reachable from an initial state of the* max/+-*automaton, with cost* -1 *at most.*

After introducing preliminary notions and recalling useful results on max/+-automata (Section 2), we briefly explain how the positivity problem can be encoded into a reachability problem (Sect. 3). Next we explain how to tackle this reachability problem using two semi-decision approaches. The first one (developed in Sect. 4) is based on a configuration space exploration using a pruning property to reduce the search. The second one (exposed in Sect. 5) uses a rewriting encoding of the problem and applies approximation techniques developed in

O. Bournez and I. Potapov (Eds.): RP 2009, LNCS 5797, pp. 79–92, 2009.

the rewriting theoretical framework. We report on experiments with the two semi-algorithms that were implemented (Sect. 6), in particular when bounding the depth of search. Section 7 contains a discussion on possible ways to tackle remaining unsolved instances and gives some perspectives before concluding in Sect. 8.

Well-structured transition systems, or WSTSs, are a general family of transition systems where general decidability results exist [17,1]. It turns out that it is possible to give to many classes of models a structure of WSTSs [18]. We want to emphasise the fact that it is not the case for max/+-automata. Consequently, thanks to the expressivity results in [7], the determinisation reachability graphs corresponding to max/+-automata do not give rise to systems sitting inside some level of the symbolic transition systems (STS) hierarchy in [26].

In a verification context, weighted (priced) systems have been studied in many recent works (see e.g.,[3,12,32,2]). The central underlying problem of these works is to compute the optimal weight of a path to reach a given configuration (from an initial configuration); the difficulties are due to timed constraints (for locations and/or transitions). In this paper, the main difficulty lies in the quantification *for all words u*.

2   Preliminaries

In this paper, Σ denotes a finite alphabet, i.e. a finite set of symbols whose elements are called *letters*. We assume that the reader is familiar with basic language theory notions as word, language, etc. In the paper, the words *weight* and *cost* are indistinctly used.

We denote by $\overline{\mathbb{Z}}$ the set $\mathbb{Z} \cup \{-\infty\}$. Addition and max-function are classically extended to $\overline{\mathbb{Z}}$ by: for every $x \in \overline{\mathbb{Z}}$, $-\infty + x = x + -\infty = -\infty$ and $\max(x, -\infty) = \max(-\infty, x) = x$.

Definition 1. *A max/+-automaton $\mathcal{A}$ over Σ is a quintuplet $\mathcal{A} = (Q, \Sigma, E, I, F)$ where Q is the finite set of states, $E \subseteq Q \times \Sigma \times \mathbb{Z} \times Q$ is the set of transitions, $I \subseteq Q$ is the set of initial states, and $F \subseteq Q$ is the set of final states. Moreover, $\mathcal{A}$ satisfies the following condition: if (p, a, c, q) and (p, a, d, q) are in E, then $c = d$.*

Figure 1 gives two examples of max/+-automata. Initial states are represented with an input arrow, and final states with a double circle.

A *path* π of a max/+-automaton $\mathcal{A}$ is a finite sequence $\pi = (p_0, a_0, c_0, q_0), (p_1, a_1, c_1, q_1), \ldots, (p_n, a_n, c_n, q_n)$ of transitions of $\mathcal{A}$ such that for every $0 \leq i < n$, $q_i = p_{i+1}$. If we add the conditions: p_0 is an initial state, q_n is a final state, then we call π a *successful path*. The *label* $lab(\pi)$ of the path π is the word $a_0 a_1 \ldots a_n$, and the *cost* of the path π is the sum of the c_i's: $\mathrm{cost}_{\mathcal{A}}(\pi) = \sum_{i=0}^{n} c_i$. The *cost* of a word u, denoted $\mathcal{A}(u)$, is the maximum of all costs of successful paths of label u: $\mathcal{A}(u) = \max\{\mathrm{cost}_{\mathcal{A}}(\pi) \mid lab(\pi) = u\}$.

Example 1. For instance, for the max/+-automaton $\mathcal{A}_{exe1}$ in Fig. 1, the word $baaab$ labels the successful paths $(q_1, b, 1, q_2), (q_2, a, 2, q_1), (q_1, a, 0, q_1), (q_1, a, 0, q_1),$

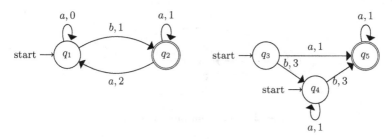

Fig. 1. max/+-automata $\mathcal{A}_{exe1}$ and $\mathcal{A}_{exe2}$

$(q_1, b, 1, q_2)$, $(q_1, b, 1, q_2), (q_2, a, 1, q_2), (q_2, a, 2, q_1), (q_1, a, 0, q_1), (q_1, b, 1, q_2)$ and $(q_1, b, 1, q_2), (q_2, a, 1, q_2), (q_2, a, 1, q_2), (q_2, a, 2, q_1)(q_1, b, 1, q_2)$. Therefore $\mathcal{A}_{exe1}$ $(baaab) = 6$.

Notice that since u is finite, there are finitely many successful paths of label u. A max/+-automaton is finitely ambiguous if there exists an integer k such that every word accepted by the automaton is the label of k successful paths, at most. In Fig. 1, $\mathcal{A}_{exe2}$ is finitely ambiguous, whereas $\mathcal{A}_{exe1}$ is not: the word $ba^n b$ is accepted by $n - 1$ different successful paths. We end this section by recalling some useful results on decision procedures for finite (integer weighted) automata exploited in this paper.

Theorem 1. *Given a* max/+-*automaton* $\mathcal{A}$, *it is undecidable to test whether for every* $u \in L(\mathcal{A})$, $\mathcal{A}(u) \geq 0$ *[31], and polynomial time decidable whether for every* $u \in L(\mathcal{A})$, $\mathcal{A}(u) \geq 0$ *if* $\mathcal{A}$ *is finitely ambiguous [25,38].*

3   Reachability Encoding

Given an max/+-automaton $\mathcal{A}$, while it is undecidable to test whether for every $u \in L(\mathcal{A})$, $\mathcal{A}(u) \geq 0$ [31], we define a determinisation-based abstraction of the model, leading to graphs for which reachability can be semi-decided. More precisely, in this section, the operational semantics of a max/+-automaton $\mathcal{A}$ over Σ is given as a determinisation reachability graph where for a given word in Σ^*, the corresponding configuration contains the information on maximal costs for reaching states of $\mathcal{A}$.

Let $\mathcal{A} = (Q, \Sigma, E, I, F)$ be a max/+-automaton. The *determinisation graph* $\mathcal{G}(\mathcal{A}) = (V, \delta, s_0, K)$ of $\mathcal{A}$ is defined by

- $V = \overline{\mathbb{Z}}^Q$, $s_0 \in V$ and $s_0(p) = 0$ if $p \in I$, and $s_0(p) = -\infty$, otherwise;
- $\delta \subset (V \times \Sigma) \times V$ is the function defined $\delta(s, a) = s'$ iff $s'(p) = \max\{s(q) + c \mid (q, a, c, p) \in E\}$, with the convention that $\max \emptyset = -\infty$;
- $K = \{s \in V \mid \exists q \in F, \ s(q) \neq -\infty \text{ and } \forall p \in F, \ s(p) < 0\} \subseteq V$.

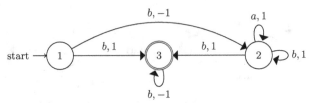

Fig. 2. max/+-automata $\mathcal{A}_{exe3}$

Example 2. Let us consider for instance the automaton $\mathcal{A}_{exe3}$ depicted in Fig. 2.

An element s of $\overline{\mathbb{Z}}^{\{1,2,3\}}$ is denoted (x, y, z) if $s(1) = x$, $s(2) = y$ and $s(3) = z$. $\mathcal{G}(\mathcal{A}_{exe3}) = (\overline{\mathbb{Z}}^{\{1,2,3\}}, \delta_{exe3}, (0, -\infty, -\infty), K_{exe3})$ with $K_{exe3} = \{(x, y, z) \mid z < 0$ and $z \neq -\infty\}$. A part of δ_{exe3} is depicted in Fig. 3 (at this stage, we are not concerned with dashed arrows).

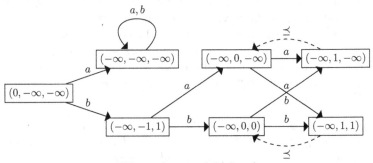

Fig. 3. A part of $\mathcal{G}(\mathcal{A}_{exe3})$

The automaton $\mathcal{A}$ is said to be *non-positive* if in $\mathcal{G}(\mathcal{A})$ there exists a path from s_0 to an element of K.

Proposition 1. *Let* $\mathcal{A} = (Q, \Sigma, E, I, F)$ *be a* max/+-*automaton. There exists* $u \in L(\mathcal{A})$ *such that* $\mathcal{A}(u) < 0$ *if and only if* $\mathcal{A}$ *is non-positive.*

Proposition 1 is a direct consequence of the following lemma. The reader familiar with max/+-automata may notice that this lemma is a direct consequence of matricial presentation of max/+-automata.

Lemma 1. *Let* $u \in \Sigma^+$, $\mathcal{A} = (Q, \Sigma, E, I, F)$ *be a* max/+-*automaton and* $\mathcal{G}(\mathcal{A}) = (V, \delta, s_0, K)$ *its determinisation graph. There is a path in* $\mathcal{G}(\mathcal{A})$ *from* s_0 *to* s *labelled by* u *if and only if for every* $p \in Q$,

$$s(p) = \max\{cost_{\mathcal{A}}(\pi) \mid \pi \text{ is a path in } \mathcal{A} \text{ from an initial state to } p\}.$$

Proof. We will prove the lemma by induction on the length of u.

Assume that $u \in \Sigma$ and that $\delta(s_0, u) = s$. By definition of δ, for every state p, $s(p) = \max\{s_0(q) + c \mid (q, a, c, p) \in E\}$. Therefore and by definition of s_0, $s(p)$

is exactly the maximal value of all transition weights from an initial state to p, proving the lemma for u's in Σ.

Assume now that the lemma is true for all words of length $k \geq 1$. Let $u \in \Sigma^{k+1}$. There exists $v \in \Sigma^k$ and $a \in \Sigma$ such that $u = va$. Let $s_1 = \delta(s_0, v)$. Each path π in $\mathcal{A}$ from an initial state to p can be decomposed into $\pi = \pi_1, (q, a, c, p)$ where π_1 is labelled by v and $(q, a, c, p) \in E$. Since $\mathrm{cost}_{\mathcal{A}}(\pi) = \mathrm{cost}_{\mathcal{A}}(\pi_1) + c$, one has

$$
\begin{aligned}
s(p) &= \max\{s_1(q) + c \mid (q, a, c, p) \in E\} \\
&= \max\{\max\{\mathrm{cost}(\pi_1) \mid \pi_1 \text{ from an initial state to } q\} + c \mid (q, a, c, p) \in E\} \\
&= \max\{\max\{\mathrm{cost}(\pi_1) + c \mid \pi_1 \text{ from an initial state to } q\} \mid (q, a, c, p) \in E\} \\
&= \max\{\mathrm{cost}(\pi_1) + c \mid \pi_1 \text{ from an initial state to } q \text{ and } (q, a, c, p) \in E\} \\
&= \max\{\mathrm{cost}_{\mathcal{A}}(\pi) \mid \pi \text{ is a path in } \mathcal{A} \text{ from an initial state to } p\}.
\end{aligned}
$$

Consequently, the lemma is true for words of Σ^{k+1}, concluding the proof.

4 State Space Exploration

We are interested in semi-deciding whether a max/+-automaton $\mathcal{A}$ is non-positive. Clearly, this is a matter of walking – by classical algorithms like depth-first search, random-walk, etc. – the determinisation graph $\mathcal{G}(\mathcal{A})$ defined above, until either a configuration in K is reached or there is an argument to prove such a configuration can no longer be found. Unfortunately, the determinisation reachability graph is generally infinite, and it is not easy to determine when it is safe to stop. Consequently, these algorithms may not terminate and can only conclude that $\mathcal{A}$ is non-positive but, when $\mathcal{G}(\mathcal{A})$ has infinitely many reachable configurations, they cannot conclude that $\mathcal{A}$ is not non-positive.

While reachability seems to a be a good tool to find configurations in K, for practical problems the determinisation graph usually has far too many configurations to calculate. To alleviate this problem, we exploit a *pruning configuration* approach. For that there is a need to introduce the relation $\preceq$ over configurations of a determinisation graph $\mathcal{G}(\mathcal{A})$ of an max/+-automaton $\mathcal{A}$. We define this relation by: $s_1 \preceq s_2$ iff for every state p in $\mathcal{A}$, $s_1(p) = -\infty$ iff $s_2(p) = -\infty$ and $s_1(p) \leq s_2(p)$ otherwise. The pruning is based on the following property.

Proposition 2. *Let $\mathcal{A} = (Q, \Sigma, E, I, F)$ be a max/+-automaton and $\mathcal{G}(A) = (V, \delta, s_0, K)$ its determinisation graph. Let $s_1, s_2 \in \overline{\mathbb{Z}}^Q$ such that $s_1 \preceq s_2$. Then if a configuration s_2' in K is reachable in $\mathcal{G}(A)$ from s_2, then there also is a configuration s_1' in K reachable from s_1.*

Proposition 2 can be proved by a direct induction using the following lemma.

Lemma 2. *Let $\mathcal{A} = (Q, \Sigma, E, I, F)$ be a max/+-automaton and $\mathcal{G}(A) = (V, \delta, s_0, K)$ its determinisation graph. Let $s_1, s_2 \in \overline{\mathbb{Z}}^Q$ such that $s_1 \preceq s_2$. Then for every letter $a \in \Sigma$, $\delta(s_1, a) \preceq \delta(s_2, a)$.*

Proof. Notice first that δ is a function defined on $V \times \Sigma$, thus $\delta(s_1, a)$ and $\delta(s_2, a)$ both exist.

Now $\delta(s_1, a)(p) = -\infty$ iff $\{q \mid s_1(q) \neq -\infty\} \cap \{q \mid \exists(q, a, c, p) \in E\} = \emptyset$. Since $s_1 \preceq s_2$, $\{q \mid s_1(q) \neq -\infty\} = \{q \mid s_2(q) \neq -\infty\}$. Therefore, $\delta(s_1, a)(p) = -\infty$ iff $\delta(s_2, a)(p) = -\infty$. Moreover, for every state p,

$$\delta(s_1, a)(p) = \max\{s_1(q) + c \mid (q, a, c, p) \in E\}$$
$$\leq \max\{s_2(q) + c \mid (q, a, c, p) \in E\}$$
$$= \delta(s_2, a)(p),$$

proving the lemma.

While bounding the depth, Proposition 2 leads to the search based algorithm depicted in Fig. 4. In this algorithm, δ and K are related to the determinisation graph of $\mathcal{A}$. Notice too that a **Return** instruction ends the execution of the algorithm. Integer k is the bound of the number of computed configurations of the determinisation graph of $\mathcal{A}$. Set C is the set of computed accessible configurations. Set L encodes configurations to explore. Line 08, the function **Get** takes an element of L: the way this function is implemented may lead to different search approaches (depth first search, breadth first search, etc.). Next the graph is classically computed but only for configurations s such that there is no $s' \in C$ such that $s' \preceq s$ (notice that $\preceq$ is reflexive). The procedure ends at Line 06 if there is no more configuration to visit: K is not reachable. The algorithm then returns 1, indicating that for all $u \in \Sigma^+$, $\mathcal{A}(u) \geq 0$. The procedure ends at Line 10 if a configuration of K is reachable. Then the algorithm returns -1 indicating there exists u such that $\mathcal{A}(u) < 0$. At Line 18, the algorithm returns 0, indicating that it cannot conclude whether $\mathcal{A}$ is non-positive or not.

For instance, let consider the max/+-automaton depicted in Fig. 2. The exploration algorithm computes the graph depicted in Fig. 3 where dashed arrows represent the $\preceq$ relation. On this example, the algorithm stops after a few steps and returns 1.

Algorithm Name: Explore
Input: $\mathcal{A}$, $k \in \mathbb{N}$
Local Variables: L, C finite sets,
Begin

01.	Compute $C := \emptyset$	10.	**Return** -1
02.	Compute s_0	11.	**EndIf**
03.	Compute $L := \{s_0\}$	12.	**If not** exists $s' \in C$ s.t. $s' \preceq s$
04.	**While** $(k \geq 0)$	13.	$C := C \cup \{s\}$
05.	**If** $C \cap K = \emptyset$ and $L = \emptyset$	14.	$L := L \cup \{\delta(s, a) \mid a \in \Sigma\}$
06.	**Return** 1	15.	**EndIf**
07.	**EndIf**	16.	$k := k - 1$
08.	**Get** $s \in L$	17.	**EndWhile**
09 .	**If** $C \cap K \neq \emptyset$	18.	**Return** 0

End

Fig. 4. Exploration algorithm

5   Rewriting Techniques Approach

Rewriting techniques are also well-suited for performing reachability analysis. In particular, reachability analysis allows verifying safety properties on critical systems: Java programs [35,16], cryptographic protocols [20] or Java Bytecode programs [9].

For the use of such techniques, rewriting semantics are defined for a given reachability problem, and the reachability analysis is performed from a rewriting point of view. Section 5.1 describes the rewriting model we use for determinisation graphs, and Section 5.2 explains how to show that an max/+-automaton is positive.

5.1   Rewriting Model for Determinisation Graphs

Focusing on the abstraction chosen in this paper, we specify the determinisation graph $\mathcal{G}(\mathcal{A})$ of a given automaton $\mathcal{A}$ as follows: its states are represented by terms and its transition relation is then compiled into rewrite rules. Integers are manipulated in their peano representations, i.e., using the constructors s (for successor), p (for predecessor) and 0. For example, 1 is represented by the term $s(0)$ and -2 by $p(p(0))$.

Thus, a configuration of a determinisation graph $\mathcal{G}(\mathcal{A})$ is specified by a term of the form $run(w_1, \ldots, w_n)$ where n is the number of states of $\mathcal{A}$, w_i is either a peano integer or $-\infty$. Considering this representation, the initial configuration $(0, -\infty, -\infty)$ of the determinisation graph in Fig. 3 is specified by the term $run(0, -\infty, -\infty)$.

The transition relation of a determinisation graph $\mathcal{G}(\mathcal{A})$ is then specified by a term rewriting system (TRS), i.e., a set of rewrite rules. The algorithm for generating such a TRS is simple. For a given max/+−automaton $\mathcal{A} = (\mathcal{Q}, \Sigma, E, I, F)$, we generate a set of rules per symbol of Σ by anticipating every possible scenario.

For instance, concerning $\mathcal{A}_{exe3}$ of Fig. 2 and the letter b, b can be read from the states 1, 2 and 3. So, a configuration of the determinisation graph when b is reading is a term of the form $run(t_1, t_2, t_3)$ where x_i's are variables, $t_1 \in \{-\infty, s(x_1), p(x_1)\}$, $t_2 \in \{-\infty, s(x_2), p(x_2)\}$ and $t_3 \in \{-\infty, s(x_3), p(x_3)\}$. For each of these terms, according to the transition relation of $\mathcal{G}(\mathcal{A}_{exe3})$, a successor term can be defined.

Example 3. For example, let $run(s(x_1), -\infty, p(x_3))$ be one of the forms mentioned right above. According to the $\mathcal{G}(\mathcal{A}_{exe3})$ transition relation, the following successor term can be set: $run(-\infty, +(s(x_1), p(0)), max(+(s(x_1), s(0)), +(p(x_3), p(0))))$. Consequently, one can define the rewrite rule

$$run(s(x_1), -\infty, p(x_3)) \rightarrow run(-\infty, +(s(x_1), p(0)), max(+(s(x_1), s(0)), +(p(x_3), p(0)))).$$

Doing so for each letter of Σ and for each form of terms, the whole transition relation can be defined as a TRS $\mathcal{R}$. In addition to these rules, those concerning the function max and the addition $+$ between two peano integers complete the set of rewrite rules.

5.2   Reachability Analysis

The rewriting model is now defined. Since we face systems whose number of states is potentially infinite, a complete and exact rewriting analysis is in general impossible. A well-suited approach as proposed in [22] is to compute an over-approximation of the reachable terms by rewriting – with a given set of rewrite rules $\mathcal{R}$ – from an initial set of terms E.

Initially, terms and subterms of terms in E are split into equivalence classes. For example, one can use tree automata to define equivalence classes where classes are actually the states of these automata. We refer the interested reader to [14,23] for more detail on tree automata and theoretical results on this topic. The technique in [22] enhances and creates new equivalence classes of terms and subterms by rewriting. If a term t is in an equivalence class C and t' is reachable by rewriting from t, then t' added into the equivalence class C. Moreover, new equivalence classes may be added if there are subterms of t' which are not in existing equivalence classes. One proceeds in this way for all equivalence classes defined.

Approximations are done by manipulating equivalence classes of terms. In [21], Genet uses equations for merging equivalence classes. Let $c = c'$ be an equation where c and c' are two patterns, i.e., two terms that may contain variables. Let also C and C' be two equivalence classes of terms built with the technique described in [22]. If there exists a solution of c in C (resp. C') and a solution of c' in C' (resp. C), then the two equivalence classes are merged.

Example 4. For example, let consider the equation $s(x) = s(s(x))$ and the equivalence classes C_0, C_1, C_2, C_3 and C_4 such that $C_i = \{s^{(i)}(0)\}$. Since $s(0)$ is in C_1 and $s(s(0))$ is in C_2, using the equation we obtain that $s(0) = s(s(0))$. Consequently, C_1 and C_2 are merged into $C_{1,2}$. The same process can be applied for C_3 and C_4. Thus, the merging of C_3 and C_4 results in the equivalence class $C_{3,4}$. Once again, $s(s(0))$ and $s(s(s(0)))$ are respectively in $C_{1,2}$ and $C_{3,4}$. Using the given equation, the process results in a single equivalence class denoted C_{1to4}. Finally, using the given equation over the five equivalence classes gives rise to only two equivalence classes: C_0 and C_{1to4}.

As soon as the set of equivalence classes is stable by equation, rewriting is performed anew, and so on. The computation stops when all equivalence classes are closed by rewriting, i.e., when a fix-point set of terms is computed. Thus, the final set of terms is an over-approximation of the set of reachable terms.

For performing a reachability analysis, we can check on the fix-point set of terms if a pattern has a solution. If no solution exists then we can conclude that no term matching such a pattern is reachable from an initial set of terms E by rewriting with the given TRS $\mathcal{R}$.

Example 5. For example, in Fig. 2, the state 3 is the final state of $\mathcal{A}_{exe3}$. From the rewriting model, if we obtain a fix-point set of terms E', we have to check whether the patter $run(x, y, p(z))$ has a solution. In the negative case, we can conclude that no path in the determinisation graph has a negative cost. And, consequently, we also conclude that for every $u \in \mathcal{L}(\mathcal{A}_{exe3})$, $\mathcal{A}_{exe3}(u) \geq 0$. Whereas in the positive case, no conclusion can be raised. Indeed, the solution of the pattern may come from a side-effect of the approximation.

Section 6 reports on the implementation and experimental results for the proposed rewriting model. Notice that the rewriting-based encoding and analysis are used when the exploration algorithm in Fig. 4 returns 0. Obviously, other rewriting models and other rewriting approximations can be defined.

6   Experiments

In order to evaluate our approaches, we randomly generate non-deterministic finite max/+-automata using the following method: given a set of states $\{1, \ldots, n\}$, for each letter a and each i, j, there is a fixed probability $p_{\text{transition}}$ to have a transition of the form (i, a, c, j). If such a transition exists, its weight is uniformly picked up between $-c_{\text{max}}$ and c_{max}. Moreover, 1 is the unique initial state, n is always a final state, and there is a fixed probability p_{final} for each other state to be final. If a generated automaton accepts the empty language, it is rejected. We have done several tests with different values of c_{max}, $p_{\text{transition}}$ and p_{final}. Table 1 reports on results obtained with $c_{\text{max}} = 3$, $p_{\text{transition}} = 0.3$ and $p_{\text{final}} = 0.1$. For each value of n from 2 to 20, we randomly generate 1000 automata. Line n is the number of states of the automata. We first run the Explore algorithm developed in Sec. 4 with $k = 10n$. Line *pos.* (resp. *neg.*) reports on the proportion of inputs when the algorithm returns 1 (resp. -1). Line *??* indicates the number of automata (out of 1000 automata generated for each n) for which the algorithm returns 0. Line *depth* reports on the average number of computed reachable configurations in the Explore algorithm (when it returns 1 or -1).

When the first algorithm gives the inconclusive results, we apply the second rewriting approximation approach to them. Experiments have been led for $n = 2, 3, 4$ and 5 using equations allowing to split integers into 13 equivalence classes: $< -5, = -5, = -4, = -3, = -2, = -1, = 0, = 1, = 2, = 3, = 4, = 5$ and > 5. For example, the equivalence class < -5 is defined by the equation $p(p(p(p(p(p(x)))))) = p(p(p(p(p(p(p(x))))))).$

Table 1 reports at line *inc.* on the number of automata that are not shown to be positive using the rewriting approximation technique among inconclusive

Table 1. Experimental results

n	2	3	4	5	6	7	8	9	10	12	14	16	18	20
pos.	0.34	0.23	0.15	0.1	0.1	0.12	0.13	0.16	0.21	0.27	0.33	0.40	0.42	0.6
neg.	0.65	0.74	0.82	0.86	0.87	0.85	0.83	0.81	0.77	0.71	0.66	0.59	0.57	0.54
depth	2.45	3.83	4.66	5.92	6.68	6.88	7.14	7.10	7.40	7.35	7.46	7.64	7.47	7.37
??	4	21	25	39	27	28	27	22	21	23	8	8	5	4
TRS	36	43	58	79	125	296	554	1068	2094	8242	32822	131130	524350	$\approx 2^{21}$
inc.	0	4	6	10	6	T	T	T	T	T	T	T	T	T

analyses from the pruning approach. The result T points out that the implementation of the rewriting approach fails to answer because of a stack overflow. This table also gives details (line TRS) about the average number of rewrite rules generated for the rewriting specifications.

7   Discussions and Perspectives

Let consider the max/+-automaton $\mathcal{A}_{exe4}$ depicted in Fig. 5. Notice that $\mathcal{A}_{exe4}$ is not non-positive.

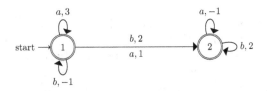

Fig. 5. max/+-automaton $\mathcal{A}_{exe3}$

For this automaton, the exploration will never end since $\mathcal{G}(\mathcal{A}_{exe4})$ has infinitely many configurations of the form $(-n, 2n)$, which are pairwise incomparable by $\preceq$. For more difficult reasons, similar to those given in [10], the approximation technique can not conclude either.

We discuss and propose several ways to handle remaining intractable cases.

7.1   Counter Systems Encoding

Presburger logic is the first order logic over $(\mathbb{Z}, +, =)$. A *n-counter-system* $\mathcal{C}$ is a tuple (Q, T, P) where Q is a finite set of *states*, P is a finite set of Presburger formulas with $2n$ free variables, and T is a finite set of elements of the form (p, φ, q) where $\varphi \in P$. For every $\varphi(x_1, \ldots, x_n, y_1, \ldots, y_n) \in P$ we define the relation $\rightarrow_\varphi$ on $\mathbb{Z}^n \times \mathbb{Z}^n$ by: $(a_1, \ldots, a_n) \rightarrow_\varphi (b_1, \ldots, b_n)$ iff $\varphi(a_1, \ldots, a_n, b_1, \ldots, b_n)$ is true. Finally, given the set $S_0 \subseteq \mathbb{Z}^n$, the set $\text{Post}^*_\mathcal{C}(S_0)$ (resp. $\text{Pre}^*_\mathcal{C}(S_0)$) is the set of $s \in \mathbb{Z}^n$ such that there exist $w = w_1 \ldots w_k \in P^*$ ($w_i \in P$), $s_0 \in S_0$ and $s_1, \ldots, s_k \in \mathbb{Z}^n$, where $s_k = s$ and for every i, $s_i \rightarrow_{w_{i+1}} s_{i+1}$ (resp. $s_{i+1} \rightarrow_{w_{i+1}} s_i$).

It is known [24] that subsets of $\mathbb{Z}^n$ that are definable by a Presburger formula with n free variables are exactly regular subsets of $(\mathbb{Z}^n, +)$. This nice property, associated with nice connections to Petri nets, has lean to many works to compute sets of the form $\text{Post}^*_\mathcal{C}(S_0)$ or $\text{Pre}^*_\mathcal{C}(S_0)$ (see [33] for a recent work with references), supported by tools as FAST [6], LASH [11] or TReX [4].

We now illustrate how to encode our problem into this model. Let $\mathcal{A} = (Q, \Sigma, E, I, F)$ be a max/+-automaton. Without loss of generality we may assume that $Q = \{1, \ldots, n\}$. We consider the function ψ from $\overline{\mathbb{Z}}^Q$ into $\mathbb{Z}^{2n}$ defined as follows: for every $s \in \overline{\mathbb{Z}}^Q$, $\psi(s)$ is the vector $(s_1, \ldots, s_{2n})$ where for every

$\varphi_b(x_1, x_2, x_3, x_4, x_5, x_6, y_1, y_2, y_3, y_4, y_5, y_6) :=$

$\quad y_1 = 0 \wedge y_4 = 1$

$\quad \wedge ((x_4 = 0 \wedge x_5 = 0 \wedge x_6 = 0) \Rightarrow (y_2 = 0 \wedge y_3 = 0 \wedge y_5 = 1 \wedge y_6 = 1))$

$\quad \wedge ((x_4 = 0 \wedge x_5 = 0 \wedge x_6 = 1) \Rightarrow (y_2 = 0 \wedge y_3 = x_3 - 1 \wedge y_5 = 0 \wedge y_6 = 1))$

$\quad \wedge ((x_4 = 0 \wedge x_5 = 1 \wedge x_6 = 0) \Rightarrow (y_2 = x_2 + 1 \wedge y_3 = x_2 + 1 \wedge y_5 = 0 \wedge y_6 = 0))$

$\quad \wedge ((x_4 = 0 \wedge x_5 = 1 \wedge x_6 = 1) \Rightarrow (y_2 = x_2 + 1 \wedge \varphi_{\max}(y_3, x_2 + 1, x_3 - 1) \wedge y_5 = 0$
$$\wedge y_6 = 0))$$

$\quad \wedge ((x_4 = 1 \wedge x_5 = 0 \wedge x_6 = 0) \Rightarrow (y_2 = x_1 - 1 \wedge y_3 = x_1 + 1 \wedge y_5 = 0 \wedge y_6 = 0))$

$\quad \wedge ((x_4 = 1 \wedge x_5 = 0 \wedge x_6 = 1) \Rightarrow (y_2 = x_1 - 1 \wedge \varphi_{\max}(y_3, x_1 + 1, x_3 - 1) \wedge y_5 = 0$
$$\wedge y_6 = 0))$$

$\quad \wedge ((x_4 = 1 \wedge x_5 = 1 \wedge x_6 = 0) \Rightarrow (\varphi_{\max}(y_2, x_1 - 1, x_2 + 1)$
$$\wedge \varphi_{\max}(y_3, x_1 + 1, x_2 + 1) \wedge y_5 = 0 \wedge y_6 = 0))$$

$\quad \wedge ((x_4 = 1 \wedge x_5 = 1 \wedge x_6 = 1) \Rightarrow \varphi_{\max}(y_2, x_1 - 1, x_2 + 1)$
$$\wedge \exists z \, (\varphi_{\max}(y_3, x_1 + 1, z) \wedge (\varphi_{\max}(z, x_2 + 1, x_3 - 1)) \wedge y_5 = 0 \wedge y_6 = 0))$$

Fig. 6. Presburger formula

$1 \leq i \leq n$, $s_i = s(i)$ and $s_{n+i} = 0$ if $s_i \in \mathbb{Z}$, and $s_i = 0$ and $s_{n+i} = 1$ otherwise. For instance if $Q = \{1, 2, 3\}$ and $s(1) = -1$, $s(2) = 3$ and $s(3) = -\infty$, then $\psi(s) = (-1, 3, 0, 0, 0, 1)$. Notice first that the max-function is Presburger definable: $z = \max(x, y)$ iff x, y, z satisfy the formula

$$\varphi_{\max}(z, x, y) := ((z = x \vee z = y) \wedge ((x \leq y) \Rightarrow z = y))$$

Writing exact formulas encoding a generic $\mathcal{A}$ is quite long. Since we do not use this approach and since our goal is just to show how to use it, we provide the encoding for the automaton $\mathcal{A}_{exe3}$. One has $\delta_{exe3}(s, b) = s'$ iff $\varphi_b(\psi(s), \psi(s'))$ is satisfied, where φ_b is depicted in Fig. 6.

In this context, the non-positivity problem is reduced either to $\mathrm{Pre}_\mathcal{C}^*(\{\psi(s_0))\}) \cap \psi(K) = \emptyset$? or, equivalently, to $\psi(s_0) \in \mathrm{Post}_\mathcal{C}^*(\{\psi(K))\})$? where $\mathcal{C}$ is the counter system encoding $\mathcal{A}$. One can also easily verify that $\psi(K)$ is Presburger definable.

7.2   Using max/+ Theory

Another way to improve the approach consists in using theoretical results on max/+-automata. For instance, a recent work [29] points out new subclasses of max/+-automata for which the positivity problem is decidable. However, the proposed constructive proof is far from being effective, and an algorithmic research has still to be done.

Another very interesting direction may be to use results of [34]: for a one-letter alphabet, many problems becomes decidable. In particular, such results can be used during the exploration of a determinisation graph. When visiting a configuration s, for each letter a, one can test with one step whether there exists $n \geq 0$ such that $\delta(s, a^n) \cap K \neq \emptyset$. It may deeply reduce the exploration for non-positive max/+-automata. Moreover, we think this approach can be used to perform a symbolic exploration of the determinisation graph: rather than visiting each accessible configuration, we would work on infinite sets of configurations similarly to the counter system encoding presented below.

8   Conclusion

We proposed to exploit abstractions and approximations to semi-decide the positivity problem over max/+-automata whose determinisation reachability graphs are infinite state systems. The positivity problem is then reduced to a reachability problem on these graphs. We developed two semi-decision procedures and explained how to conclude more often and how to do it efficiently.

The first kind of determinisation-based reachability graphs abstractions together with pruning technique gives rise to a semi-decision procedure. The experimental results on thousands of automatically generated max/+-automata show that when bounding the depth of search in the determinisation graphs, the algorithm seems to be efficient enough.

The second kind of abstractions is based on the reachability analysis through rewriting approximations as well as tree automata. The rewriting-based reachability encoding has been applied to the inconclusive cases previously obtained with the exploration algorithm.

Rewriting approximation techniques were already implemented in [5]. In the future we plan to integrate integer weighted automata based algorithms into this tool in order to treat practical applications. Obviously, other rewriting models and other rewriting approximations can be defined. Moreover, one can propose an abstraction refinement for rewriting approximations guided by the property to be verified, as in [8].

Finally, we plan to experiment with other random generators of non-deterministic finite automata, for instance using the technique developed in [37].

References

1. Abdulla, P.A., Cerans, K., Jonsson, B., Yih-Kuen, T.: General decidability theorems for infinite-state systems. In: Proc. 11th IEEE Symp. Logic in Computer Science (1996)
2. Abdulla, P.A., Mayr, R.: Minimal cost reachability/coverability in priced timed petri nets. In: de Alfaro, L. (ed.) FOSSACS. LNCS, vol. 5504, pp. 348–363. Springer, Heidelberg (2009)
3. Alur, R., La Torre, S., Pappas, G.J.: Optimal paths in weighted timed automata. In: Di Benedetto, M.D., Sangiovanni-Vincentelli, A.L. (eds.) HSCC 2001. LNCS, vol. 2034, pp. 49–62. Springer, Heidelberg (2001)
4. Annichini, A., Bouajjani, A., Sighireanu, M.: Trex: A tool for reachability analysis of complex systems. In: Berry, G., Comon, H., Finkel, A. (eds.) CAV 2001. LNCS, vol. 2102, pp. 368–372. Springer, Heidelberg (2001)

5. Balland, E., Boichut, Y., Genet, T., Moreau, P.-E.: Towards an efficient implementation of tree automata completion. In: Meseguer, J., Roşu, G. (eds.) AMAST 2008. LNCS, vol. 5140, pp. 67–82. Springer, Heidelberg (2008)
6. Bardin, S., Finkel, A., Leroux, J., Petrucci, L.: Fast: acceleration from theory to practice. STTT 10(5), 401–424 (2008)
7. Bertrand, N., Schnoebelen, P.: A short visit to the sts hierarchy. Electr. Notes Theor. Comput. Sci. 154(3), 59–69 (2006)
8. Boichut, Y., Courbis, R., Héam, P.-C., Kouchnarenko, O.: Finer is better: Abstraction refinement for rewriting approximations. In: Voronkov, A. (ed.) RTA 2008. LNCS, vol. 5117, pp. 48–62. Springer, Heidelberg (2008)
9. Boichut, Y., Genet, T., Jensen, T., Le Roux, L.: Rewriting Approximations for Fast Prototyping of Static Analyzers. In: Baader, F. (ed.) RTA 2007. LNCS, vol. 4533, pp. 48–62. Springer, Heidelberg (2007)
10. Boichut, Y., Héam, P.-C.: A theoretical limit for safety verification techniques with regular fix-point computations. Inf. Process. Lett. 108(1), 1–2 (2008)
11. Boigelot, B., Wolper, P.: Representing arithmetic constraints with finite automata: An overview. In: Stuckey, P.J. (ed.) ICLP 2002. LNCS, vol. 2401, pp. 1–19. Springer, Heidelberg (2002)
12. Bouyer, P., Brihaye, T., Bruyère, V., Raskin, J.-C.: On the optimal reachability problem of weighted timed automata. Formal Methods in System Design 31(2), 135–175 (2007)
13. Buchsbaum, A.L., Giancarlo, R., Westbrook, J.: An approximate determinization algorithm for weighted finite-state automata. Algorithmica 30(4), 503–526 (2001)
14. Comon, H., Dauchet, M., Gilleron, R., Jacquemard, F., Lugiez, D., Tison, S., Tommasi, M.: Tree automata techniques and applications (2002), http://www.grappa.univ-lille3.fr/tata/
15. Droste, M., Gastin, P.: Weighted automata and weighted logics. Theor. Comput. Sci. 380(1-2), 69–86 (2007)
16. Farzan, A., Chen, C., Meseguer, J., Rosu, G.: Formal analysis of java programs in javafan. In: Alur, R., Peled, D.A. (eds.) CAV 2004. LNCS, vol. 3114, pp. 501–505. Springer, Heidelberg (2004)
17. Finkel, A.: Reduction and covering of infinite reachability trees. Information and Computation 89(2), 144–179 (1990)
18. Finkel, A., Schnoebelen, P.: Well-structured transition systems everywhere! TCS 256(1-2), 63–92 (2001)
19. Gaubert, S.: Performance Evaluation of (max,+) Automata. IEEE Trans. on Automatic Control 40(12) (1995)
20. Genet, T., Klay, F.: Rewriting for Cryptographic Protocol Verification. In: McAllester, D. (ed.) CADE 2000. LNCS (LNAI), vol. 1831, Springer, Heidelberg (2000)
21. Genet, T.: Timbuk 3.0 : Equationnal approximations, http://www.irisa.fr/lande/genet/timbuk/
22. Genet, T.: Decidable approximations of sets of descendants and sets of normal forms. In: Nipkow, T. (ed.) RTA 1998. LNCS, vol. 1379, p. 151. Springer, Heidelberg (1998)
23. Gilleron, R., Tison, S.: Regular tree languages and rewrite systems. Fundamenta Informaticae 24, 157–175 (1995)
24. Ginsburg, S., Spanier, E.H.: Bounded regular sets. Proceedings of the American Mathematical Society 7, 1043–1049 (1966)
25. Hashiguchi, K., Ishiguro, K., Jimbo, S.: Decidability of the Equivalence Problem for Finitely Ambiguous Finance Automata. IJAC 12(3) (2002)

26. Henzinger, T.A., Majumdar, R., Raskin, J.F.: A classification of symbolic transition systems. ACM Trans. Comput. Log. 6(1), 1–32 (2005)
27. Culik II, K., von Rosenberg, P.C.: Generalized weighted finite automata based image compression. J. UCS 5(4), 227–242 (1999)
28. Katritzke, F., Merzenich, W., Thomas, M.: Enhancements of partitioning techniques for image compression using weighted finite automata. TCS 313(1), 133–144 (2004)
29. Kirsten, D., Lombardy, S.: Deciding unambiguity and sequentiality of polynomially ambiguous min-plus automata. In: STACS, pp. 589–600 (2009)
30. Klimann, I., Lombardy, S., Mairesse, J., Prieur, C.: Deciding unambiguity and sequentiality from a finitely ambiguous max-plus automaton. TCS 327(3), 349–373 (2004)
31. Krob, D.: The Equality Problem for Rational Series with Multiplicities in the Tropical Semiring is Undecidable. IJAC 4(3) (1994)
32. Larsen, K.G., Rasmussen, J.I.: Optimal reachability for multi-priced timed automata. Theor. Comput. Sci. 390(2-3), 197–213 (2008)
33. Leroux, J.: Structural presburger digit vector automata. TCS 409(3), 549–556 (2008)
34. Lombardy, S.: Sequentialization and unambiguity of (max,+) rational series over one letter. In: Gaubert, S., Loiseau, J.-J. (eds.) Workshop on max-plus Algebras and Their Applications to Discrete-event Systems, TCS, and Optimization, Prague. IFAC, Elsevier Sciences (2001)
35. Meseguer, J., Rosu, G.: Rewriting logic semantics: From language specifications to formal analysis tools. In: Basin, D., Rusinowitch, M. (eds.) IJCAR 2004. LNCS, vol. 3097, pp. 1–44. Springer, Heidelberg (2004)
36. Mohri, M., Pereira, F., Riley, M.: Weighted finite-state transducers in speech recognition. Computer Speech & Language 16(1), 69–88 (2002)
37. Tabakov, D., Vardi, M.Y.: Experimental evaluation of classical automata constructions. In: Sutcliffe, G., Voronkov, A. (eds.) LPAR 2005. LNCS, vol. 3835, pp. 396–411. Springer, Heidelberg (2005)
38. Weber, A.: Finite-valued Distance Automata. In: TCS, p. 134 (1994)

A Reduction Theorem for the Verification of Round-Based Distributed Algorithms*

Mouna Chaouch-Saad[1], Bernadette Charron-Bost[2], and Stephan Merz[3]

[1] Faculté des Sciences, Tunis, Tunisia
Mouna.Saad@fst.rnu.tn
[2] CNRS & LIX, Palaiseau, France
charron@lix.polytechnique.fr
[3] INRIA Nancy & LORIA, Nancy, France
Stephan.Merz@loria.fr

Abstract. We consider the verification of algorithms expressed in the Heard-Of Model, a round-based computational model for fault-tolerant distributed computing. Rounds in this model are communication-closed, and we show that every execution recording individual events corresponds to a coarser-grained execution based on global rounds such that the local views of all processes are identical in the two executions. This result helps us to substantially mitigate state-space explosion and verify Consensus algorithms using standard model checking techniques.

1 Introduction

Distributed algorithms are often quite subtle, both in the way they operate and in the assumptions they make. Formal verification is therefore crucial in distributed computing. Unfortunately, due to their asynchronous nature, distributed algorithms almost invariably give rise to state-space explosion, severely limiting the applicability of model checking techniques. This is particularly true for fault-tolerant algorithms whose correctness relies on elaborate failure hypotheses.

The key to overcome this problem is to make use of the inherently non-sequential nature of distributed executions and to exploit the causality relation [4] between events of the execution in order to reduce the number of executions that have to be analyzed. In this paper we study reductions that hold for distributed algorithms that are structured in *rounds*: each process first sends messages and receives messages sent for the round, and finally makes a local state transition. More specifically, Charron-Bost and Schiper [1] recently proposed the Heard-Of (HO) model, a round-based model for fault-tolerant distributed computing. We formally prove that for verifying interesting properties for algorithms in this model, it suffices to model executions as infinite sequences of global rounds. Moreover, rounds in the HO model are communication closed, hence the medium of communication can be considered as empty at the end

---

* We gratefully acknowledge support by CMCU (Comité Mixte de Coopération Universitaire) project DEFI Utique.

O. Bournez and I. Potapov (Eds.): RP 2009, LNCS 5797, pp. 93–106, 2009.

of each round, and the overall state can be represented just by the collection of local states for each process. These two observations induce reductions that go beyond well-known techniques of partial-order reduction [3,11], and that can indeed justify reductions from infinite-state to finite-state models. We validate our approach by verifying finite instances of some of the Consensus algorithms proposed in [1], using a standard explicit-state model checker.

The paper is organised as follows: Section 2 provides a short introduction to the HO model and defines executions. Section 3 proves the reduction theorem that establishes a close correspondence between the two representations of executions. We present in Section 4 some experiments on model checking Consensus algorithms in the HO model. Section 5 discusses related work and concludes.

2   The Heard-of Model for Distributed Algorithms

Computations in the HO model are organized in rounds, in which each process exchanges messages, takes a step, and then proceeds to the next round. Without any specific synchronization assumptions, processes execute rounds at their own pace. In particular, the difference between the numbers of rounds that two different processes are executing at any given moment may be arbitrarily large. Rounds are communication-closed layers in the terminology of Elrad and Francez [2]: messages are valid only for the round they were sent in. Thus the model generalizes the classical notion of synchronized rounds developed for synchronous systems [7].[1]

2.1   A Round-Based Computational Model

We suppose that we have a finite, non-empty set Π of process identifiers[2] and a set of messages M. By including a designated *empty* message in M that processes use to indicate absence of useful information, we may assume w.l.o.g. that each process sends some message to every process in Π, in each round. We denote the cardinality of Π by $N > 0$, let $\bot \notin M$ be a placeholder indicating that no message has (yet) been received, and write $M_\bot = M \cup \{\bot\}$. To each p in Π, we associate a *process specification* $Proc_p = (\Sigma_p, s_{0,p}, S_p, T_p)$ whose components are the following:

- Σ_p is the set of p's *states*, and $s_{0,p} \in \Sigma_p$ is the *initial state* of process p,
- $S_p : \mathbb{N} \times \Sigma_p \times \Pi \to M$ is the *message sending function* such that $S_p(r, s, q)$ denotes the message that p sends to q at round r, given the state s of p, and
- $T_p : \mathbb{N} \times \Sigma_p \times M_\bot^\Pi \to \Sigma_p$ is the *next-state function*: $T_p(r, s, \mu)$ yields the successor state of process p at round r, given its current state s and the partial vector $\mu = (\mu_q)_{q \in \Pi}$ of messages where μ_q indicates the message that p received from q at round r, or $\bot$ if no message was received.

The collection of process specifications $Proc_p$ is called an *algorithm* on Π.

---

[1] Communication-closedness can be ensured in asynchronous settings by buffering messages which are early, and by discarding messages which are late.

[2] When there is no risk of confusion, we simply speak of *processes* in Π.

2.2   Executions of HO Algorithms

Each process of an HO algorithm executes an infinite sequence of rounds, which are numbered consecutively, starting with round 0. At the beginning of each round r, process p first emits messages to all processes, computed according to the message sending function S_p. It then waits for messages to arrive for round r before it executes a state transition according to the next-state function T_p, based on its current state and the vector of messages received, and starts a new round. The *heard-of set* $HO(p, r)$ for p at round r is the set of processes from which p receives a message at round r.

Formally, we define executions with respect to a given *HO collection*

$$HO : \Pi \times \mathbb{N} \to 2^{\Pi}$$

that specifies, for each $p \in \Pi$ and round $r \in \mathbb{N}$, the heard-of set $HO(p, r)$. Process p proceeds to round $r+1$ when it has received messages from the processes in $HO(p, r)$. We make HO collections an explicit parameter of the definition of executions because algorithms are unlikely to work under completely arbitrary HO collections. Assumptions on the underlying system model and communication network, such as the degree of synchronism and the failure model, are formally expressed by *communication predicates* $\mathcal{P} \subseteq (\Pi \times \mathbb{N} \to 2^{\Pi})$, and the correctness of an algorithm is asserted relative to a certain communication predicate $\mathcal{P}$. As discussed in [1], standard failure models with various degrees of synchronism can be represented in this way. The weaker the communication predicate is, the more freedom the system has to provide heard-of sets, and the harder it will be to achieve coordination among processes in the corresponding failure model.

Fine-grained executions. We define two models of execution, whose relationship will be explored in Section 3. The *fine-grained model* represents events of individual processes and the way they interleave, and so faithfully models the asynchronous execution of distributed algorithms. A *configuration* of an algorithm is a tuple $(rd, st, sent, rcv, msgs)$:

-   rd, st, $sent$ and rcv are arrays indexed by processes where $rd(p) \in \mathbb{N}$, $st(p) \in \Sigma_p$, $sent \subseteq \Pi$, and $rcv(p) \in M_{\perp}^{\Pi}$ denote, for process p, its current round, its local state, the set of processes to which p has sent messages in the current round, and the partial vector of messages received;
-   $msgs \subseteq \Pi \times \mathbb{N} \times \Pi \times M$ represents the messages in transit: $(p, r, q, m) \in msgs$ if p sent message m at round r to q, but q has not yet received m.

The algorithm starts in the *initial configuration* c where $c.rd(p) = 0$, $c.st(p) = s_{0,p}$, $sent(p) = \emptyset$, and $c.rcv(p) = (q \in \Pi \mapsto \perp)$ for all $p \in \Pi$, and where no messages are in transit, i.e. $c.msgs = \emptyset$.

Configuration c' is a *successor configuration* of c if one of the following cases holds:

– Transition (c, c') represents process p *sending a message* to process q:

$$q \in \Pi \setminus c.sent(p), \quad c'.sent = c.sent\big(p := c.sent(p) \cup \{q\}\big),$$
$$c'.rd = c.rd, \qquad c'.st = c.st, \qquad c'.rcv = c.rcv,$$
$$c'.msgs = c.msgs \cup \big\{ \big(p, c.rd(p), q, S_p(c.rd(p), c.st(p), q)\big) \big\}$$

The transition is enabled if p has not yet sent a message to q during its current round. The effect of the transition is to add the message (computed according to function S_p) to the set of messages in transit and to record the fact that the message has been sent in the *sent* field of configuration c' for process p.

– Transition (c, c') represents a *message reception*: there exist $p, q \in \Pi$ and $m \in M$ such that

$$q \in HO(p, c.rd(p)), \qquad\qquad (q, c.rd(p), p, m) \in c.msgs,$$
$$c'.msgs = c.msgs \setminus \{(q, c.rd(p), p, m)\}, \quad c'.rd = c.rd, \quad c'.st = c.st,$$
$$c'.rcv = c.rcv\big(p := c.rcv(p)(q := m)\big), \quad c'.sent = c.sent.$$

The transition is enabled if q is a member of p's heard-of set for p's current round and message m is in transit from q to p for that round. The effect of the transition is to transfer the message from the set of messages in transit to the vector of messages received by p, while the rounds, process states, and *sent* fields remain unchanged.

– Transition (c, c') is a *local transition* of some process $p \in \Pi$:

$$c.sent(p) = \Pi, \qquad\qquad \operatorname{dom} c.rcv(p) = HO(p, c.rd(p)),$$
$$c'.rd = c.rd\big(p := c.rd(p) + 1\big),$$
$$c'.st = c.st\big(p := T_p(c.rd(p), c.st(p), c.rcv(p))\big),$$
$$c'.sent = c.sent(p := \emptyset), \quad c'.rcv = c.rcv\big(p := (q \in \Pi \mapsto \bot)\big),$$
$$c'.msgs = c.msgs$$

where $\operatorname{dom} c.rcv(p)$ denotes the set $\{q \in \Pi : c.rcv(p, q) \neq \bot\}$.[3] A local transition of p is enabled when p has sent messages for the current round to all processes and has received messages from precisely the processes specified by the HO collection for its current round. The configuration c' is obtained by incrementing the round number of process p, updating its local state according to the next-state function T_p, and resetting the *sent* and *rcv* fields for process p.

A *fine-grained execution* is an ω-sequence $c_0 c_1 \ldots$ of configurations such that c_0 is the initial configuration, c_{i+1} is a successor configuration of c_i for all $i \in \mathbb{N}$, and for each $p \in \Pi$ there are infinitely many $i \in \mathbb{N}$ such that (c_i, c_{i+1}) is a local transition of p. The last condition specifies a condition of (local) progress for each process; since p can execute a local transition ending round r only if it has sent messages to all processes and has received messages from all $q \in HO(p, r)$, this condition also implies the existence of sufficiently many transitions of type message sending and reception.

---

[3] We identify a function $f : A \times B \to C$ and its "curried" version $f_c : A \to (B \to C)$.

Coarse-grained executions. We now define an execution model of HO algorithms that is based on the much coarser abstraction where entire rounds are the unit of atomicity. Thus, a *coarse-grained execution* is an ω-sequence $\sigma_0\sigma_1\ldots$ where each σ_i is an array of local states $\sigma_i(p) \in \Sigma_p$ indexed by $p \in \Pi$, such that

- $\sigma_0(p) = s_{0,p}$ is the initial state of p, for all $p \in \Pi$, and
- at every step, *all* processes make a transition according to their next-state function and the HO collection: for all $p \in \Pi$ and all $r \in \mathbb{N}$,

$$\sigma_{r+1}(p) = T_p\big(r, \sigma_r(p), rcvd(p, r)\big)$$

$$\text{where} \quad rcvd(p, r) = \Big(q \in \Pi \mapsto \begin{cases} S_q(r, \sigma_r(q), p) & \text{if } q \in HO(p, r) \\ \bot & \text{otherwise} \end{cases}\Big).$$

In words, the state $\sigma_{r+1}(p)$ is computed according to the next-state function T_p (at the current round r) from the state $\sigma_r(p)$, and the vector of messages that p receives at round r according to the HO collection. A step of a coarse-grained execution encapsulates a move by each process; because messages can be received only in the rounds for which they have been sent, there is no need to represent messages in transit.

Variations. We made some choices in the above definitions. For example, all message sending transitions for a process in a given round could be grouped into a single transition, and possibly even combined with the local transitions, yielding an intermediate granularity of executions. Another alternative would be to define executions without fixing the HO collection in advance. Instead, a local transition in the fine-grained model could occur at any point after the process has sent all messages for the current round. In the coarse-grained model, the HO sets $HO(p, r)$ would be chosen non-deterministically. Indeed, the representation of HO algorithms in TLA$^+$ presented in Section 4 is defined in such a way.

Charron-Bost and Schiper [1] define a variant of HO algorithms called *co-ordinated* HO algorithms, whose message-sending functions S_p and transition relations T_p depend on an additional parameter indicating the process that p believes to be the coordinator of the current round. Correspondingly, executions are defined in this variant with respect to a HO collection as well as an assignment of coordinators $Coord(p, r) \in \Pi$ per process and round.

The reduction theorem presented in the following section can be adapted to any of these alternative definitions. It also extends to non-deterministic settings where each process has a set of possible initial states and a next-state relation instead of a next-state function. The only essential requirement is that processes react only to messages intended for the round they are currently executing.

3   A Reduction Theorem for HO Algorithms

We now present our main theorem, which asserts, informally, that in the HO model, the fine-grained and coarse-grained execution semantics are indistinguishable from the point of view of any process.

3.1   Relating the Two Models of Execution

Given a (fine-grained or coarse-grained) execution ρ and a process $q \in \Pi$, we define the q-*view* ρ^q of ρ for process q as the sequence of local states that q assumes in ρ. More precisely, for a fine-grained execution $\xi = c_0 c_1 \ldots$, we define

$$\xi^q = c_0.st(q) \; c_1.st(q) \; \ldots.$$

For a coarse-grained execution $\sigma = \sigma_0 \sigma_1 \ldots$, the q-view is simply

$$\sigma^q = \sigma_0(q) \, \sigma_1(q) \, \cdots$$

Any two executions ρ_1 and ρ_2 can be compared with respect to the views that they generate for the processes in Π. We say that two executions ρ_1 and ρ_2 are q-*equivalent* (for $q \in \Pi$) if $\rho_1^q \simeq \rho_2^q$ where $\simeq$ denotes *stuttering equivalence* [5], i.e. if their q-views agree up to finite repetitions of states. We call ρ_1 and ρ_2 *locally equivalent*, written $\rho_1 \approx \rho_2$, if they are q-equivalent for all $q \in \Pi$.

The following theorem asserts that fine-grained executions do not generate any more local views of an algorithm than coarse-grained ones.

Theorem 1. *For any fine-grained execution $\xi = c_0 c_1 \ldots$ of an HO algorithm for some HO collection $\big(HO(p, r)\big)_{p \in \Pi, r \in \mathbb{N}}$, there exists a coarse-grained execution σ of the same algorithm for the same HO collection such that $\sigma \approx \xi$.*

Proof (sketch). Given execution $\xi = c_0 c_1 \ldots$ and some process $p \in \Pi$, let $\ell_0^p = 0$ and for $n > 0$, $\ell_n^p = k + 1$ if (c_k, c_{k+1}) is the n-th local transition of p in ξ; remember that every process p performs infinitely many local transitions in a fine-grained execution. By the definition of fine-grained executions, round numbers and local states of p change only during local transitions. It follows that $c_i.rd(p) = c_{\ell_n^p}.rd(p) = n$ and $c_i.st(p) = c_{\ell_n^p}.st(p)$ for all $n \in \mathbb{N}$ and all $\ell_n^p \leq i < \ell_{n+1}^p$.

We will now show that the sequence $\sigma = \sigma_0 \sigma_1 \ldots$ defined by

$$\sigma_n = \big(p \in \Pi \mapsto c_{\ell_n^p}.st(p)\big)$$

is a coarse-grained execution of the same algorithm for the given HO collection HO. By the observations above, this definition of σ ensures that $\sigma^p \simeq \xi^p$ for all $p \in \Pi$, and therefore $\sigma \approx \xi$.

To show the initialization condition, it suffices to observe that

$$\sigma_0(p) = c_{\ell_0^p}.st(p) = c_0.st(p) = s_{0,p}$$

is the initial state for all $p \in \Pi$. It remains to show that for all $p \in \Pi$ and $n \in \mathbb{N}$, we have $\sigma_{n+1}(p) = T_p\big(n, \sigma_n(p), rcvd(p, n)\big)$.

By induction on the definition of fine-grained executions, it is easy to verify the following invariants, for all $n, r \in \mathbb{N}$, $m \in M$, and $p, q \in \Pi$:

- If $(p, r, q, m) \in c_n.msgs$ then $m = S_p(r, c_{\ell_r^p}.st(p), q)$: any message for round r in transit from p to q was computed according to p's send function for round r, based on p's local state at (the beginning of) round r.

- If $r = c_n.rd(q)$ and $\perp \neq m = c_n.rcv(q,p)$ then $m = S_p(r, c_{\ell_r^p}.st(p), q)$: any message received by q from p for round r was computed according to p's send function for round r, based on p's local state at (the beginning of) round r.

For $n \in \mathbb{N}$, consider now the transition of process p from round n to round $n+1$ in execution ξ, i.e. the transition $(c_{\ell_{n+1}^p - 1}, c_{\ell_{n+1}^p})$. For simplicity of notation, we write $c = c_{\ell_{n+1}^p - 1}$ and $c' = c_{\ell_{n+1}^p}$. From the observations about round numbers and local states we know that $c.rd(p) = n$, $c.st(p) = c_{\ell_n^p}.st(p)$, and $c'.rd(p) = n+1$. Since (c, c') is a local transition of p, we have dom $c.rcv(p) = HO(p,n)$, and in particular $c.rcv(p,q) = \perp$ iff $q \in \Pi \setminus HO(p,n)$. Moreover, the second invariant above implies that

$$c.rcv(p,q) = S_q(n, c_{\ell_n^q}.st(q), p)$$

for all $q \in HO(p,n)$. Altogether this means $c.rcv(p) = rcvd(p,n)$.

Using the fact that $c'.st(p) = T_p(c.rd(p), c.st(p), c.rcv(p))$ and rewriting with the above equalities, we obtain that

$$\sigma_{n+1}(p) = c'.st(p) = T_p(n, \sigma_n(p), rcvd(p,n)),$$

which completes the proof. $\square$

We note in passing that the converse of Theorem 1 is true almost trivially.

Theorem 2. *For any coarse-grained execution σ of an HO algorithm for some HO collection $(HO(p,r))_{p \in \Pi, r \in \mathbb{N}}$, there exists a fine-grained execution ξ of the same algorithm for the same HO collection such that $\xi \approx \sigma$.*

Proof (sketch). Given a coarse-grained execution σ, it is easy to construct a corresponding fine-grained execution where processes execute rounds in lock-step, first sending all messages, then receiving the messages according to the HO sets $HO(p,r)$ and finally performing their respective local transitions. $\square$

3.2   Application: Verification of Local Properties

Theorem 1 can be used to verify linear-time properties of HO algorithms that are expressed in terms of local views of processes, and that are insensitive to specific interleavings. More formally, we say that a property P is *local* if for any (coarse- or fine-grained) executions ρ_1 and ρ_2 such that $\rho_1 \approx \rho_2$ we have $\rho_1 \models P$ iff $\rho_2 \models P$.[4]

Corollary 3. *If P is a local property and $\sigma \models P$ holds for all coarse-grained executions σ of an algorithm, then $\xi \models P$ also holds for all fine-grained executions ξ of the same algorithm.*

---

[4] As usual, $\rho \models P$ means that P is satisfied by execution ρ.

Proof. Let ξ be some fine-grained execution (over some HO collection), then Theorem 1 yields a coarse-grained execution σ (over the same HO collection) such that $\sigma \approx \xi$. By assumption, we must have $\sigma \models P$, and since P is local, this implies $\xi \models P$. $\qquad\square$

Having to verify a given property just for all coarse-grained executions represents a significant reduction because coarse-grained executions afford a simpler representation of the system state, and because fewer (types of) transitions must be considered.

Corollary 3 is useful in practice if typical correctness properties are indeed local. Observe that local properties must be stuttering invariant [8], by the definition of local equivalence $\approx$ of executions. Moreover, their satisfaction should not depend on the specific interleaving of process transitions. As a trivial example for a non-local property, suppose that each process $p \in \Pi$ maintains a counter of its current round in the variable rnd_p. Then any coarse-grained execution by definition satisfies the LTL formula

$$\bigwedge_{p,q \in \Pi} \Box(rnd_p = rnd_q) \tag{1}$$

asserting that all processes execute the same round at any moment; this formula obviously does not hold for fine-grained executions.

In the following we indicate a sufficient syntactic criterion for determining when a formula of LTL-X, i.e. linear-time temporal logic without the next-time operator expresses a local property.[5] We assume that the set of (flexible) state variables that appear in formulas is of the form $\mathcal{V} = \bigcup_{p \in \Pi} \mathcal{V}_p$ where $\mathcal{V}_p \cap \mathcal{V}_q = \emptyset$ for different processes $p \neq q$, and such that any state $s \in \Sigma_p$ of a process $p \in \Pi$ uniquely determines the values of $\mathcal{V}_p$.

We say that a formula φ is a *p-formula*, for $p \in \Pi$, if it contains only state variables from $\mathcal{V}_p$. It is easy to see that p-formulas are local properties, as are first-order combinations of p-formulas, for possibly different processes $p \in \Pi$. However, temporal combinations of p-formulas are in general not local because they can express the simulaneity of local states of different processes, or assert temporal relations between states of processes, and the formula (1) is a typical example since variables of different processes appear in the scope of a temporal operator.

3.3   Consensus as a Local Property

We argue that local properties express many interesting correctness properties of distributed algorithms. As a concrete and important example, consider the specification of the Consensus problem [7]. We assume that the state variables $\mathcal{V}_p$ include variables x_p and $decide_p$. The intuitive idea is that at the beginning of an execution the variable x_p holds the initial value of process p. Variable

---

[5] LTL-X formulas are stuttering invariant [8]; our criterion carries over to the logic TLA considered in Section 4 because LTL-X is a sublogic of TLA.

$decide_p$, initially *null*, represents the decision taken by process p in the sense that $decide_p$ is updated to the value $v \neq null$ when process p decides value v.

The Consensus problem is specified by the conjunction of the following formulas of LTL-X, which are all local according to the criterion introduced in Section 3.2.

Integrity. The integrity property asserts that decision values must be among the initial values (possibly of some other process). This property is expressed by the following first-order combination of p-formulas:

$$\forall v : v \neq null \wedge \left(\bigvee_{p \in \Pi} \Diamond(decide_p = v) \right) \Rightarrow \bigvee_{q \in \Pi} x_q = v.$$

Irrevocability. A process that has decided must never change its decision value. This property is expressed by the following p-formula, for all $p \in \Pi$:

$$\forall v : v \neq null \Rightarrow \Box\big(decide_p = v \Rightarrow \Box(decide_p = v)\big)$$

Agreement. The core correctness property of Consensus algorithms requires that if any two processes decide, they decide on the same value. Again, this can be expressed as a first-order combination of p-formulas:

$$\begin{aligned}
\forall v, w : \quad & v \neq null \wedge w \neq null \\
& \wedge \bigvee_{p,q \in \Pi} \big(\Diamond(decide_p = v) \wedge \Diamond(decide_q = w)\big) \\
& \Rightarrow v = w.
\end{aligned}$$

Termination. The preceding properties are all safety properties. The final property required by Consensus is that all (non-faulty) processes eventually decide. Because the HO model does not flag processes as being faulty [1], this property is simply expressed by the following p-formula, for all $p \in \Pi$:

$$\Diamond(decide_p \neq null).$$

4   Model Checking HO Algorithms

We validate the effectiveness of our reduction-based approach to verification by verifying finite instances of Consensus algorithms in the HO model.

Exploiting Corollary 3, we model coarse-grained executions of HO algorithms in TLA$^+$ [6]. We instantiate this generic model for some of the Consensus algorithms that are discussed in [1], and use the TLA$^+$ model checker TLC [12] for verification. In this work, we do not aim at utmost efficiency, but prefer the high level of abstraction offered by TLA$^+$ that lets us obtain readable models, close to the mathematical description of HO algorithms in Section 2.

However, model checking even finite instances of these algorithms would be impossible in the fine-grained execution model: the model would have to include round numbers and therefore be infinite-state. Even if we artificially imposed bounds on round numbers (abandoning the verification of liveness properties), state explosion would make verification impractical.

────────────────────── MODULE *HeardOf* ──────────────────────
EXTENDS *Naturals*
CONSTANTS *Proc, State, Msg, roundsPerPhase, Start*(_)*, Send*(_, _, _, _)*, Trans*(_, _, _, _)
VARIABLES *round, state, heardof*

$$Init \triangleq \land round = 0$$
$$\land state = [p \in Proc \mapsto Start(p)]$$
$$\land heardof = [p \in Proc \mapsto \{\}]$$

$Step(HO) \triangleq$ LET $rcvd(p) \triangleq \{\langle q, Send(q, round, state[p], p)\rangle : q \in HO[p]\}$
  IN $\land round' = (round + 1)\%roundsPerPhase$
    $\land state' = [p \in Proc \mapsto Trans(p, round, state[p], rcvd(p))]$
    $\land heardof' = HO$

$Next \triangleq \exists HO \in [Proc \to \text{SUBSET } Proc] : Step(HO)$

$vars \triangleq \langle round, state, heardof \rangle$

$NoSplit(HO) \triangleq \forall p, q \in Proc : HO[p] \cap HO[q] \neq \{\}$

$NextNoSplit \triangleq \exists HO \in [Proc \to \text{SUBSET } Proc] : NoSplit(HO) \land Step(HO)$

$Uniform(HO) \triangleq \exists S \in \text{SUBSET } Proc : S \neq \{\} \land HO = [q \in Proc \mapsto S]$

Fig. 1. Generic TLA$^+$ module for HO algorithms

4.1 A Generic TLA$^+$ Model for HO Algorithms

We begin by introducing a generic representation of coarse-grained executions of HO algorithms in TLA$^+$. It is similar to the semantic presentation in Section 2.2, except for two differences that help us obtain finite-state models. The first difference concerns round numbers, which are formally a parameter of the functions S_p and T_p. Many actual algorithms do not refer to the absolute round number, but are organized in *phases*, where a phase consists of a fixed finite number of rounds. Therefore, the functions S_p and T_p only depend on the current round number relative to the phase number, and it suffices to count rounds modulo the number of rounds per phase. The second difference, already indicated at the end of Section 2.2, is to choose assignments of HO sets to processes non-deterministically for each step of the algorithm instead of fixing them in advance.

A generic TLA$^+$ module that represents HO algorithms appears in Fig. 1. It begins by importing the standard TLA$^+$ module for arithmetic over natural numbers and then declares the constant and variable parameters of the module: the sets *Proc, State* and *Msg* represent processes, process states, and messages. Parameter *roundsPerPhase* indicates the number of rounds per phase. The parameters *Start, Send,* and *Trans* will be instantianted to specify the behavior of concrete algorithms: for each $p \in Proc$, the predicate $Start(p)$ characterizes the initial state of p, $Send(p, r, s, q)$ yields the message that process p sends to process q at round r of a phase, given p's current local state s, and $Trans(p, r, s, rcvd)$ computes the next state of p at round r, given p's local state s and the partial vector $rcvd$ of messages received, which we represent as a set of pairs $\langle q, m \rangle$ indicating that m was received from q.

A round of the system is represented by three variables: *round* indicates the number of the current step modulo *roundsPerPhase*, *state* is an array[6] of local states per process, and *heardof* records the HO assignment of the preceding transition (its initial value is chosen arbitrarily). This auxiliary variable serves to express communication predicates.

With this understanding, the initialization and next-state predicates closely follow the definition of coarse-grained executions in Section 2.2. The predicates *NoSplit* and *Uniform* are two examples of formulas serving to express communication predicates. Action *NextNoSplit* defines a variant of the next-state relation that enforces non-split rounds. The remaining communication predicates appearing in [1] can be defined in a similar way.

4.2   Modeling and Verifying Concrete HO Algorithms in TLA$^+$

Charron-Bost and Schiper [1] propose several Consensus algorithms in the HO model. As an example of how these can be encoded in our TLA$^+$ framework, a specification of their *OneThirdRule* algorithm appears in Fig. 2. The module declares the constant parameter N (the number of processes) and *defines* the sets *Proc* and *Msg*: we arbitrarily specify that each process $p \in 1..N$ proposes $10 * p$ as its initial value. Next, the module defines the remaining constant parameters of module *HeardOf*. Phases of *OneThirdRule* consist of only one round. Process states are represented as records with two fields x and *decide*, whose initialization is obvious. At each round, each process sends its current x field to all processes. The next-state function is defined as follows: if a process has received messages from more than 2/3 of all processes, it updates its x field to the smallest most frequently received value (cf. definition of *min*). If it has received some value v from more than 2/3 of all processes, then it also updates its *decide* field to v.

Charron-Bost and Schiper show that the algorithm *OneThirdRule* always achieves the integrity, irrevocability, and agreement properties, and that it guarantees termination for runs that eventually execute some uniform round for "sufficiently large" heard-of sets. We express these properties in TLA and use TLC to verify these theorems. Observe that we have expressed the correctness properties in module *OneThirdRule* using different, but equivalent formulas than those given in Section 3.3. In particular, TLC checks *Validity*, *Agreement*, and *Irrevocability* as state and transition invariants while computing the state space, which is more efficient than verifying arbitrary temporal formulas. Because the concept of local properties is a semantic one, it is independent of the particular syntactic formulation of the property, and we can apply Corollary 3 to deduce that the formulas are also satisfied by fine-grained executions.

Figure 3 gives the number of generated and distinct states and the running time of TLC for verifying these properties, as well as for the somewhat more complicated *UniformVoting* algorithm that is encoded in TLA$^+$ in a similar way. Measurements were taken on an Intel® 2.16GHz Core Duo® laptop with 2GB RAM running Mac OSX 10.5. It is apparent that the non-deterministic

---

[6] TLA$^+$ uses square brackets to denote functions.

───────────────────── MODULE *OneThirdRule* ─────────────────────

EXTENDS *Naturals, FiniteSet*
CONSTANTS N
VARIABLES *round, state, heardof*

──

$roundsPerPhase \triangleq 1$

$Proc \triangleq 1 .. N$

$InitValue(p) \triangleq 10 * p$

$Value \triangleq \{InitValue(p) : p \in Proc\}$

$Msg \triangleq Value$

$null \triangleq 0$

$ValueOrNull \triangleq Value \cup \{null\}$

──

$State \triangleq [x : Value, decide : ValueOrNull]$

$Init(p) \triangleq [x \mapsto InitValue(p), decide \mapsto null]$

$Send(p, r, s, q) \triangleq s.x$

$Trans(p, r, s, rcvd) \triangleq$
  IF $Cardinality(rcvd) > (2 * N) \div 3$
    THEN LET $Freq(v) \triangleq Cardinality(\{q \in Proc : \langle q, v \rangle \in rcvd\})$
           $MFR(v) \triangleq \forall w \in Value : Freq(w) \leq Freq(v)$
           $min \triangleq$ CHOOSE $v \in Value : MFR(v) \wedge (\forall w \in Value : MFR(w) \Rightarrow v \leq w)$
           $willDecide \triangleq \exists v \in Value : Freq(v) > (2 * N) \div 3$
      IN $[x \mapsto min,$
          $decide \mapsto$ IF $willDecide$ THEN CHOOSE $v \in Value : Freq(v) > (2 * N) \div 3$
                ELSE $s.decide]$
    ELSE s

──

INSTANCE *HeardOf*

$Safety \triangleq Init \wedge \Box[Next]_{vars}$

$Liveness \triangleq \Diamond(Uniform(heardof) \wedge Cardinality(heardof) > (2 * N) \div 3)$

$Integrity \triangleq \forall p \in Proc : \Box(state[p].decide \in \{null\} \cup \{InitValue(p) : p \in Proc\})$

$Irrevocability \triangleq \forall p \in Proc : \Box[state[p].decide = null]_{state[p].decide}$

$Agreement \triangleq \forall p, q \in Proc : \Box(state[p].decide \neq null \wedge state[q].decide \neq null$
                         $\Rightarrow state[p].decide = state[q].decide)$

$Termination \triangleq \forall p \in Proc : \Diamond(state[p].decide \neq null)$

──

THEOREM $Safety \Rightarrow Integrity \wedge Irrevocability \wedge Agreement$
THEOREM $Safety \wedge Liveness \Rightarrow Termination$

Fig. 2. TLA$^+$ specification of the algorithm *OneThirdRule*

choice of a collection of heard-of sets at every transition induces a combinatorial
explosion in the number of successor states that are generated, although many of
them are identical (cf. the low number of distinct states generated by the model
checker). As mentioned above, we regard these results just as an indication of the
feasibility of the approach; there is ample room for improvement using standard

	OneThirdRule		UniformVoting	
	$N = 3$	$N = 4$	$N = 3$	$N = 4$
states	5633	9,830,401	21351	15,865,770
distinct	11	150	122	887
time (s)	2.26	1546	13.8	1330

Fig. 3. Results of verification with TLC

optimization techniques or symbolic model checking. In particular, we did not apply symmetry reduction, except for identifying states that differ only in the value of the auxiliary variable *heardof*.

5   Conclusion

The main contribution of this paper is the precise statement and the proof of a reduction theorem for algorithms expressed in the HO model. The key ingredient for obtaining the reduction theorem is the fact that the HO model relies on communication-closed rounds. For this reason, the local transition (of a fine-grained execution) in which process p passes from round r to round $r + 1$ is causally independent of all transitions of processes at rounds $r' > r$. Hence, executions can be rearranged into coarse-grained executions whose unit of atomicity is that of global rounds of all processes, without changing the local observations of any process.

As a corollary to the reduction theorem, we obtain a method for applying standard model checking algorithms for the verification of local properties of distributed algorithms, that is, properties whose satisfaction only depends on local views of processes. We have shown that this class of properties contains the correctness properties of Consensus algorithms. Specifically, we have been able to verify (finite instance of) some Consensus algorithms proposed in [1], which would be impossible in a standard, fine-grained representation of executions.

Tsuchiya and Schiper [9,10] applied symbolic and bounded model checkers to verify Consensus algorithms in the HO model over coarse-grained runs. However, they do not explain why the verification of coarse-grained models is sufficient. Our contribution can be understood as a formal justification of their models; we also delimit the applicability of the approach by introducing the notion of local properties.

In future work, we intend to further validate this approach by verifying more distributed algorithms. We are also interested in syntactic criteria (beyond the basic one presented in Section 3.3) for determining if a temporal formula expresses a local property. A longer-term goal would be to have model checkers apply this kind of reduction automatically whenever the user attempts to verify a local property of a distributed algorithm.

References

1. Charron-Bost, B., Schiper, A.: The Heard-Of model: Computing in distributed systems with benign failures. In: Distributed Computing (to appear, 2009)
2. Elrad, T., Francez, N.: Decomposition of distributed programs into communication-closed layers. Science of Computer Programming 2(3) (April 1982)
3. Godefroid, P.: Partial-Order Methods for the Verification of Concurrent Systems. LNCS, vol. 1032. Springer, Heidelberg (1996)
4. Lamport, L.: Time, clocks, and the ordering of events in a distributed system. Communications of the ACM 21(7), 558–565 (1978)
5. Lamport, L.: What good is temporal logic? In: Mason, R.E.A. (ed.) Information Processing 83: Proceedings of the IFIP 9th World Congress, September 1983, pp. 657–668. IFIP, North-Holland, Paris (1983)
6. Lamport, L.: Specifying Systems. Addison-Wesley, Boston (2002)
7. Lynch, N.A.: Distributed Algorithms. Morgan Kaufmann, San Francisco (1996)
8. Peled, D., Wilke, T.: Stutter-invariant temporal properties are expressible without the next-time operator. Inf. Proc. Letters 63(5), 243–246 (1997)
9. Tsuchiya, T., Schiper, A.: Model checking of consensus algorithms. In: 26th IEEE Symp. Reliable Distributed Systems (SRDS 2007), pp. 137–148. IEEE Computer Society, Beijing (2007)
10. Tsuchiya, T., Schiper, A.: Using bounded model checking to verify consensus algorithms. In: Taubenfeld, G. (ed.) DISC 2008. LNCS, vol. 5218, pp. 466–480. Springer, Heidelberg (2008)
11. Valmari, A.: The state explosion problem. In: Reisig, W., Rozenberg, G. (eds.) APN 1998. LNCS, vol. 1491, pp. 429–528. Springer, Heidelberg (1998)
12. Yu, Y., Manolios, P., Lamport, L.: Model checking TLA+ specifications. In: Pierre, L., Kropf, T. (eds.) CHARME 1999. LNCS, vol. 1703, pp. 54–66. Springer, Heidelberg (1999)

Computable CTL^* for Discrete-Time and Continuous-Space Dynamic Systems*

Pieter Collins and Ivan S. Zapreev

Centrum Wiskunde & Informatica,
Science Park 123, 1098 XG Amsterdam
Pieter.Collins@cwi.nl, I.Zapreev@cwl.nl

Abstract. The CTL^* model-checking problem is thoroughly studied and is fully understood for finite and countable state spaces. Yet, in most models arising in the sciences and engineering the system's sate space is uncountable. Then, the standard computability and complexity theory is inapplicable but the semantics of CTL^* has to be in some sense computable to allow for model-checking algorithms that are implementable on digital computers. To tackle this problem, we consider discrete-time continuous-space dynamic systems for which we study the computability of the standard semantics of CTL^* and provide a variant thereof computable in the sense of Type-2 Theory of Effectivity.

Keywords: Computability, Model Checking, CTL^*, Dynamic Systems.

1 Introduction

Dynamic systems are widely applied for modelling and analysis in physiology, biology, chemistry and engineering. The high-profile and safety-critical nature of such applications has resulted in a large amount of work on formal methods for dynamic systems: mathematical logics, computational methods, formal verification, and etc. In our work, we focus on the verification approach called model checking, and its computability aspects.

In general, given a formal model $\mathcal{M}$ of a system design, along with a specification formula ϕ that represents a desired system property, model checking uses exhaustive state-space exploration to answer the question: "Does the model $\mathcal{M}$ satisfy the property ϕ?". This question is typically put as a formula $\mathcal{M} \models \phi$, that uses a satisfiability relation $\models$.

Model checking is expected to be fully automated, i. e. implementable using digital computers. Yet, the semantics of the formula ϕ verified on the model $\mathcal{M}$ does not always lead to a computable (decidable) model-checking algorithm. Therefore, it is not only important to realise which system properties are practically verifiable, but also to provide a semantics for logical formalisms that ensures computability (semi-decidability) of the induced model-checking procedures.

---

* This research was supported by the Nederlandse Organisatie voor Wetenschappelijk Onderzoek (NWO) Vidi grant 639.032.408.

O. Bournez and I. Potapov (Eds.): RP 2009, LNCS 5797, pp. 107–119, 2009.

For dynamic systems, the system model $\mathcal{M}$ can be described using various formalisms, such as time-automaton, hybrid automaton, differential equation, differential inclusion and others. The system property ϕ is typically described using a logical formalism, some sort of temporal or modal logic, such as LTL [16], CTL [4], CTL^* [10], or propositional modal/temporal μ-calculus [11]. Typical system properties, that need to be verified, are *reachability* – "Does the system reach the certain set of goal states?" – and *repeated reachability* – "Does the system return to the set of goal states infinitely often?".

The state-spaces of dynamic systems are typically not only infinite, but also continuous (e. g. $\mathbb{R}^n$) or hybrid (e. g. a product of $\mathbb{R}^n$ and $\mathbb{Z}$). The model-checking algorithms, require computation of system's reachable states (images or pre-images of sets of states under the system's evolution function), as well as computation of union, intersection of sets, and testing them for inclusions. This is the key place where the computability aspect comes into play, because for having effective model-checking procedures these sets and operations have to be in some sense computable, i. e. effectively implementable on digital computers.

The ordinary computability and complexity theory is not powerful enough to express the computability of real-valued functions and therefore sets of any continuous or hybrid domain. Thus, to decide on whether this or that model-checking algorithm is computable, we have to use a more powerful approach. Our choice is Type-2 Theory of Effectivity (TTE) [20] which defines computability based on Turing machines with finite and infinite input/output sequences. This theory has been already applied to analysis of computability of reachable sets of control systems in [6], and was used for providing computable semantics of CSL for discrete-time continuous-space dynamic systems (DTCSDSs) in [8,9].

In this paper, we devise a computable semantics for model checking CTL^* on DTCSDSs. This work is motivated by the fact that CTL^* is a strict super set of CTL and LTL, which allows it to express more interesting reachability properties by combining linear- and branching-time semantics. Our computable semantics for CTL^* is topological, for other topological logics see, e. g., [12,1]. Similar to [1], we require state and paths formulae to result in (respectively) open sets of states and paths. To achieve this, unlike [1], we do not use Alexandrov spaces, which are quite restrictive, but rather provide computable and open (under) approximations for the sets satisfying the formulae. Due to (necessary) choices, our computable semantics does not preserve the Law of Excluded Middle. In particular, the formulae containing negation, henceforth or release operators can be true (on the given model) but not computably verifiable. Yet, we argue that the provided computable semantics is optimal. Note that, the results of this paper are another step towards providing a computable logic for hybrid systems.

This paper is organised as follows: Section 2 contains the preliminary material. Section 3 outlines the computable semantics of CTL. In Section 4, we discuss computability of CTL^* in its original semantics and devise a computable one. This sketches the approximate model-checking algorithms that can be implemented on digital computers. Section 5 provides a theorem that allows to propagate quantifiers inside CTL^* formulae. This can be used for optimising model-checking procedures. Section 6 concludes.

2 Preliminaries

TTE relies on topological spaces and thus we begin with Section 2.1 that recalls some of the important aspects thereof. Since DTCSDS models are typically expressed by multivalued maps, in Section 2.2 we discuss continuity aspects of the latter ones and provide some of their properties. Further, in Section 2.3, we talk about TTE and computability of various sets/functions. After, in Section 2.4, we recall the standard semantics of CTL and CTL^* on Kripke structures.

2.1 Topological Spaces

A *topological space* is a pair $T = (X, \tau)$ where X is an arbitrary set and $\tau \subseteq 2^X$ is such that: $\emptyset, X \in \tau$, $\forall U_1, U_2 \in \tau \Rightarrow U_1 \cap U_2 \in \tau$, and $\forall \mathbb{U} \subseteq \tau \Rightarrow \bigcup_{U \in \mathbb{U}} U \in \tau$. For a topological space T, elements of τ are called *open* and their complements in X are called *closed*. Let $x \in X$ and $B \subseteq X$ then B is a *neighbourhood* of point x if there exists an open set $U \in \tau$ such that $x \in U \subseteq B$. Let $B \subseteq X$ and $\mathbb{U} \subseteq \tau$ then $\mathbb{U}$ is an *open cover* of B if $B \subseteq \bigcup_{U \in \mathbb{U}} U$. Let $S \subseteq X$, then the set $\mathrm{Int}(S) = \cup \{U | U \subseteq S \wedge U \in \tau\}$ is called the *interior* of S and $\mathrm{Cl}(S) = \cap \{A | S \subseteq A \wedge A \text{ is closed}\}$ is called the *closure* of S. A set $C \subseteq X$ is *compact* iff every open cover of C has a finite sub cover. A subset of X is *pre-compact* iff its closure is compact. For a topological space we have $\mathcal{O}$ – a set of open, $\mathcal{A}$ – a set of closed, and $\mathcal{K}$ – a set of compact sets.

Let $T = (X, \tau)$ be a topological space. Then $\beta \subseteq \tau$ is a *base* of the topology τ if every element of τ can be represented as a union of elements from β. A topological space is called *second countable* if its topology has a countable base. A *Hausdorff space (T_2 space)* is a topological space such that $\forall x, y \in X$ where $x \neq y$ there exist $U_x, U_y \in \tau$ such that $x \in U_x$, $y \in U_y$ and $U_x \cap U_y = \emptyset$.

A *path space* is a topological space (X^ω, τ^ω) where X^ω is the countable Cartesian product of X. Let $\sigma \in X^\omega$ and $\sigma = s_0 s_1 s_2 \dots$ then $\forall i \in \mathbb{N}$ we define the *canonical projection* $p_i : X^\omega \to X$ such that $p_i(\sigma) = s_i$. Let τ be a topology on X, and any $U^\omega \in \tau^\omega$ be a countable (or finite) union of finite intersection of sets from $B^\omega := \{B_U^\omega \subseteq X^\omega | \exists n \in \mathbb{N} : ((\forall i < n : p_i(B_U^\omega) \in \tau) \wedge (\forall i \geq n : p_i(B_U^\omega) = X))\}$. Then, τ^ω is called a *product topology* on X^ω (induced by τ). The product topology is the coarsest topology for which all the projections p_i are continuous, and in addition every p_i is an open-valued map. Note that, if (X, τ) is second-countable Hausdorff space then (X^ω, τ^ω) with the product topology τ^ω is also second-countable and Hausdorff.

2.2 DTCSDSs and Multivalued Maps

We consider discrete-time continuous-space dynamic systems (DTCSDSs) for which the state-space is continuous and the time domain is discrete (the system state changes at discrete time points). In system theory, dynamic systems are given by functions $f : X \times U \to X$, where X is the state space, and U can either represent control or system noise. These functions are typically converted into multivalued maps $F : X \rightrightarrows X$ such that $F(x) = f(x, U)$.

A *multivalued map* $F : X \rightrightarrows Y$, also known as multivalued function or multifunction, is a total relation on $X \times Y$. If we define $F(S) = \{F(x) \,|\, x \in S\}$ for $S \subseteq X$ then F can be seen as a function $F : X \to 2^Y$. This last definition is more convenient when we want to talk about function composition. For example, for two multivalued maps $F : X \rightrightarrows Y$, $G : Y \rightrightarrows Z$ and their composition $G \circ F$ we have $G \circ F : X \rightrightarrows Z$ and thus for any $x \in X$ we can simply write $G \circ F(x) = G(F(x))$. A *weak preimage* of F on $B \subseteq Y$ is $F^{-1}(B) = \{x \in X \,:\, F(x) \cap B \neq \emptyset\}$ and a *strong preimage* is $F^{\Leftarrow}(B) = \{x \in X \,:\, F(x) \subseteq B\}$. The notion of *continuity* for multivalued maps is an extension of continuity for the regular functions. Let us only note that, for a continuous multivalued map F and an open set $U \subseteq Y$ the pre-images $F^{-1}(U)$ and $F^{\Leftarrow}(U)$ are open sets.

2.3   Type-2 Theory of Effectivity (TTE)

TTE [20], as well as are regular computability theory, is based on Turing machines. The difference is that TTE (type-2) machines allow for infinite computations. In particular they can accept infinite inputs and produce infinite outputs. The computability is first defined on type-2 machines and then is extended to arbitrary functions, sets and their elements by means of notations and representations.

Let M be a type-2 machine with a fixed finite alphabet Σ, $k \geq 0$ input tapes, one output tape and $Y_i \in \{\Sigma^*, \Sigma^\omega\}$ where $i \in 0, \ldots, k$. Then, a (partial) string function $f_M : Y_1 \times \ldots \times Y_k \to Y_0$ is *computable* iff it is realised by a type-2 machine M. The latter means that for $y_i \in Y_i$ we have $f_M(y_1, \ldots, y_k) = y_0 \in \Sigma^*$ iff M halts on input $(y_1, \ldots, y_k)$ with y_0 on the output tape and $f_M(y_1, \ldots, y_k) = y_0 \in \Sigma^\omega$ iff M computes forever on input $(y_1, \ldots, y_k)$ and writes y_0 to the output.

The computability on Σ^* and Σ^ω is generalised by means of notations and representations. A *notation* of set X is a partial surjective function $\nu : \Sigma^* \to X$ and a *representation* is a partial surjective function $\delta : \Sigma^\omega \to X$. These functions encode elements of the domain X into strings and sequences.

A computable Hausdorff space is a tuple (X, τ, β, ν) such that (X, τ) is a second-countable Hausdorff (T_2) space; β is a countable base of τ consisting of pre-compact open sets; ν is a notation of β; we take effectivity properties in [3] (Lemma 2.3) as axioms; and assume that $Cl : \beta \to \mathcal{K}$ is computable. Let us have a computable Hausdorff space, the Sierpinski space $\mathcal{S}$ and a continuous function $F : X \to X$. Then the following operations are *computable* (continuous): countable union as $\mathcal{O} \times \mathcal{O} \to \mathcal{O}$, complement as $\mathcal{O} \to \mathcal{A}$, subset operation as $\mathcal{K} \times \mathcal{O} \to \mathcal{S}$, the $F^{-1}(.)$ and $F^{\Leftarrow}(.)$ as $\mathcal{O} \to \mathcal{O}$. The following operations are known to be *uncomputable*: closure as $\mathcal{O} \to \mathcal{A}$, interior as $\mathcal{A} \to \mathcal{O}$.

2.4   Standard Semantics of CTL and CTL^*

CTL and CTL^* are typically interpreted over Kripke structures. A Kripke structure M is a tuple (S, I, R, L) where S is a countable set of states; $I \subseteq S$ is a set of initial states; $R \subseteq S \times S$ is a transition relation such that $\forall s \in S, \exists s' \in S : (s, s') \in R$; AP is a finite set of atomic propositions; and $L : S \to 2^{AP}$ is an labelling function. A path in M is an infinite sequence of states $s_0 s_1 s_2 \ldots$ such that $\forall i \geq 0 : (s_i, s_{i+1}) \in R$. A set of paths starting in state s is denoted as $Paths_M(s)$. For a path $\sigma \in Paths_M(s)$, where $\sigma = s_0 s_1 s_2 \ldots$, for any $j \geq 0$ we denote $\sigma_j := s_j s_{j+1} s_{j+2} \ldots$, and $\sigma[j] := s_j$.

Computational Tree Logic (CTL) [4] is divided into *state formulae*: $\Phi ::= p \mid \neg \Phi \mid \Phi \wedge \Phi \mid \forall \phi \mid \exists \phi$, and *path formulae*: $\phi ::= \mathcal{X} \Phi \mid \Phi \mathcal{U} \Phi \mid \Phi \mathcal{R} \Phi$. The state formulae have the following semantics: $s \models p$ iff $p \in L(s)$; $s \models \neg \Phi$ iff $\neg (s \models \Phi)$; $s \models \Phi \wedge \Psi$ iff $(s \models \Phi) \wedge (s \models \Psi)$; $s \models \exists \phi$ iff $\exists \sigma \in Paths_M(s) : \sigma \models \phi$; $s \models \forall \phi$ iff $\forall \sigma \in Paths_M(s) : \sigma \models \phi$. The semantics of path formulae is as follows: $\sigma \models \mathcal{X} \Phi$ iff $\sigma[1] \models \Phi$; $\sigma \models \Phi \mathcal{U} \Psi$ iff $\exists j \geq 0 : (\sigma[j] \models \Psi \wedge \forall 0 \leq i < j : \sigma[i] \models \Phi)$; $\sigma \models \Psi \mathcal{R} \Phi$ iff $(\forall i \geq 0 : \sigma[i] \models \Phi) \vee (\exists j \geq 0 : (\sigma[j] \models \Psi \wedge \forall 0 \leq i \leq j : \sigma[i] \models \Phi))$.

Branching Temporal Logic (CTL*) [10] is a combination of *LTL* [16] and *CTL*, it's syntax is defined by *state formulae*: $\Phi ::= p \mid \neg \Phi \mid \Phi \wedge \Phi \mid \forall \phi \mid \exists \phi$ and *path formulae*: $\phi ::= \Phi \mid \neg \phi \mid \phi \wedge \phi \mid \mathcal{X} \phi \mid \phi \mathcal{U} \phi \mid \psi \mathcal{R} \phi$. The semantics of the state formulae is the same as for *CTL*, the semantics of path formulae is as follows: $\sigma \models \Phi$ iff $\sigma[0] \models \Phi$; $\sigma \models \neg \phi$ iff $\neg (\sigma \models \phi)$; $\sigma \models \phi \wedge \psi$ iff $(\sigma \models \phi) \wedge (\sigma \models \psi)$; $\sigma \models \mathcal{X} \phi$ iff $\sigma_1 \models \phi$; $\sigma \models \phi \mathcal{U} \psi$ iff $\exists j \geq 0 : (\sigma_j \models \psi \wedge \forall 0 \leq i < j : \sigma_i \models \phi)$; $\sigma \models \psi \mathcal{R} \phi$ iff $(\forall i \geq 0 : \sigma_i \models \phi) \vee (\exists j \geq 0 : (\sigma_j \models \psi \wedge \forall 0 \leq i \leq j : \sigma_i \models \phi))$.

Remarks: In *CTL* and *CTL**, path formulae can only be used as sub formulae. For a state formula Φ, we denote $Sat(\Phi) := \{s \in S | s \models \Phi\}$. In the following, we often identify Φ with the set $U_\Phi := Sat(\Phi)$, and use the following (standard) abbreviations: *(i)* $\forall i \in \mathbb{N}$ we define $\mathcal{X}^i \phi := \underbrace{\mathcal{X} \ldots \mathcal{X}}_{i \text{ times}} \phi$ and $\mathcal{X}^0 \phi := \phi$; *(ii)* $\mathcal{E} \psi :=$ *true* $\mathcal{U}$ ψ; *(ii)* $\mathcal{G} \phi :=$ *false* $\mathcal{R}$ ϕ. The temporal operators have the following names: $\mathcal{X}$ – *next*, $\mathcal{E}$ – *eventually*, $\mathcal{U}$ – *until*, $\mathcal{G}$ – *henceforth*, and $\mathcal{R}$ – *release*.

3   Computable Semantics for *CTL*

In this section, we briefly outline and motivate the computable semantics of *CTL*, cf. [8,9], for the (extended) DTCSDS model given below.

Definition 1. *A discrete-time continuous-space control system (DTCSDS) is a tuple $M = (T, F, L)$ where: $T = (X, \tau, \beta, \nu)$ is a computable Hausdorff space; $F \in \mathrm{C}(X, X)$ is a multivalued map which defines the system's evolution; and $L : X \to 2^{AP}$ is a labelling function where AP is a finite set of atomic propositions. For any $p \in AP$ and $x \in X$ we have that (respectively) $Sat(p) \in \tau$ and $L(x) = \{p \in AP | x \in Sat(p)\}$.*

Notice that, elements of AP identify trivial properties of system states, and each property is represented by an open set. This is necessary for reflecting the topological aspects of the: *(i)* computability theory, cf. Section 2.3; *(ii)* hybrid systems, cf. Section 5 of [15]; and *(iii)* logics for hybrid systems, cf. [12,1].

According to Section 2.4, for a model M, a set of initial states $I \subseteq X$, and a CTL formula Φ, proving $M, I \models \Phi$ is equivalent to showing that $I \subseteq Sat(\Phi)$.

In our case, M is a DTCSDS and if $\Phi := p \in AP$ then we need to check that I is a subset of an open set $Sat\,(p)$. The latter, cf. Section 2.3, is known to be computably verifiable only if I is compact. Thus, to make requirements on I uniform, and $M, I \models \Phi$ computable for any $\Phi \in CTL$: *(i)* we should only consider sets I that are compact; and *(ii)* we expect $Sat\,(\Phi)$ to be open.

The last condition of the previous paragraph turns out to be problematic. If $Sat\,(\Phi)$ is open then $Sat\,(\neg\Phi)$ is closed[1] and thus $I \subseteq Sat\,(\neg\Phi)$ is uncomputable. Defining $Sat\,(\neg\Phi) := Int\,(X \setminus Sat\,(\Phi))$, as in [1], results in $Sat\,(\neg\Phi)$ being open, but, cf. Section 2.3, uncomputable. Therefore, we suggest to transform every CTL formula into an equivalent CTL formula in the *negation normal form* (NNF). The latter is always possible, see e.g. [17]. In NNF, negations can only prefix atomic propositions and thus one can make sure that the representations of their open sets are good enough to make negations computable.

To summarize, for $M, I \models \Phi$ to be computably verifiable we require that: *(i)* I is compact; *(ii)* Φ is in NNF; *(iii)* any $p \in AP$, such that $\neg p$ occurs in Φ, has a representation of $Sat\,(p)$ such that $Int\,(X \setminus Sat\,(p))$ is computable; and *(iv)* F is such that $F^{-1}(U)$ and $F^{\Leftarrow}(U)$ for $U \in \tau$ are computable. Under these conditions, Eq. 1 to 9 provide the computable semantics for CTL. Here, we only consider the universal quantifier because accounting for the existential one boils down to using the weak preimage F^{-1} in place of the strong preimage $F^{\Leftarrow}$. In the following, $Sat'\,(\Phi)$ is either the open set of states satisfying the formula Φ, i.e. $Sat'\,(\Phi) = Sat\,(\Phi)$, or it is its open under approximation, i.e. $Sat'\,(\Phi) \subset Sat\,(\Phi)$.

$$Sat'\,(p) := U_p \text{ then } I \subseteq Sat'\,(p) \Leftrightarrow I \models p \tag{1}$$

$$Sat'\,(\neg p) := Int\,(X \setminus U_p),\ I \subseteq Sat'\,(\neg p) \Rightarrow I \models \neg p \tag{2}$$

$$Sat'\,(\Phi \vee \Psi) := U_\Phi \cup U_\psi,\ I \subseteq Sat'\,(\Phi \vee \Psi) \Leftrightarrow I \models \Phi \vee \Psi \tag{3}$$

$$Sat'\,(\Phi \wedge \Psi) := U_\Phi \cap U_\psi,\ I \subseteq Sat'\,(\Phi \wedge \Psi) \Leftrightarrow I \models \Phi \wedge \Psi \tag{4}$$

$$Sat'\,(\forall\,(\mathcal{X}\Phi)) := F^{\Leftarrow}(U_\Phi),\ I \subseteq Sat'\,(\forall\,(\mathcal{X}\Phi)) \Leftrightarrow I \models \forall\,(\mathcal{X}\Phi) \tag{5}$$

$$Sat'\,(\forall\,(\mathcal{E}\Psi)) := \bigcup_{n=0}^{\infty} S_n,\ S_0 = U_\Psi,$$

$$\forall n \geq 1:\ S_n = F^{\Leftarrow}(\bigcup_{i=0}^{n-1} S_i),\ I \subseteq Sat'\,(\forall\,(\mathcal{E}\Psi)) \Leftrightarrow I \models \forall\,(\mathcal{E}\Psi) \tag{6}$$

$$Sat'\,(\forall\,(\Phi\,\mathcal{U}\,\Psi)) := \bigcup_{n=0}^{\infty} S'_n,\ S'_0 = U_\Psi,$$

$$\forall n \geq 1:\ S'_n = F^{\Leftarrow}(\bigcup_{i=0}^{n-1} S'_i) \cap U_\Phi,\ I \subseteq Sat'\,(\forall\,(\Phi\,\mathcal{U}\,\Psi)) \Leftrightarrow I \models \forall\,(\Phi\,\mathcal{U}\,\Psi) \tag{7}$$

$$Sat'\,(\forall\,(\mathcal{G}\Phi)) := \bigcup \{B_r \in \tau | Cl\,(B_r) \subseteq U_\Phi \wedge$$

$$Cl\,(B_r) \subseteq F^{\Leftarrow}(B_r)\},\ I \subseteq Sat'\,(\forall\,(\mathcal{G}\Phi)) \Rightarrow I \models \forall\,(\mathcal{G}\Phi) \tag{8}$$

$$Sat'\,(\forall\,(\Psi\,\mathcal{R}\,\Phi)) := \bigcup \{B_r \in \tau | Cl\,(B_r) \subseteq U_\Phi \wedge (Cl\,(B_r) \subseteq U_\Psi$$

$$\vee Cl\,(B_r) \subseteq F^{\Leftarrow}(B_r \cup (U_\Psi \cap U_\Phi)))\},\ I \subseteq Sat'\,(\forall\,(\Psi\,\mathcal{R}\,\Phi)) \Rightarrow I \models \forall\,(\Psi\,\mathcal{R}\,\Phi) \tag{9}$$

---

[1] Note that, if $\Phi \in CTL$ then $\neg\Phi$ is a valid CTL formula too.

In most cases the standard semantics of CTL results in an open and computable set of states that satisfy the formula. The exceptions are the *negation* (Eq. 2), *henceforth* (Eq. 8), and *release* (Eq. 9) operators for which the resulting sets of states are closed. Thus, we had to alter their semantics by using open and computable under approximations for the sets of states satisfying the formulae. Note that, in Eq. 8 to 9, each $B_r \in \tau$ is a finite union of open rational boxes in X and so $Cl(B_r)$ is compact and computable.

A consequence of the (necessary) choice of semantics of the negation, henceforth and release operators is that there can be no proof by contradiction (the Law of Excluded Middle does not hold). In other words, if Φ contains at least one of these operators, the fact that Φ does not hold (on M, I) does not imply that the negation of Φ holds. Moreover, such a Φ can be true but not computably verifiable. It only remains to note that our choice for the computable semantics is optimal with respect to the considered DTCSDS model. The case of the negation operator was discussed earlier, for the henceforth operator the optimality was shown in [7], and the semantics for the release operator is the combination of the semantics of the henceforth and until operators.

4   Computable Semantics for CTL^*

Since CTL^* extends CTL, we use the same DTCSDS model as before and derive the conditions on the set of initial states I, the evolution function F, and the logical formula. Let us have a DTCSDS model M with a continuous evolution function F, such that $F^{-1}(U)$ with $F^{\Leftarrow}(U)$ are computable for any $U \in \tau$, and a compact initial set of states $I \subseteq X$. Then, any CTL^* formula (in NNF) that we might need to verify is given by the syntax $\Phi ::= \neg p \mid \Phi \wedge \Phi \mid \Phi \vee \Phi \mid \forall \phi \mid \exists \phi$, where ϕ is an arbitrary path formula. Here, $\neg p$, $\Phi \wedge \Phi$, and $\Phi \vee \Phi$ inherit the computable semantics of CTL, but for $\forall \phi$ and $\exists \phi$ we need another approach. This is because, ϕ can be a conjunction or disjunction of other path formulae, or an until (release) formula acting on other path formulae, and etc. The latter implies that we need to work with the set of paths satisfying the formula ϕ.

Let $Paths_M : X \to X^\omega$ be the multivalued map that maps the set of initial states I into the set of system paths starting in I; and $Paths(\phi) \subseteq X^\omega$ be the set of all paths satisfying the path formula ϕ (regardless to the system-evolution function F). Then, if $Paths(\phi)$ is an open set of paths in X^ω equipped with the product topology τ^ω, we can define the computable semantics for $\forall \phi$ and $\exists \phi$ as follows:

$$Sat'(\exists \phi) := Paths_M^{-1}(Paths(\phi)),\ I \subseteq Sat'(\forall \phi) \Leftrightarrow M, I \models \exists \phi, \quad (10)$$

$$Sat'(\forall \phi) := Paths_M^{\Leftarrow}(Paths(\phi)),\ I \subseteq Sat'(\forall \phi) \Leftrightarrow M, I \models \forall \phi. \quad (11)$$

Notice that, since F is continuous and $F^{-1}(U)$ with $F^{\Leftarrow}(U)$ are computable for any $U \in \tau$, it follows from [6,19] that $Paths_M$ is continuous and $Paths_M^{-1}(U^\omega)$ with $Paths_M^{\Leftarrow}(U^\omega)$ are computable for any $U^\omega \in \tau^\omega$. Also, by the definition of DTCSDS and the results of Section 2.1, (X^ω, τ^ω) is a computable Hausdorff space and thus we can use the computability results of Section 2.3.

Now, to complete the computable semantics for CTL^*, it suffices to consider each possible path formula $\phi \in CTL^*$ and to either show that it results in an open and computable set of paths $Paths(\phi)$, or to provide its open approximation $Paths'(\phi)$ such that:

$$Paths_M^{-1}\left(Paths'(\phi)\right) \subset Paths_M^{-1}\left(Paths(\phi)\right) \text{ and}$$
$$Paths_M^{\Leftarrow}\left(Paths'(\phi)\right) \subset Paths_M^{\Leftarrow}\left(Paths(\phi)\right). \tag{12}$$

In the latter case, Eq. 10 to 11 will turn into implications (from left to right), the same as it was for CTL, e. g., cf. Eq. 8. Further, we assume that ψ, ψ_1, and ψ_2 are such that $Paths(\psi)$, $Paths(\psi_1)$, and $Paths(\psi_2)$ are open and computable.

Paths of $\phi := \Phi$: Since Φ is a state formula, from Section 3 we know that we can compute an open set $Sat'(\Phi) \subseteq Sat(\Phi)$. Then, we define:

$$Paths'(\Phi) := (Sat'(\Phi), X, \ldots, X \ldots), \tag{13}$$

which is open in X^ω, is trivially computable, and $Paths'(\Phi) \subseteq Paths(\Phi)$.

Paths of $\phi := \psi_1 \vee \psi_2$: Since $Paths(\psi_1)$, and $Paths(\psi_2)$ are open, we get:

$$Paths(\psi_1 \vee \psi_2) := Paths(\psi_1) \cup Paths(\psi_2), \tag{14}$$

which is open and computable.

Paths of $\phi := \psi_1 \wedge \psi_2$: Similar to the previous case,

$$Paths(\psi_1 \wedge \psi_2) := Paths(\psi_1) \cap Paths(\psi_2), \tag{15}$$

is open and computable.

Paths of $\phi := \mathcal{X}\psi$: In this case we get:

$$Paths(\mathcal{X}\psi) := X \times Paths(\psi), \tag{16}$$

i. e. the Cartesian product of X and an open set $Paths(\psi)$. Clearly, $Paths(\mathcal{X}\psi)$ is open and computable.

Paths of $\phi := \mathcal{E}\psi$: Remember that, by the standard semantics, for any path σ, we have that $\sigma \models \mathcal{E}\psi$ is equivalent to $\sigma \models \bigvee_{i\in\mathbb{N}} \mathcal{X}^i \psi$. Therefore, we obtain:

$$Paths(\mathcal{E}\psi) := \bigcup_{i\in\mathbb{N}} U_i^\psi, \text{ where } U_i^\psi := \underbrace{(X, \ldots, X)}_{i \text{ times}} \times Paths(\psi). \tag{17}$$

Clearly, $\forall i \in \mathbb{N}$ the set U_i^ψ is open and computable, and thus $Paths(\mathcal{E}\psi)$ is open and computable (as a countable union of open sets).

Paths of $\phi := \psi_1 \, \mathcal{U} \, \psi_2$: Remember that, for any path σ, $\sigma \models \psi_1 \, \mathcal{U} \, \psi_2$ is equivalent to $\sigma \models \bigvee_{i\in\mathbb{N},i>0}\left(\bigwedge_{j\in\mathbb{N},j<i} \mathcal{X}^j\psi_1 \wedge \mathcal{X}^i\psi_2\right) \vee \psi_2$. This implies:

$$Paths(\psi_1 \, \mathcal{U} \, \psi_2) := \bigcup_{i\in\mathbb{N},i>0}\left(\bigcap_{j\in\mathbb{N},j<i} U_j^{\psi_1} \cap U_i^{\psi_2}\right) \cup Paths(\psi_2). \tag{18}$$

Clearly, $\forall i \in \mathbb{N}$ the set $\bigcap_{j \in \mathbb{N}, j<i} U_j^{\psi_1} \cap U_i^{\psi_2}$ is open and computable as a finite intersection of open sets. Thus, $Paths\,(\psi_1\,\mathcal{U}\,\psi_2)$ is open and computable as a countable union of open sets.

Paths of $\phi := \mathcal{G}\psi$: As before, following the standard semantics we get that $Paths\,(\mathcal{G}\psi) := \bigcap_{i \in \mathbb{N}} U_i^\psi$. Being a countable intersection of open sets, $Paths\,(\mathcal{G}\psi)$ is neither open nor computable. Moreover, constructing an open *under* approximation of $Paths\,(\mathcal{G}\psi)$ is generally *impossible*. Consider $p \in AP$ and $U_p := Sat\,(p) \neq X$, then $Paths\,(\mathcal{G}p) := (U_p, \ldots, U_p, \ldots)$. Since $\forall i \in \mathbb{N}$ we have that $p_i\,(Paths\,(\mathcal{G}p)) = U_p \neq X$, we deduce that $Int\,(Paths\,(\mathcal{G}p)) = \emptyset$.

From the above, to construct an *open* set $Paths'\,(\mathcal{G}\psi)$, we can only restrict finite prefixes of the considered paths. We also need to ensure that any extension of the path prefix does satisfy $\mathcal{G}\psi$. Formally, we should consider paths σ such that $\exists i \in \mathbb{N}$ for which $\sigma\,[i] \models \forall \mathcal{G}\psi$ and $\forall j < i : \sigma_j \models \psi$. This is only possible if we account for the system model. As a conservative approximation, we suggest:

$$Paths'\,(\mathcal{G}\psi) := \bigcup \{B_r^\omega \in B^\omega | \exists i \in \mathbb{N} : Cl\,(p_i\,(B_r^\omega)) \subseteq Sat\,(\forall \mathcal{G}\forall \psi) \wedge$$
$$\forall j < i : Cl\,(Sh_j\,(B_r^\omega)) \subseteq Paths\,(\psi)\}. \qquad (19)$$

In Eq. 19, we substitute $\forall \mathcal{G}\psi$ with the equivalent formula $\forall \mathcal{G}\forall \psi$, cf. Eq. 26 of Th. 1 in Section 5; and use $Sh_j : X^\omega \to X^\omega$ – the *shift map* that removes the first j components of its argument. Sh_j is computable and open-valued. B_r^ω is a finite union of open rational balls in (X^ω, τ^ω). Note that, $Cl\,(Sh_j\,(B_r^\omega))$ is computed componentwise, and since $Sh_j\,(B_r^\omega)$ is open in X^ω, we only need to compute the closure of finitely many components. Moreover, there are only finitely many i such that $p_i\,(Sh_j\,(B_r^\omega)) \neq X$. For such i we have that $p_i\,(Cl\,(Sh_j\,(B_r^\omega)))$ is compact. The latter ensures computability of $Cl\,(Sh_j\,(B_r^\omega)) \subseteq Paths\,(\psi)$.

Clearly, $Paths'\,(\mathcal{G}\psi)$ is open and computable, but it is restrictive because it excludes the paths σ such that $\sigma \models \mathcal{G}\psi$ and $\neg \exists i \in \mathbb{N} : \sigma\,[i] \models \forall \mathcal{G}\psi$. Still, we conjecture that the given choice for $Paths'\,(\mathcal{G}\psi)$ is optimal for any approach that propagates the sets of paths through the formulae. This is because $Paths'\,(\mathcal{G}\psi)$ has to contain *only* paths satisfying $\mathcal{G}\psi$ and to be an open set, i. e. we can not put conditions on any but finite path prefixes. Also, this semantics is optimal for verifying $\forall \mathcal{G}\psi$ because the latter is equivalent to $\forall \mathcal{G}\forall \psi$.

For specific formulae, there can be approximations that are better than the one given by Eq. 19. For example, in case of $\exists \mathcal{G}\psi$ we suggest to use:

$$Paths'\,(\mathcal{G}\psi) := \bigcup \{B_r^\omega \in B^\omega | \exists n \in \mathbb{N} : \forall i \leq n : Cl\,(Sh_i\,(B_r^\omega)) \subseteq Paths\,(\psi) \wedge$$
$$Cl\,(p_i\,(B_r^\omega)) \subseteq F^{-1}(\textstyle\bigcup_{j \leq n} p_j\,(B_r^\omega))\}, \qquad (20)$$

where $\forall \sigma \in Paths'\,(\mathcal{G}\psi)$ there exists $n \in \mathbb{N}$ for which $\forall i \leq n : \sigma_i \models \psi$ and $\exists \sigma' \in Paths\,(\sigma\,[n]) : \sigma' \models \mathcal{G}\psi$. Since $Paths'\,(\mathcal{G}\psi)$ contains paths with suffixes violating $\mathcal{G}\psi$, it can *only* be used for verifying $\exists \mathcal{G}\psi$. Note that, the set given by Eq. 19 satisfies Eq. 12 but the set given by Eq. 20 only satisfies its first part.

Paths of $\phi := \psi_2\,\mathcal{R}\,\psi_1$: Remember that, for any path σ we have that $\sigma \models \psi_2\,\mathcal{R}\,\psi_1$ is equivalent to $\sigma \models \mathcal{G}\psi_1 \vee (\psi_1\,\mathcal{U}\,(\psi_1 \wedge \psi_2))$. This, by Eq. 14, Eq. 15,

Eq. 18, and Eq. 19 gives us a computable and open approximation:

$$\textit{Paths}' \, (\psi_2 \, \mathcal{R} \, \psi_1) := \textit{Paths}' \, (\mathcal{G}\psi_1) \cup \textit{Paths} \, (\psi_1 \, \mathcal{U} \, (\psi_1 \wedge \psi_2)) \tag{21}$$

5   Equivalences and Implications for CTL^*

It is a well known fact, see e.g. [18], that in traditional setting the complexity of CTL and CTL^* (LTL) model checking are (respectively) *P-complete* and *PSPACE-complete*. Also on DTCSDSs, when model checking CTL we need to work with the open sets of system states, whereas for CTL^* we have to (also) operate with the open sets of paths. The latter are harder to represent and certainly require a significant storage space. This is why, in our opinion, it can be beneficial to be able to, when possible, substitute path formulae with state formulae that are either equivalent to or imply the former ones. This subject is related to the formulae-equivalence problems for CTL, LTL and CTL^* studied in, e.g., [10,14,5,13]. In order to avoid complications, in the following we restrict to a simple propagation of universal and existential quantifiers inside the CTL^* formulae. To our knowledge, the following results are not explicitly present in the current literature on CTL^*.

Below, we assume the *standard* semantics of CTL^* on Kripke structures, as outlined in Section 2.4. Before we present Th. 1 that summarises our results, let us define what it means for one CTL^* to imply another or to be equivalent.

Definition 2. *Let Φ and Ψ be two state formulae, then Φ implies Ψ ($\Phi \Rightarrow \Psi$) iff for any M and I: $M, I \models \Phi \Rightarrow M, I \models \Psi$; Φ and Ψ are equivalent ($\Phi \equiv \Psi$) iff $\Phi \Rightarrow \Psi$ and $\Psi \Rightarrow \Phi$.*

Theorem 1. *Let $\phi, \psi \in CTL^*$ be path formulae then the diagrams commute:*

$$\forall (\phi \vee \psi) \overset{\not\Leftarrow}{\Leftarrow} \forall\phi \vee \forall\psi \qquad\qquad \forall (\phi \wedge \psi) \equiv \forall\phi \wedge \forall\psi$$
$$\Downarrow\not\Uparrow \qquad\qquad \Downarrow\not\Uparrow \qquad\qquad\qquad \Downarrow\not\Uparrow \qquad\qquad \Downarrow\not\Uparrow$$
$$\exists (\phi \vee \psi) \equiv \exists\phi \vee \exists\psi \quad (22) \qquad\qquad \exists (\phi \wedge \psi) \overset{\not\Rightarrow}{\Rightarrow} \exists\phi \wedge \exists\psi \quad (23)$$

$$\forall\mathcal{X}\phi \equiv \forall\mathcal{X}\forall\phi \qquad \forall\mathcal{E}\phi \overset{\not\Leftarrow}{\Leftarrow} \forall\mathcal{E}\forall\phi \qquad \forall\mathcal{G}\phi \equiv \forall\mathcal{G}\forall\phi$$
$$\Downarrow\not\Uparrow \qquad \Downarrow\not\Uparrow \qquad\qquad \Downarrow\not\Uparrow \qquad \Downarrow\not\Uparrow \qquad\qquad \Downarrow\not\Uparrow \qquad \Downarrow\not\Uparrow$$
$$\exists\mathcal{X}\phi \equiv \exists\mathcal{X}\exists\phi \quad (24) \qquad \exists\mathcal{E}\phi \equiv \exists\mathcal{E}\exists\phi \quad (25) \qquad \exists\mathcal{G}\phi \overset{\not\Rightarrow}{\Rightarrow} \exists\mathcal{G}\exists\phi \quad (26)$$

$$\forall (\phi\,\mathcal{U}\,\psi) \overset{\not\Leftarrow}{\Leftarrow} \forall (\forall\phi\,\mathcal{U}\,\forall\psi) \qquad\qquad \forall (\psi\,\mathcal{R}\,\phi) \overset{\not\Leftarrow}{\Leftarrow} \forall (\forall\psi\,\mathcal{R}\,\forall\phi)$$
$$\Downarrow\not\Uparrow \qquad\qquad \Downarrow\not\Uparrow \qquad\qquad\qquad \Downarrow\not\Uparrow \qquad\qquad \Downarrow\not\Uparrow$$
$$\exists (\phi\,\mathcal{U}\,\psi) \overset{\not\Rightarrow}{\Rightarrow} \exists (\exists\phi\,\mathcal{U}\,\exists\psi) \quad (27) \qquad \exists (\psi\,\mathcal{R}\,\phi) \overset{\not\Rightarrow}{\Rightarrow} \exists (\exists\psi\,\mathcal{R}\,\exists\phi) \quad (28)$$

Proof (Sketch). First we prove Eq. 22 to 24, and then Eq. 25 to 28 follow as simple consequences. Note that, for each equation, implications from the row with the universal quantifiers to the row with the existential ones are trivial. Without loss of generality, we will assume that the set of initial states $I := \{s\}$.

- **Eq. 22:**
  - $\forall(\phi\vee\psi) \not\Leftarrow \forall\phi\vee\forall\psi$: Clearly, there exists a model M and an initial state s such that $Paths_M(s) = \{\sigma',\sigma''\}$ where $\sigma'\models\phi$ and $\sigma''\models\psi$. Then it is easy to see that $M,s\models\forall(\phi\vee\psi)$ but $M,s\not\models\forall\phi\vee\forall\psi$.
  - $\forall(\phi\vee\psi)\Leftarrow\forall\phi\vee\forall\psi$: $M,s\models\forall\phi\vee\forall\psi\Leftrightarrow(\exists\sigma'\in Paths_M(s):\sigma'\models\psi)\vee(\exists\sigma''\in Paths_M(s):\sigma''\models\phi)\Longrightarrow(\exists\sigma'\in Paths_M(s):\sigma'\models\phi\vee\psi)\vee(\exists\sigma''\in Paths_M(s):\sigma''\models\phi\vee\psi)\Leftrightarrow M,s\models\forall(\phi\vee\psi)$.
  - $\exists(\phi\vee\psi)\equiv\exists\phi\vee\exists\psi$: $M,s\models\exists(\phi\vee\psi)\Leftrightarrow\exists\sigma\in Paths_M(s):\sigma\models\phi\vee\psi\Leftrightarrow\exists\sigma\in Paths_M(s):\sigma\models\phi\vee\sigma\models\psi\Leftrightarrow(\exists\sigma'\in Paths_M(s):\sigma'\models\phi\vee\sigma'\models\psi)\vee(\exists\sigma''\in Paths_M(s):\sigma''\models\phi\vee\sigma''\models\psi)\Leftrightarrow(\exists\sigma'\in Paths_M(s):\sigma'\models\phi)\vee(\exists\sigma''\in Paths_M(s):\sigma''\models\psi)\Leftrightarrow M,s\models\exists\phi\vee\exists\psi$.
- **Eq. 23:** Follows from Eq. 22 by negating the diagram and taking into account that: $\neg(\forall(\phi\vee\psi))\equiv\exists(\neg\phi\wedge\neg\psi)$, $\neg(\forall\phi\vee\forall\psi)\equiv\exists\neg\phi\wedge\exists\neg\psi$, $\neg(\exists(\phi\vee\psi))\equiv\forall(\neg\phi\wedge\neg\psi)$, $\neg(\exists\phi\vee\exists\psi)\equiv\forall\neg\phi\wedge\forall\neg\psi$.
- **Eq. 24:** For an arbitrary model M and $I:=\{s\}$:
  - $\forall\mathcal{X}\phi\equiv\forall\mathcal{X}\forall\phi$: Notice that $M,s\models\forall\mathcal{X}\forall\phi\Leftrightarrow\forall\sigma\in Paths_M(s):\sigma[1]\models\forall\phi\Leftrightarrow\forall\sigma\in Paths_M(s)\forall\sigma'\in Paths_M(\sigma[1]):\sigma'\models\phi$ and $M,s\models\forall\mathcal{X}\phi\Leftrightarrow\forall\sigma''\in Paths_M(s):\sigma''\models\mathcal{X}\phi\Leftrightarrow\forall\sigma''\in Paths_M(s):\sigma''_1\models\phi$. Clearly, $\forall\mathcal{X}\forall\phi\Rightarrow\forall\mathcal{X}\phi$ because if $M,s\models\forall\mathcal{X}\forall\phi$ then for any $\sigma''\in Paths_M(s)$ we have that $\sigma''_1\in Paths_M(\sigma[1])$ for some $\sigma\in Paths_M(s)$. At the same time $\forall\sigma\in Paths_M(s)\forall\sigma'\in Paths_M(\sigma[1]):\sigma'\models\phi$ and thus $\sigma''_1\models\phi$.

    Clearly, $\forall\mathcal{X}\phi\Rightarrow\forall\mathcal{X}\forall\phi$. Let $M,s\models\forall\mathcal{X}\phi$ and we chose any $\sigma\in Paths_M(s)$ and consider the set $\{\sigma''_1|\sigma''[1]=\sigma[1]$, and $\sigma''\in Paths_M(s)\}$. Notice that this set equals to $Paths_M(\sigma[1])$ and for any σ' from the set we have that $\sigma'\models\phi$, because for any $\sigma''\in Paths_M(s)$ we have that $\sigma''_1\models\phi$.
  - $\exists\mathcal{X}\phi\equiv\exists\mathcal{X}\exists\phi$: Follows from $\forall\mathcal{X}\phi\equiv\forall\mathcal{X}\forall\phi$ by the fact that $\neg\forall\mathcal{X}\phi\equiv\exists\mathcal{X}\neg\phi$ and $\neg\forall\mathcal{X}\forall\phi\equiv\exists\mathcal{X}\exists\neg\phi$.

Before we proceed, let us notice that Eq. 22 (Eq. 23) holds for any countable disjunction (conjunction) of formulae.

- **Eq. 25:** Follows from Eq. 22 and Eq. 24 by the fact that for any model path σ we have $\sigma\models\mathcal{E}\phi$ iff $\sigma\models\bigvee_{i\in\mathbb{N}}\mathcal{X}^i\phi$.
- **Eq. 26:** Follows from Eq. 23 and Eq. 24 by the fact that for any model path σ we have $\sigma\models\mathcal{G}\phi$ iff $\sigma\models\bigwedge_{i\in\mathbb{N}}\mathcal{X}^i\phi$.
- **Eq. 27:** Follows from Eq. 22 to 24 by the fact that for any model path σ we have $\sigma\models\phi\,\mathcal{U}\,\psi$ iff $\sigma\models\bigvee_{i\in\mathbb{N}}\left(\bigwedge_{j\in\mathbb{N},j<i}\mathcal{X}^i\phi\wedge\mathcal{X}^i\psi\right)$.
- **Eq. 28:** Follows from Eq. 22 to 24 by the fact that for any model path σ we have $\sigma\models\psi\,\mathcal{R}\,\phi$ iff $\sigma\models\left(\bigwedge_{i\in\mathbb{N}}\mathcal{X}^i\phi\right)\vee\bigvee_{j\in\mathbb{N}}\left(\bigwedge_{k\in\mathbb{N},k\leq j}\mathcal{X}^k\phi\wedge\mathcal{X}^j\psi\right)$.

Remarks: *(i)* In some cases implications of Th. 1 can turn into equivalences. E. g., in Eq. 22 we get $\forall(\phi\vee\psi)\equiv\forall\phi\vee\forall\psi$ in case ϕ or ψ are state formulae. Remember that, in CTL^* a state formula can be also seen as a path formula, cf. Section 2.4. *(ii)* Th. 1 assumes the standard semantics of CTL^* on Kripke structures which are defined for countable state spaces. In our case, the state space (X) can be uncountable, but this does not restrict the applicability of Th. 1, because the standard semantics, cf. Section 2.4, *does not* account for the cardinality of X.

6 Concluding Remarks

CTL^* is a well known temporal logics that combines the power of linear-time (LTL) and branching-time (CTL) semantics. In this work we focused on model checking CTL^* on discrete-time continuous-space control systems (DTCSDSs) and in particular on computability aspects thereof. Due to continuity of the state space, regular computability and complexity theory is not applicable and therefore we resort to the (more powerful) Type Two Effectivity theory (TTE).

Here, we extended the work on computable semantics for CTL, cf. [8,9], by providing a computable semantics for CTL^* on DTCSDS models. For the latter, the main computability requirements stays intact. I.e., we assume a compact set of initial states I, a CTL^* formula Φ in its negation normal form, and a system model $M = (T, F, L)$, where T is a computable Hausdorff space and a continuous map F is such that $F^{-1}(U)$ and $F^{\Leftarrow}(U)$ are computable for any open U. Also, the atomic propositions of Φ that are prefixed with negations, must have representations that allow for computing interiors of their complements.

Similar to CTL, the linear-time semantics of the formulae including negation, henceforth or release operators required changes in their interpretation. The biggest challenge there was to provide computable and open approximation for the sets of paths satisfying the formula. In the constructed computable semantics of CTL^*, if Φ contains the aforementioned operators then Φ can be true (on M, I) but not computably verifiable. Note that, if the formula holds in the computable semantics, then it also holds in the original one. Since, in the traditional setting, the complexity of CTL and CTL^* (LTL) model checking are (respectively) *P-complete* and *PSPACE-complete*, we have also provided a set of implication that allow to propagate quantifies inside the CTL^* formulae.

In the future, we plan to tackle the computability of the automata-based model checking of CTL^* on DTCSDSs, and to extend the present approach towards computable model checking of CTL^* on systems with hybrid time and space domains. Also, we intend to implement these model checking algorithms in a framework for reachability analysis of hybrid systems called Ariadne [2].

References

1. Artemov, S.N., Davoren, J.M., Nerode, A.: Logic, Topological Semantics and Hybrid Systems. In: International Conference on Decision and Control, CDC 1997, vol. 1, pp. 698–701. IEEE Press, Los Alamitos (1997)
2. Balluchi, A., Casagrande, A., Collins, P., Ferrari, A., Villa, T., Sangiovanni-Vincentelli, A.L.: Ariadne: A Framework for Reachability Analysis of Hybrid Automata. In: Symposium on Mathematical Theory of Networks and Systems (MTNS 2006), Kyoto, Japan (July 2006) (to appear)

3. Brattka, V., Presser, G.: Computability on subsets of metric spaces. Theoretical Computer Science 305(1-3), 43–76 (2003)
4. Clarke, E.M., Emerson, E.A., Sistla, A.P.: Automatic verification of finite-state concurrent systems using temporal logic specifications. AMC Transactions On Programming Languages And Systems 8(2), 244–263 (1986)
5. Clarke, E.M., Draghicescu, I.A.: Expressibility Results for Linear-Time and Branching-Time Logics. In: Linear Time, Branching Time and Partial Order in Logics and Models for Concurrency, School/Workshop, London, UK, pp. 428–437. Springer, Heidelberg (1989)
6. Collins, P.: Continuity and computability of reachable sets. Theoretical Computer Science 341(1), 162–195 (2005)
7. Collins, P.: Optimal Semicomputable Approximations to Reachable and Invariant Sets. Theory of Computing Systems 41(1), 33–48 (2007)
8. Collins, P.J., Zapreev, I.S.: Computable CTL for Discrete-Time and Continuous-Space Dynamic Systems. Technical Report MAS-E0903, MAS, Centrum Wiskunde & Informatica (2009), http://www.cwi.nl/ftp/CWIreports/MAS/MAS-E0903.pdf
9. Collins, P.J., Zapreev, I.S.: Computable CTL for Discrete-Time and Continuous-Space Dynamic Systems. In: Computability in Europe, CiE (2009), To be published in a local pre-conference proceedings volume
10. Emerson, E.A., Halpern, J.Y.: "sometimes" and "Not Never" Revisited: On Branching versus Linear Time Temporal Logic. Journal of the Association for Computing Machinery (ACM) 33(1), 151–178 (1986)
11. Kozen, D.: Results on the propositional μ-calculus. Research Report RC 10133 (44981), IBM Research Division, August 1983, p. 42 (1983)
12. Kremer, P., Mints, G.: Dynamic topological logic. Annals of Pure and Applied Logic 131(1–3), 133–158 (2005)
13. Kupferman, O., Vardi, M.Y.: From Linear Time to Branching Time. ACM Transactions on Computational Logic (TOCL) 6(2), 273–294 (2005)
14. Maidl, M.: The Common Fragment of CTL and LTL. In: Annual Symposium on Foundations of Computer Science, FOCS 2000, pp. 643–652. IEEE Computer Society, Los Alamitos (2000)
15. Nerode, A., Kohn, W.: Models for Hybrid Systems: Automata, Topologies, Controllability, Observability. In: Grossman, R.L., Ravn, A.P., Rischel, H., Nerode, A. (eds.) HS 1991 and HS 1992. LNCS, vol. 736, pp. 317–356. Springer, Heidelberg (1993)
16. Pnueli, A.: The Temporal Semantics of Concurrent Programs. In: Kahn, G. (ed.) Semantics of Concurrent Computation. LNCS, vol. 70, pp. 1–20. Springer, Heidelberg (1979)
17. Schneider, K.: Verification of Reactive Systems: Formal Methods and Algorithms. In: TTCSS. Springer, Heidelberg (2004)
18. Schnoebelen, P.: The complexity of temporal logic model checking. In: Balbiani, P., Suzuki, N.-Y., Wolter, F., Zakharyaschev, M. (eds.) Selected Papers from the 4th Workshop on Advances in Modal Logics (AiML), Toulouse, France, pp. 393–436. King's College Publication (2003)
19. Spreen, D.: On the Continuity of Effective Multifunctions. Electron. Notes Theor. Comput. Sci. 221, 271–286 (2008)
20. Weihrauch, K.: Computable Analysis: An Introduction. Springer, New York (2000)

An Undecidable Permutation
of the Natural Numbers

Eero Lehtonen*

Department of Information Technology,
University of Turku,
FI-20014 Turku Finland
elleht@utu.fi

Abstract. In this paper, an undecidability result concerning a permutation of natural numbers is presented. More precisely, it is shown that for a certain piecewise defined permutation, which consists of five affine transformations, it is undecidable whether a given number belongs to a finite cycle or not.

Keywords: Permutation, Undecidability.

1 Introduction

The aim of this paper is to present an undecidability result concerning the cycle structure of a permutation of natural numbers, and to introduce a method to easily obtain such results. This work continues that of [1], where the main motivation was to study the undecidability of generalized Collatz-like problems (after Lothar Collatz, cf. [2]) such as the $3n + 1$ problem and the one more carefully investigated in this paper, the *Collatz's original problem* [2]. The first undecidability result concerning the generalized $3n + 1$ problem was proven in [3], where it was shown that one can simulate the computation of a Minsky machine by a set of periodically piecewise linear functions. In contrast to [3], in [1] and the current paper, a different coding is used and the functions to be iterated are given in an explicit form.

To avoid ambiguity in notations we define the natural numbers to be the set $\mathbb{N} = \{1, 2, 3 \ldots\}$.

Problem 1 (Collatz's original problem). Let a permutation $f : \mathbb{N} \to \mathbb{N}$ be defined as

$$f(n) = \begin{cases} (4/3)n - (1/3), & \text{if } n \in 3\mathbb{N} - 2 \\ (4/3)n + (1/3), & \text{if } n \in 3\mathbb{N} - 1 \\ (2/3)n + 0, & \text{if } n \in 3\mathbb{N}. \end{cases} \tag{1}$$

Is the cycle of this permutation containing $n = 8$ finite?

As a generalization of this problem we give

---

* This work was partly supported by the Nokia Foundation Scholarship.

O. Bournez and I. Potapov (Eds.): RP 2009, LNCS 5797, pp. 120–126, 2009.

Problem 2. Let $f : \mathbb{N} \to \mathbb{N}$ be a given permutation of the form

$$f(n) = \begin{cases} a_1 n + b_1, & \text{if } n \in A_1 \\ a_2 n + b_2, & \text{if } n \in A_2 \\ \vdots \\ a_\nu n + b_\nu, & \text{if } n \in A_\nu \end{cases} \tag{2}$$

where all a_i, b_i are rational numbers and the sets A_i are recursive, i.e., the relation $n \in A_i$ is effectively decidable for all i.

For a given $n \in \mathbb{N}$, decide whether or not there exists a $k \in \mathbb{N}$ such that

$$f^{(k)}(n) = \underbrace{f \circ f \circ \cdots \circ f}_{k \text{ times}}(n) = n. \tag{3}$$

In other words, decide whether or not n belongs to a finite cycle of f.

Remark 1. Problem 2 is a reachability problem: For a given permutation f and a natural number n it is asked, whether or not a finite cycle is reached when iterating f on n.

We shall now, using a variation of the coding introduced in [1], try to find a permutation f with minimal possible number of affine transformations (and hence, recursive sets A_i) for which Problem 2 is undecidable. The main result of this paper is

Theorem 1. *There are recursive sets $A_1, A_2, \ldots, A_5$ forming a partition of $\mathbb{N}$ such that*

$$f(n) = \begin{cases} (1/2)n - (1/2), & \text{if } n \in A_1 \\ (1/2)n + 0, & \text{if } n \in A_2 \\ (1/2)n + (1/2), & \text{if } n \in A_3 \\ 2n + 0, & \text{if } n \in A_4 \\ 2n + 1 & \text{if } n \in A_5 \end{cases} \tag{4}$$

is a permutation for which Problem 2 is undecidable.

Proof (Sketch). We define the recursive sets A_i such that for natural numbers n of specific type, iterating f on n corresponds to a computation of a Turing machine $\mathcal{M}$ on a specific input. This is called *ascending computation*. At the same time we define *descending computation* (iterating f^{-1} on n) which collides with the ascending one if and only if $\mathcal{M}$ halts on the given input. If $\mathcal{M}$ does not halt,

$$\lim_{k \to \infty} f^{(k)}(n) = \lim_{k \to \infty} (f^{-1})^{(k)}(n) = \infty.$$

Finally, for numbers n which do not correspond to a description of a Turing machine and its input, we define $f(n)$ such that f is a permutation of natural numbers. The claim follows from undecidability of the halting problem [4]. $\square$

2   Encoding a Turing Machine's Computation

We assume familiarity with Turing Machines and their configurations and computations as in [5]. Let us assume that the input and tape alphabets coincide, hence $\Sigma = \Gamma = \{a_1, \ldots, a_k\}$. The set of states is denoted by $Q = \{q_1, \ldots, q_m\}$ and it is assumed that q_1 and q_2 are the initial and final state, correspondingly ($q_1 \neq q_2$). The list of transition rules of a Turing machine is $\Delta = [u_1 \vdash v_1, \ldots, u_n \vdash v_n]$. Now, such a Turing machine can be given as a triplet

$$\mathcal{M} = [[a_1, \ldots, a_k], [q_1, q_2, \ldots, q_m], [u_1 \vdash v_1, \ldots, u_n \vdash v_n]] \qquad (5)$$

Definition 1. *Define a coding $\mathcal{C}'$ to the binary alphabet $\{0, 1\}$ as follows:*

$$\mathcal{C}'(a_i) = 0^{5+i}1 \ \ for \ i = 1, \ldots, k \quad and \quad \mathcal{C}'(q_i) = 0^{5+k+i}1 \ \ for \ i = 1, \ldots, m,$$

where the codewords are regarded as words in $\{0, 1\}^*$. We also need codings for the special letters:

$$\mathcal{C}'(\ [\ ) = 01 \qquad \mathcal{C}'(\ ]\ ) = 001 \qquad \mathcal{C}'(\ ,\ ) = 0^31$$
$$\mathcal{C}'(\vdash) = 0^41 \qquad \qquad \mathcal{C}'(\ \#\ ) = 0^51$$

For words $w = w_1 w_2 \cdots w_l$ whose length is greater than one, define

$$\mathcal{C}'(w) = \mathcal{C}'(w_1)\mathcal{C}'(w_2)\cdots\mathcal{C}'(w_l).$$

Note, that this applies to the elements of the list Δ. In the same way we now know how to encode the description (5) of a Turing machine.

Finally, for any word w whose encoding has been defined, let

$$\mathcal{C}(w) = 1\,\mathcal{C}'(w). \qquad (6)$$

Notice that the codewords $\mathcal{C}'(x)$, for a Turing machine given in the form (5), form a prefix code. Moreover, there is a bijective correspondence between the codewords $\mathcal{C}(w)$ and the natural numbers. In the following we write $n = \mathcal{C}(w)$, where $n \in \mathbb{N}$ is the natural number corresponding to the binary representation $\mathcal{C}(w)$.

3   Defining the Computing Part of the Permutation

Consider natural numbers n of the form

$$n = \mathcal{C}(\mathcal{M}\#w_0\#w_1\#\cdots\#w_l\#\alpha), \qquad (7)$$

where $\mathcal{M}$ is any Turing machine given in the form (5), w_0 is a legal initial ID of $\mathcal{M}$, $w_0 \vdash w_1 \vdash \cdots \vdash w_l$, α does not contain the special letter $\#$, and n (as a word) is a prefix of $\mathcal{C}(\mathcal{M}\#w_0\#w_1\#\cdots\#w_l\#w_{l+1}\#)$, where $w_l \vdash w_{l+1}$.

In the form (7) we allow $l = 0$ and α to be the empty word. Notice that $w_0 \vdash^* w_l$ is not a halting computation in (7). We define the *set of ascending computations* S_A to be

$$S_A = \{n \in \mathbb{N} \mid n \text{ is of the form (7)}\}. \tag{8}$$

Next, let $n \in S_A$ and define $f(n)$ to be the unique number whose binary representation is one bit longer than that of the number n and for which

$$f(n) \text{ is a prefix of } C(\mathcal{M}\#w_0\#w_1\#\cdots\#w_l\#w_{l+1}\#).$$

Thus, $f(n)$ continues the encoding of the computation by one bit and $f(n) = 2n + 0$ or $f(n) = 2n + 1$.

Next we define the *set of descending computations* S_D. In the following $n = w1$ means that the binary representation of n is of the form $w1$ where $w \in \{0, 1\}^*$. Let

$$S_D = \{n = w1 \mid w0 \in S_A\} \cup \{n = w10 \mid w01 \in S_A, w01 \neq C(\mathcal{M}\#w_0\#)\}, \tag{9}$$

Notice, that there is an obvious injection from S_D to S_A and that $S_A \cap S_D = \emptyset$.

Let $n \in S_D$ be of the form $n = w1$.

Suppose first that

$$w = C(\mathcal{M}\#w_0\#),$$

where w_0 is an initial ID of a Turing machine $\mathcal{M}$. Define then

$$f(n) = (1/2)(n - 1) = (1/2)n - 1/2,$$

i.e., $f(n) = w \in S_A$. Suppose then that

$$C(\mathcal{M}\#w_0\#) \text{ is a proper prefix of } w.$$

For this case we define $f(n) = (1/2)(n + 1) = (1/2)n + 1/2$. Now the binary representation of $f(n)$ is one bit shorter than that of n and it is either of the form $f(n) = w1$ or $f(n) = w10$, where $w0 \in S_A$ or $w01 \in S_A$, correspondingly. In both cases, $f(n) \in S_D$.

Let then $n \in S_D$ be of the form $n = w10$. Define $f(n) = (1/2)n + 0$, i.e. $f(n) = w1$. Since $w01 \in S_A$ it follows that $w0 \in S_A$ and thus $f(n) = w1 \in S_D$.

Finally we define the *set of halted computations* S_H in the obvious way:

$$S_H = \{n = C(\mathcal{M}\#w_0\#\cdots\#w_h\#) \mid w_0 \vdash^* w_h \text{ is a halting computation of } \mathcal{M}\}. \tag{10}$$

If $n \in S_H$, define $f(n) = (1/2)(n + 1) = (1/2)n + (1/2)$. Then $f(n)$ is of the form $f(n) = w1$ where $w0 \in S_A$ and therefore $f(n) \in S_D$.

Now we are ready to present two Lemmata. For them, denote

$$S = S_A \cup S_D \cup S_H. \tag{11}$$

Lemma 1. *Let $wx \in S$, where $w \in \{0,1\}^*$ and $x \in \{0,1\}$. Then exactly one of the following is true:*

(i) $wx = C(\mathcal{M}\#w_0\#)$ for some $\mathcal{M}$ and its initial ID w_0
(ii) $w \in S$

Proof. If $wx = C(\mathcal{M}\#w_0\#)$ it follows from the unique decodability of C that $w \notin S$.

Suppose then that wx is not of the form (i) and let $wx \in S_A \cup S_H$. Then wx is a possibly incomplete encoding of a computation of Turing machine and also $w \in S_A \subset S$.

Suppose that $wx \in S_D$. If $x = 1$ it follows that $w0 \in S_A$ and by the same argumentation as above $w \in S_A \subset S$. If $x = 0$ we can write $wx = u10$ where $u01 \in S_A$, and thus $u0 \in S_A$ from which it follows that $w = u1 \in S_D \subset S$.   □

Lemma 2. *The function $f : S \to S$ is bijective.*

Proof. Let $n = C(\mathcal{M}\#w_0\#)$ for some $\mathcal{M}$ and its initial ID w_0. We defined $f(n)$ to be the number whose binary representation extends the encoding of the computation of $\mathcal{M}$ on w_0 by one bit. In the same way $f^{(i)}(n) \in S_A$, for $i \geq 2$, until possibly for some $k \in \mathbb{N}$

$$f^{(k)}(n) = C(\mathcal{M}\#w_0\#\cdots\#w_h\#) \in S_H,$$

where $w_0 \vdash^* w_h$ is a halting computation of $\mathcal{M}$.

On the other hand, the preimage of f on n is uniquely determined as

$$f^{-1}(n) = C(\mathcal{M}\#w_0\#)1 \in S_D.$$

Iterating f^{-1} on n results in $(f^{-1})^{(i)}(n) \in S_D$ until possibly for some $k \in \mathbb{N}$

$$(f^{-1})^{(k)}(n) = C(\mathcal{M}\#w_0\#\cdots\#w_h\#) \in S_H.$$

Thus, for fixed n, iterating f and f^{-1} results in a finite or infinite cycle.

Since a computation of a Turing machine is deterministic, cycles with different initial IDs do not intersect.   □

Notice that in the definition of $f : S \to S$ we used all the affine transformations $(1/2)n\pm(1/2)$, $(1/2)n+0$, $2n+0$ and $2n+1$ and that each of these transformations corresponds to a recursive set A_i.

4   Extending f

We could extend f to a permutation of natural numbers by defining $f(n) = n$ whenever $n \notin S$. However, as our motivation was to find a permutation f with as few affine transformations as possible, we reuse these transformations when extending f.

Extending f is done by iterating the following procedure:

Algorithm 1

1. Let $I = \emptyset$
2. Let $n \in \mathbb{N}$ be the smallest integer for which $f(n)$ has not been defined. Set $I = I \cup \{n\}$.
3. Let the binary representation of n be w
4. Define $T_n = w1^* \cup w0^+$
5. If $m \in w1^*$, define $f(m) = w1$ (i.e. $f(m) = 2m + 1$)
6. If $m \in w0^+$, define $f(m) = w$ (i.e. $f(m) = (1/2)m + 0$)
7. Go to 2.

Let us denote $T = \mathbb{N} \setminus S$. To show that Algorithm 1 truly extends $f : S \to S$ to a bijection $f : \mathbb{N} \to \mathbb{N}$, we must prove that

(i) $T = \bigcup_{i \in I} T_i$
(ii) $f(T) = T$ and
(iii) $f : T \to T$ is bijective.

The following auxiliary result is clear by the definition of the Algorithm 1.

Lemma 3. *Let $i \in I$ and $wx \in T_i$. Then either $w \in T_i$ or $i = wx$.*

As corollary we obtain

Corollary 1. *Let $i, j \in I$ be different. Then $T_i \cap T_j = \emptyset$.*

Proof. Suppose $T_i \cap T_j \neq \emptyset$. By Lemma 3, $i \in T_j$ or $j \in T_i$, which is impossible. $\square$

Lemma 4. *The sets T_i form a partition of T.*

Proof. From Algorithm 1 it directly follows that

$$T \subset \bigcup_{i \in I} T_i.$$

Suppose there is a $v \in T_i \cap S$ for some $i \in I$. Then by Lemmata 1 and 3 there must exist $w \in T_i$ and $x \in \{0, 1\}$ such that

$$wx = \mathcal{C}(\mathcal{M} \# w_0 \#) \in T_i \cap S. \tag{12}$$

From the definition of the coding $\mathcal{C}$ we see that wx ends to bits 0001 (i.e. $x = 1$) and thus the three final bits of w are 000. Let us write $w = w'00$. Clearly $w'0 \notin S$, and thus $w'0$ and w belong both in the same set T_i. But then $w1 \notin T_i$, a contradiction.

Therefore $T = \bigcup T_i$ and by Corollary 1 the sets T_i are disjoint. $\square$

Now we are ready to present the final Lemma needed for the proof of Theorem 1.

Lemma 5. *Let $f : \mathbb{N} \to \mathbb{N}$ be the extension of $f : S \to S$ as described in Algorithm 1. Then $f(T) = T$ and $f : T \to T$ is bijective.*

Proof. By Algorithm 1, each set T_i corresponds to iterations of f and f^{-1} on $i \notin S$. Thus f is bijective on each T_i. From Lemma 4 it follows that f is bijective on T. $\square$

5   Proof of the Main Theorem

Finally, we are ready to present the proof of the Theorem 1.

Proof (Theorem 1). By Lemmata 2 and 5 we have constructed a recursive permutation $f : \mathbb{N} \to \mathbb{N}$ of the form

$$f(n) = \begin{cases} (1/2)n - (1/2), & \text{if } n \in A_1 \\ (1/2)n + 0, & \text{if } n \in A_2 \\ (1/2)n + (1/2), & \text{if } n \in A_3 \\ 2n + 0, & \text{if } n \in A_4 \\ 2n + 1 & \text{if } n \in A_5 \end{cases} \tag{13}$$

where the sets A_i are recursive.

If $n \in \mathbb{N}$ is of the form

$$n = \mathcal{C}(\mathcal{M}\#w_0\#) \tag{14}$$

for some Turing machine $\mathcal{M}$ and its initial ID w_0, the following is true:

There exists a $k > 1$ for which $f^{(k)}(n) = n \iff \mathcal{M}$ halts

on its initial ID w_0.

The claim follows from the undecidability of the halting problem [4].   □

6   Conclusion

In this paper, a coding of a Turing machine's computation introduced in [1] was used to derive an undecidability result concerning the cycle structure of a certain explicitly defined permutation. The main motivation was to present an effective way to attain such results, but also to find an undecidable permutation with as few affine transformations as possible. A future task is to further investigate the number of affine transformations for which the general cycle structure problem remains undecidable. Such an analysis may prove to be valuable when analyzing the decidability of such iterative permutation problems as the Collatz's original problem.

References

1. Lehtonen, E.: Two Undecidable Variants of Collatz's Problems. In: Theoretical Computer Science (2008), doi:10.1016/j.tcs.2008.08.029
2. Lagarias, J.C.: The $3x+1$ problem and its generalizations. Amer. Math. Monthly 92, 3–23 (1985)
3. Conway, J.H.: Unpredictable Iterations. In: Proceedings of the 1972 Number Theory Conference, pp. 49–52. University of Colorado, Colorado (1972)
4. Turing, A.: On computable numbers, with an application to the Entscheidungsproblem. In: Proceedings of the London Mathematical Society, vol. 2, p. 42 (1936)
5. Rozenberg, G., Salomaa, A.: Handbook of Formal Languages, vol. 1, pp. 177–179. Springer, Heidelberg (1997)

Forward Analysis of Dynamic Network of Pushdown Systems Is Easier without Order

Denis Lugiez

LIF UMR 6166 Aix-Marseille Université- CNRS, France
denis.lugiez@lif.univ-mrs.fr

Abstract. Dynamic networks of Pushdown Systems (*PDN* in short) have been introduced to perform static analysis of concurrent programs that may spawn threads dynamically. In this model the set of successors of a regular set of configurations can be non-regular, making forward analysis of these models difficult. We refine the model by adding the associative-commutative properties of parallel composition, and we define Presburger weighted tree automata, an extension of weighted automata and tree automata, that accept the set of successors of a regular set of configurations. This allows forward analysis of *PDN* since these automata have a decidable emptiness problem and are closed under intersection.

Introduction

Dynamic networks of Pushdown Systems is a model of concurrent programs that models thread generation and it has been introduced for performing the static analysis of these programs [BMOT05]. This model follows from a stream of works that have advocated the use of automata techniques for the static analysis of programs for more and more complex problems (from intraprocedural analysis to interprocedural concurrent analysis [EP00]) and more and more complex models, from Pushdown system [AB97, RSJ03] to Process Algebra [LS98] and networks of Pushdown systems [BMOT05], possibly involving data structure or synchronization [KG07, LMOW09]. In [BMOT05], the authors consider pushdown processes that can generate new processes yielding configurations that are sets of ordered unranked trees and they prove that the set of predecessors of a regular set of configurations is regular where regularity refers to hedge automata, the standard extension of tree automata to unranked trees. However, the set of successors can be non-regular making forward analysis of such systems difficult. In this paper we enrich this model by assuming that parallel composition is an associative-commutative operation therefore the execution of threads generated by some pushdown system is independent of their order. This amounts to considering configurations that are *unranked unordered trees* and using the notion of regularity relying on Presburger automata [ZL06]. The main result of the paper is to prove that the set of successors of a regular set of configurations can be non-regular but is accepted by a Presburger weighted tree automaton, an extension of Presburger automata which enjoy properties that allow to perform forward

O. Bournez and I. Potapov (Eds.): RP 2009, LNCS 5797, pp. 127–140, 2009.

analysis. The regularity of the set of predecessors of a regular set is derived from the results of [BMOT05].

Section 1 presents the basic definitions while section 2 introduces pushdown systems and dynamic networks of pushdown systems. Regular sets are defined in section 4. Weighted word and tree automata are defined in section 5 and the computation of the set of successors is explained in section 6. Section 7 shows how to derive the computation of the set of predecessors using known results.

1   Analysis of Transition Systems

A system S is given by an (infinite) set of configurations $\bar{C}$ and a transition relation $\rightarrow_S$ between configurations. Configurations are formal objects (words, trees,...) that describe the current state of the system, and the dynamic behavior of the systems is given by the relation $\rightarrow_S$ which is usually defined by a finite set of transition rules R. The reflexive transitive closure of $\rightarrow_S$ is denoted as $\xrightarrow{*}_S$.

The set of *successors* of a configuration c is the set $Post^*_S(c) = \{c' \in \bar{C} \mid c \xrightarrow{*}_S c'\}$ and for $L \subseteq \bar{C}$, $Post^*_S(L) = \bigcup_{c \in L} Post^*_S(c)$.

The set of *predecessors* of a configuration c is the set $Pred^*_S(c) = \{c' \in \bar{C} \mid c' \xrightarrow{*}_S c\}$ and for $L \subseteq \bar{C}$, $Pred^*_S(L) = \bigcup_{c \in L} Pred^*_S(c)$.

The set of all possible initial configurations $Init$ and the set of all bad configurations Bad can be infinite and are usually defined as regular language for some appropriate notion of regularity. *Backward analysis* tests the emptiness of the set $Init \cap Pred^*_S(Bad)$ when *Forward analysis* test the emptiness of the set $Post^*_S(Init) \cap Bad$. These analyzes allow to detect statically if the execution of the system S can lead to an error state. When the languages $Init$ and Bad are regular for a good notion of regularity (that enjoys closure under boolean operations and decision of emptiness) the feasibility of these analysis boils down to proving that $Pred^*_S(L)$ and $Post^*_S(L)$ are regular when L is regular and providing effective constructions of devices accepting these sets. Forward analysis is more difficult than backward analysis, since regularity is easier to preserve under inverse image that under direct image, see [CDJ+99] for regular tree languages.

In the following, we drop the subscript S of $\rightarrow_S, \ldots$ when S is clear from the context.

2   Pushdown Systems

A *pushdown system* (PDS in short, see [AB97] for details) P is a triple (Q, Σ, R) where Q is a finite set of states, Σ is a finite stack alphabet and R is a finite set of transition rules $qa \rightarrow q'\gamma$ with $q, q' \in Q, a \in \Sigma, \gamma \in \Sigma^*$ (also called rewrite rules in the following). The set $\bar{C}$ of configurations is the set of words qw with $q \in Q$ (the state), $w \in \Sigma^*$ (the content of the stack). The transition relation $\rightarrow$ between configurations is the relation defined by $qw \rightarrow q'w'$ iff $w = aw''$ and $w' = \gamma.w''$ and there is a rule $qa \rightarrow q'\gamma \in R$. The relation $\xrightarrow{*}$ is exactly the prefix rewriting relation defined by the set of rewrite rules R.

Let P be a pushdown system, let $L \subseteq \bar{C}$ be a regular language. Then $Pred^*(L)$ and $Post^*(L)$ are regular languages, see [Buc64, AB97]. Actually the relation R on pairs of configurations defined by qw $\mathcal{R}$ $q'w'$ iff $qw \xrightarrow{*} q'w'$ is a rational relation (see [Cau00] for extensions).

The following proposition states that the successors of a regular set of configurations L is the set of successors of a unique configuration, provided that (i) we extend the initial PDS with new states and new rules (ii) that we keep only the configurations that corresponds to the initial alphabet.

Proposition 1. *Let $P = (Q, \Sigma, R)$ be a PDS and $L \subseteq \bar{C}$ be a regular language. There exists $P' = (Q \cup Q', \Sigma \cup \{\$\}, R \cup R')$ such that $Post^*_P(L) = Post^*_{P'}(q_0\$) \cap \bar{C}$ where $q_0 \in Q'$.*

3   Dynamic Network of Pushdown Systems

Dynamic networks of pushdown processes [BMOT05] generalize PDS since (i) a configuration may have several PDS running in parallel, (ii)a transition rule of a PDS not only changes the state and stack of the PDS, but may also spawn one (or more) new PDS which is a son of the process. There is no limitation in the creation of processes (a process has an arbitrary number of sons) and in the recursion depth for process creation (each process may create sons that can create sons, ...). Therefore configurations are isomorphic to unranked tree-like structures. In this work, we enrich the original model by assuming that parallel composition is associative-commutative, hence trees are also unordered.

3.1   Configurations

The set PDN of configurations and the set $PDN_\parallel$ of parallel configurations are defined by:

$$PDN \ni c \quad ::= qw \quad | \quad qw(c_\parallel) \quad q \in Q, w \in \Sigma^*$$
$$PDN_\parallel \ni c_\parallel ::= c_1 \parallel \ldots \parallel c_n \quad n \geq 1, \forall i = 1, \ldots, n \ c_i \in PDN$$

The parallel composition is independent of the order of its arguments and the equality $\equiv$ between configurations is defined by:

$$qw \equiv q'w' \qquad \text{iff } q = q' \text{ and } w = w'$$
$$qw(c_1 \parallel \ldots \parallel c_n) \equiv q'w'(c'_1 \parallel \ldots \parallel c'_m) \text{ iff } q = q', \ w = w', \ n = m \text{ and } \exists \sigma$$
$$\text{permutation of } \{1, \ldots, n\}$$
$$\text{such that } c_i \equiv c'_{\sigma(i)}$$

The set $Sub(t)$ of process subterms of a configuration t is defined by:

$$Sub(qw) = \{qw\}$$
$$Sub(qw(t_1 \parallel \ldots \parallel t_n)) = \{qw(t_1 \parallel \ldots \parallel t_n)\} \cup \{qw\} \cup \bigcup_{i=1,\ldots,n} Sub(t_i)$$

A context $C[\square]$ is a configuration t where some process subterm is replaced by the symbol $\square$. The notation $C[s]$ denotes the configuration obtained by replacing $\square$ by $s \in PDN$.

Example 1. Let $c = qa(qaa \parallel q'b \parallel q(qa \parallel q'b))$. We can draw this configuration as sets of vertical lines (each one being a PDS) combined in a tree-like structure:

$$
\begin{array}{ccccccc}
 & & & q & & & \\
 & & & a & & & \\
q & \parallel & q' & \parallel & & q & \\
a & & b & & q & \parallel & q' \\
a & & & & a & & b
\end{array}
$$

We have $Sub(c) = \{q, qa, qaa, q'b, qb\}$ and $c \equiv qa(q'b \parallel q(q'b \parallel qa) \parallel qaa)$.

3.2   PDN and Their Transition Relation

A *dynamic network of pushdown systems* (*PDN* in short) P is a triple (Q, Σ, R) where Q is a finite set of states, Σ is a finite alphabet and R is a finite set of rules of the form $qa \rightarrow q'\gamma$ or $qa \rightarrow q'\gamma(q_1\gamma_1 \parallel \ldots \parallel q_n\gamma_n)$ where w, γ, γ_i for $i = 1, \ldots, n$ belong to Σ^*, q, q', q_i for $i = 1, \ldots, n$ belong to Q.

The relation $\rightarrow$ induced by a *PDN* P on pairs of configurations is defined by:

- if $qa \rightarrow q'\gamma \in R$ then $qaw \rightarrow q'\gamma w$
- if $qa \rightarrow q'\gamma(q_1\gamma_1 \parallel \ldots \parallel q_n\gamma_n) \in R$ then
    - $qaw \rightarrow q\gamma w(q_1\gamma_1 \parallel \ldots \parallel q_n\gamma_n)$
    - $qaw(s_1 \parallel \ldots \parallel s_m) \rightarrow q\gamma w(q_1\gamma_1 \parallel \ldots \parallel q_n\gamma_n \parallel s_1 \ldots \parallel s_m)$
- if $c \rightarrow c'$ *and* $c \in Sub(t)$, then $t = C[c] \rightarrow t' = C[c']$,
- if $c \rightarrow c'$ *and* $\bar{c} \equiv c, \bar{c'} \equiv c'$ then $\bar{c} \rightarrow \bar{c'}$.

Example 2. Let $P = (\{q, q'\}, \{a, b\}, \{qa \rightarrow q'bb, q'b \rightarrow qab(qa \parallel qa)\})$ be a *PDN*. A sequence of transition of P is:

$$
qa \rightarrow q'bb \rightarrow qabb(qa \parallel qa) \rightarrow qabb(qa \parallel q'bb) \rightarrow q'bbb(qa \parallel q'bb)
$$
$$
\rightarrow qabbb(qa \parallel qa \parallel qa \parallel q'bb) \rightarrow qabbb(qa \parallel qa \parallel qa \parallel qabb(qa \parallel qa))
$$

The next proposition states that a transition sequence can be done by applying transition rules to the top *PDS* first and then on the arguments of parallel compositions.

Proposition 2. $qw(t_1 \parallel \ldots \parallel t_n) \xrightarrow{*} q'w'(u_1 \parallel \ldots \parallel u_m)$ *iff* $qw(t_1 \parallel \ldots \parallel t_n) \xrightarrow{*} q'w'(t_1 \parallel \ldots \parallel t_n \parallel s_{n+1} \parallel \ldots \parallel s_m) \xrightarrow{*} q'w'(u_1 \parallel \ldots \parallel u_m)$

4   Regular Sets of PDN Configurations

Configurations as Unranked-Unordered Trees. A configuration is isomorphic to a unranked-unordered tree on the signature $\{\#\} \cup Q \cup \Sigma \cup \{\parallel\}$ where $\#$ is a constant, each symbol in Q or Σ is now a monadic symbol. The symbol $\parallel$

is a permutative variadic symbol. The operations $c2t$ and $t2c$ that transform a configuration c into a tree t and conversely are defined as follows:

$$c2t(qa_1\ldots a_n) = q(a_1(\ldots(a_n(\#))))$$
$$c2t(qa_1\ldots a_n(c_1 \parallel \ldots \parallel c_n)) = q(a_1(\ldots(a_n(\parallel (c2t(c_1),\ldots,c2t(c_n)))))))$$
$$t2c(q(a_1(\ldots(a_n(\#))))) = qa_1\ldots a_n$$
$$t2c(q(a_1(\ldots(a_n(\parallel (c_1,\ldots,c_n)))))) = qa_1\ldots a_n(t2c(c_1) \parallel \ldots \parallel t2c(c_n))$$

Since the transformation is one-to-one, results and definitions stated for configurations are valid for trees. The set of nodes of a tree t is denoted by $Nodes(t)$. A symbol of $\{\#\} \cup Q \cup \Sigma \cup \{\parallel\}$ is attached to each node N of t.

Top-down Presburger Automata. Presburger arithmetic is the first-order theory of $\mathbb{N}, +, =$ and a Presburger formula is a formula of this theory. For instance $\exists z \; x = y + 2z + 1$ is a Presburger formula in the free variables x, y. This theory is decidable [Pre29].

Word regular languages and tree regular languages have been generalized for unranked-unordered trees using Presburger automata [ZL06]. To fit the framework of PDN, we slightly change the definitions and we use *top-down automata* instead of bottom-up automata.

Definition 1. *A top-down Presburger automaton is a tuple $\mathcal{A} = (S, S_I, R \cup R')$ where:*

- $S = \{s_1,\ldots,s_p\}$ *is a finite ordered set of states,*
- $S_I \subseteq S$ *is a set of initial states,*
- R *is a set of rules of the form $s \to \#$ or $s \xrightarrow{a} s'$ where $s, s' \in S, a \in Q \cup \Sigma$,*
- R' *is a set of rules $s \xrightarrow{\parallel} \phi(x_1,\ldots,x_p)$ where $\phi(x_1,\ldots,x_p)$ is a Presburger formula.*

A run of the automaton $\mathcal{A}$ on a tree t is a labelling $r : Nodes(t) \to S$ of the nodes of t by the states of S such that:

- if $\#$ is the symbol of a node N of t, the node N is labelled by $r(N) = s$ such that the rule $s \to \#$ belongs to R,
- if $a \in Q \cup \Sigma$ is the symbol of a node N of t, and N is labelled by $r(N) = s$, then the unique son of N is labelled by s' such that the rule $s \xrightarrow{a} s'$ belongs to R,
- if $\parallel$ is the symbol of a node N of t and N is labelled by $r(N) = s$, if $N_1,\ldots,N_m$ are the sons of N, then
  - (i) there are n_1 sons of N labelled by s_1, $\ldots$, n_p sons of N labelled by s_p,
  - (ii) there is a rule $s \xrightarrow{\parallel} \phi(x_1,\ldots,x_p) \in R'$, such that $\phi(n_1,\ldots,n_p)$ is true.

A tree t is accepted by $\mathcal{A}$ is there is a run of $\mathcal{A}$ on t such that the root of t is labelled by a state $s \in S_I$. $L(\mathcal{A})$ is the set of trees accepted by $\mathcal{A}$ and a language L is a *Presburger regular language* iff $L = L(\mathcal{A})$. When there is no ambiguity, we say regular language. By construction $t \in L(\mathcal{A})$ and $s \equiv t$ implies $s \in L(\mathcal{A})$.

Example 3. Let $q(a^*(\ldots))$ denote any $q(a^n(\ldots))$ for $n \geq 0$. The set of trees of the form $q(a^*(\| (q(b^*(\#)), \ldots, q(b^*(\#)), q(c^*(\#)), \ldots, q(c^*(\#))))$ where the parallel operator $\|$ has as many arguments $qb^*\#$ as arguments $qc^*\#$ is a Presburger regular language since it is accepted by the automaton $\mathcal{A} = (S = \{s_a, s_b, s_c\}$,
$S_I = \{s_a\}, R = \{s_a \xrightarrow{q} s_a; s_a \xrightarrow{a} s_a; s_b \xrightarrow{q} s_b; s_b \xrightarrow{b} s_b; s_c \xrightarrow{q} s_c; s_c \xrightarrow{c} s_c; s_b \rightarrow \#; s_c \rightarrow \#\}$,
$R' = \{s_a \xrightarrow{\|} x_a = 0 \wedge x_b = x_c\})$ with x_a the variable for s_a, x_b the variable for s_b, x_c the variable for s_c.

The usual constructions on tree automata can be adapted to Presburger automata which yields the decidability of the emptiness of $L(\mathcal{A})$ and the closure of regular languages under boolean operations, see [ZL06].

5   Weighted Automata for PDN

5.1   Semilinear Sets

Let $b \in \mathbb{N}^m$, let $P = \{p_1, \ldots, p_n\}$ be a finite subset of $\mathbb{N}^m$. The linear set $L(b, \mathcal{P})$ of $\mathbb{N}^m$ is the set $\{b + \Sigma_{i=1}^{i=n} \lambda_i p_i \mid \lambda_i \in \mathbb{N}\}$. A semilinear set is a finite union of linear sets. The $+$ operation on subsets of $\mathbb{N}^m$ is defined by $L + M = \{x + y \mid x \in L, y \in M\}$. The $*$ operation is defined by $L^* = \cup_{n \geq 0} L^n$ where $L^0 = \{(0, \ldots, 0)\}$, and $L^n = \underbrace{L + \ldots + L}_{n}$.

The set of rational expressions of semilinear sets is defined by

$$Rat ::= L \mid Rat + Rat \mid Rat \cup Rat \mid Rat^*$$

where L denotes any semilinear set. A rational expression $R \in Rat$ denotes a set $[R]$ inductively defined by

$$[L] = L \ [R + R'] = [R] + [R'] \ [R \cup R'] = [R] \cup [R'] \ [R^*] = [R]^*$$

Proposition 3. *Let $R \in Rat$. Then there exists a effectively constructible semilinear set L such that $L = [R]$.*

Semilinear sets and Presburger arithmetic are closely related since the set of valuations such that a formula $\phi(x_1, \ldots, x_p)$ is true is an effectively constructible semilinear set [GS66]. From now on, for the sake of simplicity, a semilinear set which contains only one element c will be written c (instead of $\{c\}$).

5.2   Presburger Weighted Word Automata

A semiring is a structure $(K, \oplus, \otimes, 0, 1)$ such that (i) $K, \oplus$ is a commutative monoid with 0 as neutral element, (ii) $K, \otimes$ is a monoid with 1 as neutral element, (iii) $x \otimes (y \oplus z) = (y \oplus z) \otimes x = x \otimes y \oplus x \otimes z$ (iv) for all $x \in K$, $0 \otimes x = x \otimes 0 = 0$. Let $\mathcal{SL}_m$ be the set of semilinear sets of $\mathbb{N}^m$. Then $\mathcal{S}_m = (\mathcal{SL}_m, \cup, +, \emptyset, (0, \ldots, 0))$ is a semiring. Weighted automata are word automata where the transitions are labelled by element of a semiring K. Weighted automata have already used for PDS analysis provided that K satisfies additional properties [RSJ03]. Presburger weighted automata have a similar definition, but the semiring for labels is S_p and the definition of the transition relation is slightly distinct from the standard one. Furthermore, these automata enjoy particular properties used in the reachability analysis of PDN.

Definition 2. *A Presburger weighted word automaton is an automaton $\mathcal{A} = (S, s_0, S_F, S_m, \Delta)$ where S_F is a set of pairs (s, ϕ_s) with $s \in S, \phi_s \in S_m, \Delta \subseteq S \times \Sigma \cup \{\epsilon\} \times S_m \times S$. A transition rule of Δ is denoted $s \xrightarrow{a,C} s'$.*

The transition relation $\xrightarrow{*}_{\mathcal{A}} \subseteq \Sigma^* \times \mathbb{N}^m \times S$ is inductively defined by:

- $\epsilon, (0, \ldots, 0) \xrightarrow{*}_{\mathcal{A}} s_0$
- *if $w, c \xrightarrow{*}_{\mathcal{A}} s$ then $wa, c + c' \xrightarrow{*}_{\mathcal{A}} s'$ if there is a rule $s \xrightarrow{a,C'} s'$ with $c' \in C'$,*
- *if $w, c \xrightarrow{*}_{\mathcal{A}} s$ then $w, c + c' \xrightarrow{*}_{\mathcal{A}} s'$ if there is a rule $s \xrightarrow{\epsilon,C'} s'$ with $c' \in C'$*

A pair w, c is accepted iff $w, c \xrightarrow{*}_{\mathcal{A}} s$ and $c \in \phi_s$ for $(s, \phi_s) \in S_F$. The language accepted by $\mathcal{A}$ is the set $L(\mathcal{A})$ of pairs accepted by $\mathcal{A}$.

Proposition 4. *For each $s \in S$, the set $L(s_0, s) = \{c \in SL_m \mid \exists w \in \Sigma^* \; w, c \xrightarrow{*}_{\mathcal{A}} s\}$ is an effectively computable semilinear set.*

Proof. By proposition 3, and the property that the set of words reaching a state of a (usual) word automaton can be described by a rational expression, we get that for each state s of a Presburger weighted word automaton, we can compute the semilinear set $L(s_0, s) = \{c \in SL_m \mid \exists w \in \Sigma^* \; w, c \xrightarrow{*}_{\mathcal{A}} s\}$.

Presburger weighted word automaton enjoy other properties that they share with Presburger weighted tree automaton and they are given in the next section.

5.3   Presburger Weighted Tree Automata

Presburger weighted tree automata are designed to accept set of trees corresponding to configurations of PDN.

Definition 3. *A top-down Presburger weighted tree automaton is a tuple $\mathcal{A} = (S, S_I, S_m, R \cup R')$ where:*

- $S = \{s_1, \ldots, s_p\}$ *is a finite set of states,*
- $S_I \subseteq S$ *is a set of initial states,*
- S_m *is the semiring $(SL_m, \cup, +, \emptyset, (0, \ldots, 0))$,*
- R *is a set of rules of the form $s \xrightarrow{(0,\ldots,0)} \#$ or $s \xrightarrow{a,C} s'$ where $s, s' \in S, a \in Q \cup \Sigma \cup \{\epsilon\}, C \in S_n$*
- R' *is a set of rules $s \xrightarrow{\|} \phi(x_1, \ldots, x_p, y_1, \ldots, y_m)$ where $\phi(x_1, \ldots, x_p, y_1, \ldots, y_m)$ is a Presburger formula.*

A run of the automaton $\mathcal{A}$ on a tree t is a labelling $r : Nodes(t) \times \mathbb{N}^m \to S$ of the nodes of t by the states of S and weights of $\mathbb{N}^m$ such that:

- if $\#$ is the symbol of a node N of t, the node N is labelled by $r(N) = (s,(0,\ldots,0))$ such that the rule $s \overset{(0,\ldots,0)}{\rightarrow} \#$ belongs to R.
- if $a \in Q \cup \Sigma$ is the symbol of a node N of t, and N is labelled by $r(N) = (s,c)$, then the unique son of N is labelled by $(s',c+c')$ s.t. there is a sequence of rules

$$\bar{s}_1 \overset{\epsilon,\bar{C}_1}{\rightarrow} \bar{s}_2, \ldots, \bar{s}_{i-1} \overset{\epsilon,\bar{C}_{i-1}}{\rightarrow} \bar{s}_i, \bar{s}_i \overset{a,\bar{C}_{i+1}}{\rightarrow} \bar{s}_{i+1}, \bar{s}_{i+2} \overset{\epsilon,\bar{C}_{i+2}}{\rightarrow} \bar{s}_{i+3}, \ldots, \bar{s}_l \overset{\epsilon,\bar{C}_l}{\rightarrow} \bar{s}_{l+1}$$

where $\bar{s}_1 = s$, $\bar{s}_{l+1} = s'$ and $c' = c_1 + \ldots + c_l$ with $c_i \in \bar{C}_i$ for $i = 1,\ldots,l$.
- if $\|$ is the symbol of a node N of t and N is labelled by $r(N)=(s,(c_1,\ldots,c_m))$, if $N_1,\ldots,N_n$ are the sons of N, then
  - (i) there are n_1 sons of N labelled by $(s_1,(0,\ldots,0))$, $\ldots$, n_p sons of N labelled by $(s_p,(0,\ldots,0))$,
  - (ii) there is a rule $s \overset{\|}{\rightarrow} \phi(x_1,\ldots,x_p,y_1,\ldots,y_m) \in R'$, such that $\phi(n_1,\ldots,n_p,c_1,\ldots,c_m)$ is true.

$L(\mathcal{A})$ denotes the set of trees accepted by $\mathcal{A}$.

Example 4. Let $q(b^+(\#))$ denote $q(b^n(\#))$ with $n > 0$. The (non-regular) language $\mathcal{L}^1$

$$\{q(a^{n+1}(\| \underbrace{(q(b^+(\#)),\ldots,q(b^+(\#))}_{n}, \underbrace{q(c^+(\#)),\ldots q(c^+(\#))}_{n})))) \mid n \geq 1\}$$

is accepted by the Presburger weighted tree automaton $\mathcal{A} = (S, S_I, S_2, R \cup R')$
$S = \{s_0, s_{qa}, s_{qb}, s_{qc}, s_a, s_b, s_c, s_q\}$, $S_I = \{s_0\}$,
$R = \{s_0 \overset{q,(0,0)}{\rightarrow} s_q, s_q \overset{a,(0,0)}{\rightarrow} s_a, s_a \overset{a,(1,1)}{\rightarrow} s_a, s_{qb} \overset{q,(0,0)}{\rightarrow} s_b, s_b \overset{b,(0,0)}{\rightarrow} s_b,$
$\qquad s_b \overset{(0,0)}{\rightarrow} \#, s_{qc} \overset{q,(0,0)}{\rightarrow} s_c, s_c \overset{q,(0,0)}{\rightarrow} s_c, s_c \overset{(0,0)}{\rightarrow} \#, \}$,
$R' = \{s_a \rightarrow x_{s_{qb}} = y_1 \wedge x_{s_{qc}} = y_2 \wedge \bigwedge_{s \neq s_{qb}, s_{qc}} x_s = 0$

The tree $q(a(a(a(\| (q(b(\#)), q(c(c(\#))))))))$ is accepted by a run that labels the node $\|$ by $(s_a, (1,1))$, the first son of this node by s_{qb}, the second son by s_{qc}. The tree $q(a(a(a(\| (q(b(\#)), q(c(\#)))))))$ is not accepted since the node $\|$ can be labelled only by $(s_a, (2,2))$ and this node has only two sons (when two nodes labelled by s_{qb} and two nodes labelled by s_{qc} are required).

A main feature of these automata is that the non-regular behavior is generated by weight computations. The Presburger formula of rules of R' can only be used to add additional regular constraints. For instance, we can build an automaton accepting the subset of $\mathcal{L}$ corresponding to even values of n by replacing in $\mathcal{A}$ the rule $s_a \rightarrow x_{s_{qb}} = y_1 \wedge x_{s_{qc}} = y_2 \wedge \bigwedge_{s \neq s_{qb}, s_{qc}} x_s = 0$ by the rule $s_a \rightarrow x_{s_{qb}} = y_1 \wedge x_{s_{qc}} = y_2 \wedge x_{s_{qb}}\%2 = 0 \wedge x_{s_{qc}}\%2 = 0 \wedge \bigwedge_{s \neq s_{qb}, s_{qc}} x_s = 0$.

Proposition 5. Let $\mathcal{A}$ be a Presburger weighted tree automaton.

---

[1] non-regularity follows from a straightforward pumping argument.

- *There is a Presburger weighted tree automaton $\mathcal{B}$ without ϵ-rules such that $L(\mathcal{B}) = L(\mathcal{A})$.*
- *It is decidable if $L(\mathcal{A})$ is empty.*
- *Let $\mathcal{B}$ be a Presburger weighted tree automaton. Then there is a effectively computable Presburger weighted tree automaton $\bar{C}$ s.t. $L(\bar{C}) = L(\mathcal{A}) \cap L(\mathcal{B})$.*

Since a Presburger tree automaton can be seen as a Presburger weighted tree automaton by taking $S_p = \mathbb{N}$ and replacing rules $s \xrightarrow{a} s'$ by $s \xrightarrow{a,0} s'$, we can build a Presburger weighted tree automaton that accepts $L(\mathcal{A}) \cap L(\mathcal{B})$ for $\mathcal{B}$ a Presburger tree automaton.

6   Forward Analysis of *PDN*

In this section we consider a *PDN* $P = (Q, \Sigma, R)$. From now on, we say that a subset $L \subseteq \bar{C}$ is a regular language iff the set $c2t(L) = \{t \mid t = c2t(c), c \in L\}$ is a regular tree language.

Example 5. Let $P = (\{q\}, \{a, b, c\}, \{qa \rightarrow qaa \parallel (qb, qc), qb \rightarrow qbb, qc \rightarrow qcc\})$. The automaton of example 4 accepts $Post^*(qa)$.

Since this example has a non-regular set of successors, we get:

Proposition 6. *$Post^*(L)$ can be a non-regular language.*

Our goal is to show that Presburger weighted tree automata can be used to perform the forward analysis of *PDN*.

6.1   Preliminary Computations

The two following propositions are used to simplify the computation of $Post^*(L)$ since it reduces the computation of the successors of a regular language to the computation of the successor of a single configuration.

Proposition 7. *Let $L \subseteq \bar{C}$ be a regular language. There exists $P' = (Q \cup Q', \Sigma \cup \{\$\}, R \cup R')$ such that $Post^*_P(L) = Post^*_{P'}(q_0\$) \cap \bar{C}$ where $q_0 \in Q'$.*

Proof. Let $\mathcal{A} = (S, S_I, \mathcal{R} \cup \mathcal{R}')$ be a top-down Presburger weighted tree automaton accepting $L' = c2t(L)$. For each rule $s \xrightarrow{\parallel} \phi(x_1, \ldots, x_p) \in \mathcal{R}'$ we compute the semilinear set $L_{s,\phi}$ of $\mathbb{N}^m$ equivalent to $\phi(x_1, \ldots, x_p)$. This set a finite union of linear sets $SL^i_{s,\phi} = L(b_i, P_i)$.

The *PDN* P' performs transitions that generates L and then performs the transitions of P. In the first computations, the new symbol $\$$ is used to prevent the application of rules of P.

- The set Q' is defined as follows:
  - Q' contains a starting state q_0,
  - Q' contains states $q_0(s)$ for each $s \in S$,

- if $s, s' \in S$ then Q' contains $q(s, s')$,
- if $SL_{s', \phi} = L(b, \mathcal{P})$ is a semilinear set corresponding to a rule $s' \xrightarrow{\parallel} \phi(x_1, \ldots, x_p)$, then Q' contains $q(s, s', \parallel, \phi)$ and $q(s, s', \parallel, L(b, \mathcal{P}))$.

- The set of rules R' is defined by:
  - P' chooses non-deterministically to generate a configuration qw or a configuration $qw(\ldots \parallel \ldots \parallel \ldots)$. Furthermore a run of $\mathcal{A}$ labels the root of $c2t(qw)$ by s and the last node of qw by s'. The corresponding rules are:
    * $q_0\$ \to q(s, s')\$$, meaning: generate qw such that a run of $\mathcal{A}$ labels the root of $q(w(\#))$ by s,
    * $q_0\$ \to q(s, s', \parallel, \phi)\$$, meaning: generate $qw(t_1 \parallel \ldots \parallel t_m)$ such that a run of $\mathcal{A}$ labels the root of $q(w(\parallel (c2t(t_1), \ldots, c2t(t_m)))$ by s,
  - P' performs the same choice as above, but from $q_0(s)$ instead of q_0 ($q_0\$$ occurs at the root of the initial configuration, when $q_0(s)\$$ is an initial configuration of an argument of a parallel composition)
    * $q_0(s)\$ \to q(s, s')\$$,
    * $q_0(s)\$ \to q(s, s', \parallel, \phi)\$$,
  - $q(s, s')\$ \to q(s, s'')\$a$ if $s'' \xrightarrow{a} s' \in \mathcal{R}$. This rule simulates the automaton rule $s'' \xrightarrow{a} s'$ in a backward way (which amounts to using a left-linear grammar as in the proof of proposition 1).
  - $q(s, s')\$ \to q$ if $\mathcal{A}$ contains a rule $s \xrightarrow{q} s'$.
  - if there is an automaton rule $s' \xrightarrow{\parallel} \phi(x_1, \ldots, x_p)$ and $L(b, \mathcal{P})$ is a linear set associated to ϕ, the PDN P simulates the generation of the parallel composition with the rules:
    * $q(s, s', \parallel, \phi)\$ \to q(s, s', \parallel, L(b, \mathcal{P}))\$(q_0(S)^b)$
    * $q(s, s', \parallel, L(b, \mathcal{P}))\$ \to q(s, s', \parallel, L(b, \mathcal{P}))\$(q_0(S)^p)$ for any $p \in \mathcal{P}$,
    where $q_0(S)^b$ for $b = (b_1, \ldots, b_p)$ (resp. $q_0(S)^p$ for $p = (p_1, \ldots, p_p)$) denotes $q_0(s_1)\$^{b_1} \parallel \ldots \parallel q_0(s_p)\$^{b_p}$ (resp. $q_0(s_1)\$^{p_1} \parallel \ldots \parallel q(s_p)\$^{p_p}$) where (i) $q(s_i)\$^l$ is a parallel composition of l instances $q(s_i)\$$, (ii)$S = \{s_1, \ldots, s_p\}$.
- $q(s, s', \parallel, L(b, \mathcal{P}))\$ \to q(s, s')\$$. This rules stops the generation of arguments of $\parallel$ and starts the generation of the PDS string like part qw.

If a transition of R is applied to a configuration c and a rule of R' is applied later, we can exchange the order of application and get the same result. Therefore the transition relation $\twoheadrightarrow_{P'}$ can be defined as $\twoheadrightarrow P \circ \twoheadrightarrow_{R'}$ since the rules of R and R' are independent.

A routine structural induction on c proves that $c \in Post_{R'}^*(q_0\$) \cap \bar{C}$ iff $c \in L$.
□

Proposition 8. *There exists a PDN P' such that (i) $Post_P^*(qa) = Post_{P'}^*(qa) \cap \bar{C}$ (ii) the rules of P' have the form $qa \to q'$ or $qa \to q'bc$ or $qa \to q'(q_1a_1 \parallel \ldots \parallel q_m a_m)$ or $qa \to q'bc(q_1a_1 \parallel \ldots \parallel q_m a_m)$*

Proof. (Sketch) A rule $qa \to q'a_1 \ldots a_n$ is replaced by rules $qa \to \bar{q}_{n-1}\#a_n$, $\bar{q}_{n-1}\# \to \bar{q}_{n-2}a_{n-1}, \ldots, \bar{q}_1\# \to q'a_1$ where the $\bar{q}_i$ and $\#$ are new symbols. This can be done for all rules and yields a new PDN P' of size linear in the size of the initial PDN P. By construction $Post_P^*(L) = Post_{P'}^*(L) \cap \bar{C}$.   □

From now on, we assume that the rules of $P = (Q, \Sigma, R)$ have the required form. Let $m = |Q||\Sigma|$ and let us define an ordering of the set $\{qa \mid q \in Q, a \in \Sigma\}$. The ith element qa in this ordering is identified with the tuple $C(qa) = (0, \ldots, 0, 1, 0, \ldots, 0)$. Therefore a parallel composition $c_\| = q_{i_1}a_{j_1} \| \ldots \| q_{i_k}a_{j_k}$ can be identified with a tuple $C(c_\|)$ of $\mathbb{N}^m$.

Example 6. For the PDN of example 2, $m = 4$ and assuming the ordering $qa, qb, q'a, q'b$, if $c_\| = qa \| qb \| q'b$ then $C(c_\|) = (1, 1, 0, 1)$ and if $c'_\| = qa \| qb \| qb$ then $C(c'_\|) = (1, 2, 0, 0)$.

The transition relation $\leadsto$ is the head-rewriting relation generated by R, i.e. it is the restriction of $\to$ to the initial PDS of a configuration $qw(c_\|)$ (i.e. no transition is applied to any element of $c_\|$). The reflexive transitive closure of $\leadsto$ is denoted by $\overset{*}{\leadsto}$.

The next construction is inspired by [BMOT05] and computes, for each q, a, q', a context-free grammar G such that the Parikh mapping of $L(G)$ is the set $\{C(c_\|) \mid qa \overset{*}{\leadsto} q'(c_\|)\}$, i.e. it describes all possible parallel compositions generated by transitions $qa \overset{*}{\leadsto} q'(\ldots)$. In the following, we shall identify the ith element qa to the letter l_i of a alphabet $\{l_1, \ldots, l_m\}$. We recall that the *Parikh mapping* $\#_P(w)$ of a word w is the tuple $(n_1, \ldots, n_m) \in \mathbb{N}^m$ such that the letter l_1 has n_1 occurrences in $w, \ldots$, the letter l_m has n_m occurrences in w.

- The set of terminals is $\{l_1, \ldots, l_m\}$. In the following, $w(c_\|)$ denotes the word $l_1^{n_1} \cdot \ldots \cdot l_m^{n_m}$ where $(n_1, \ldots, n_m) = C(c_\|)$ (with notation $l_i^0 = \epsilon$).
- The set of non-terminals is $\{X_{q,a,q'} \mid q, q' \in Q, a \in \Sigma\}$. They generate the words c such that $\#_P(c) = C(c_\|)$ iff $qa \overset{*}{\to} q'(c_\|)$.
- The set Δ of production rules is defined by
  - if $qa \to q'(c_\|)$ belongs to R, then $X_{q,a,q'} \to w(c_\|) \in \Delta$,
  - if $qa \to \bar{q}bc(c_\|)$ belongs to R, then $X_{q,a,q'} \to X_{\bar{q},b,\bar{q}}X_{\bar{q},c,q'}w(c_\|) \in \Delta$,

Proposition 9. *For all $q, q'' \in Q, a \in \Sigma$, $X_{q,a,q'} \overset{*}{\to}_G w$ iff $qa \overset{*}{\leadsto} q'(c_\|)$ and $C(c_\|) = \#_P(w)$*

Parikh's theorem and the results of [VSS05] yield that:

Proposition 10. *The set $C(q, a, q') = \{C(c_\|) \mid qa \overset{*}{\leadsto} q'(c_\|)\}$ is a semilinear set and an existential Presburger arithmetic formula defining $C(q, a, q')$ can be computed in polynomial time.*

6.2 A Presburger Weighted Tree Automaton Accepting $Post^*(qa)$

Firstly, we construct a Presburger weighted word automaton $\mathcal{A}=(S, s_0, S_F, S_m, \Delta)$ such that $qa \overset{*}{\leadsto} q'w(c_\parallel)$ iff $\mathcal{A}$ accepts $q'w, C(c_\parallel)$.

- $S = \{s_0\} \cup \{s_q \mid q \in Q\} \cup \{s_{qa} \mid s \in Q, a \in \Sigma\}$, $S_F = \{(s_{qa}, \mathbb{N}^m)\}$. States s_q are reached by the unique word q, and the state s_{qa} is reached by all pairs $q'w, c$ such that $qa \overset{*}{\leadsto} q'w(c_\parallel)$ and $c = C(c_\parallel)$.
- Δ is defined by:
    - $s_0 \overset{q,(0,\ldots,0)}{\to} s_q \in \Delta$ and $s_q \overset{a,(0,\ldots,0)}{\to} s_{qa} \in \Delta$
    - if R contains the rule $qa \to q'(c_\parallel)$ then $s_{q'} \overset{\epsilon,C(c_\parallel)}{\to} s_{qa} \in \Delta$
    - if R contains the rule $qa \to q'bc(c_\parallel)$ then
        * $s_{q'b} \overset{c,C(c_\parallel)}{\to} s_{qa} \in \Delta$
        * $s_{q''c} \overset{\epsilon,C(q',b,q'')+C(c_\parallel)}{\to} s_{qa} \in \Delta$

Proposition 11. $\bar{q}w, c \overset{*}{\to}_{\mathcal{A}} s_{qa}$ iff $qa \overset{*}{\leadsto} \bar{q}w(c_\parallel)$ and $c = C(c_\parallel)$.

The previous Presburger weighted word automaton $\mathcal{A} = (S, s_0, S_F, S_m, R)$ is extended into a Presburger weighted tree automaton $\mathcal{B} = (S_\mathcal{B}, S_I, S_m, R_\mathcal{B})$ that accepts $Post^*(q_0a_0)$, where $m = |\{s_0^{qa} \mid q \in Q, a \in \Sigma\}|$ and:

- $S_\mathcal{B} = \{s^{qa} \mid s \in S, qa \in Q \times \Sigma\}$ and $S_I = \{s_0^{q_0a_0}\}$.
    The states of $S_\mathcal{B}$ are ordered $s_1, \ldots, s_m$ where $s_i = s_0^{qa}$ iff i is the ith component of S_p (i.e. corresponds to the ordering of qa defined for the word automaton $\mathcal{A}$).
- The set of rules contains
    - $s^{qa} \overset{a,C}{\to} s'^{qa}$ if $\mathcal{A}$ contains the rule $s \overset{a,C}{\to} s'$.
    - $s_{qa}^{qa} \overset{(0,\ldots,0)}{\to} \#$, this rule is needed since $\mathcal{A}$ deals with words and $\mathcal{B}$ deals with trees,
    - $s_{qa}^{qa} \overset{\parallel}{\to} \bigwedge_{i=1}^{i=m} x_i = y_i \wedge \bigwedge_{i>m} x_i = 0$

The last rule states that the arguments of a parallel composition are exactly the processes generated by the $q(w((\ldots)$ part above this parallel composition.

Proposition 12. $L(\mathcal{B}) = c2t(Post^*(qa))$.

Proof. The proof is by structural induction on the tree structure. We prove that for any q, a, $t \in c2t(Post^*(qa))$ iff t is accepted by $\mathcal{A}$ with a run labelling the root of t by s_0^{qa}.

Base case. $t = q(w(\#))$. By definition of the rules of $\mathcal{B}$ and adapting the proof of proposition 11, we have $q(w(\#)) \in Post^*(qa)$ iff a run of $\mathcal{B}$ labels $q(w(\#))$ by $(s_0^{qa}, (0, \ldots, 0))$ at the root and q_{qa}^{qa} at the leaf.

Induction step. $t = q(w(\parallel (t_1, \ldots, t_n)))$. We assume that there is a run of $\mathcal{B}$ labelling each t_i by $(s_0^{qa}, (0, \ldots, 0))$ iff $t_i \in Post^*(qa)$. By definition of the rules of $\mathcal{B}$ and adapting the proof of proposition 11, $qa \overset{*}{\leadsto} t2c(t)$ there is a rule labelling the node $\parallel$ by $(s_{qa}^{qa}, (n_1, \ldots, n_m))$ such that n_1 trees t_i are labelled by $(s_0^{(qa)_1}, (0, \ldots, 0)), \ldots, n_m$ trees t_i are labelled by $(s_0^{(qa)_m}, (0, \ldots, 0))$ and no t_i is labelled by another state (where $(qa)_i$ denotes the ith element in the sequence of qa's. By induction hypothesis, there is a run of $\mathcal{B}$ accepting t iff $t2c(t) \in Post^*(qa)$. $\square$

This proposition and the properties of Presburger weighted trees automata yield that forward analysis of PDN is decidable.

7   Backward Analysis of PDN

To perform backward analysis, we can rely on the results of [BMOT05] that state that $Pred^*(L)$ is accepted by a hedge automaton and the fact that the closure of a regular hedge language under commutativity is a Presburger regular language. Therefore we get:

Proposition 13. *The set $Pred^*(L)$ is accepted by a Presburger tree automaton.*

Conclusion

We have enriched the model of dynamic networks of pushdown systems by taking parallel composition as an associative-commutative operator. Using a new class of tree automata we have been able to do forward and backward analysis. Forward analysis involves an exponential blowup in the construction of the PDN of proposition 7 since a semilinear set equivalent to a Presburger formula can be exponentially larger. This problem occurs usually when switching from a word framework to a natural number framework. Some interesting questions remain and will be further investigating. The first one is to extend our result using constrained rules as in [BMOT05] or regular tree language constraints [LMOW09] that could allow to extend the model with synchronization via nested locks. These extensions can be easily done for arithmetic constraints (in rules or in Presburger tree automata rules) which are combinations of inequalities $x_i > c_i, x_i < c_i$ or moduli equations $x_i \equiv c_i(k_i)$ but allowing any Presburger arithmetic constraint makes the analysis much more difficult and it is nt clear if the reachability problem is still decidable. Another one is to consider a flat model where all parallel compositions are at the same level and a configuration is now a set of PDS instead of a tree.

References

[AB97]   Maler, O., Bouajjani, A., Esparza, J.: Reachability analysis of pushdown automata: Application to model-checking. In: Mazurkiewicz, A., Winkowski, J. (eds.) CONCUR 1997. LNCS, vol. 1243, pp. 14–25. Springer, Heidelberg (1997)

[BMOT05]   Bouajjani, A., Müller-Olm, M., Touili, T.: Regular symbolic analysis of dynamic networks of pushdown systems. In: Abadi, M., de Alfaro, L. (eds.) CONCUR 2005. LNCS, vol. 3653, pp. 473–487. Springer, Heidelberg (2005)

[Buc64]   Buchi, R.: Regular canonical systems. Archiv fur Matematische Logik und Grundlagenforschung 6(91-111) (1964)

[Cau00]   Caucal, D.: On word rewriting systems having a rational derivation. In: Tiuryn, J. (ed.) FOSSACS 2000. LNCS, vol. 1784, pp. 48–62. Springer, Heidelberg (2000)

[CDJ^{+}99]   Comon, H., Dauchet, M., Jacquemard, F., Loeding, C., Lugiez, D., Tison, S., Tommasi, M.: Tree Automata on their application. Freeware Book (1999), http://tata.gforge.inria.fr/

[EP00]   Esparza, J., Podelski, A.: Efficient algorithms for pre* and post* on interprocedural parallel flow graphs. In: POPL, pp. 1–11 (2000)

[GS66]   Ginsburg, S., Spanier, E.: Semigroups, Presburger formulas and languages. Pacific Journal of Mathematics 16, 285–296 (1966)

[KG07]   Kahlon, V., Gupta, A.: On the analysis of interacting pushdown systems. In: Hofmann, M., Felleisen, M. (eds.) POPL, pp. 303–314. ACM, New York (2007)

[LMOW09]   Lammich, P., Müller-Olm, M., Wenner, A.: Predecessor sets of dynamic pushdown networks with tree-regular constraints. In: Proc. of 21st Conf. on Computer Aided Verification (CAV). LNCS, vol. 5643, pp. 525–539. Springer, Heidelberg (2009)

[LS98]   Lugiez, D., Schnoebelen, P.: The regular viewpoint on pa-processes. In: Sangiorgi, D., de Simone, R. (eds.) CONCUR 1998. LNCS, vol. 1466, pp. 50–66. Springer, Heidelberg (1998)

[Pre29]   Presburger, M.: Uber die vollstandigkeit eines gewissen system der arithmetik ganzer zahlen in welchem die addition als einzige operation hervortritt. In: Comptes Rendus du I Congres des Mathematiciens des Pays Slaves, Warszawa (1929)

[RSJ03]   Reps, T.W., Schwoon, S., Jha, S.: Weighted pushdown systems and their application to interprocedural dataflow analysis. In: Cousot, R. (ed.) SAS 2003. LNCS, vol. 2694, pp. 189–213. Springer, Heidelberg (2003)

[VSS05]   Verma, K.N., Seidl, H., Schwentick, T.: On the complexity of equational horn clauses. In: Nieuwenhuis, R. (ed.) CADE 2005. LNCS, vol. 3632, pp. 337–352. Springer, Heidelberg (2005)

[ZL06]   Zilio, S.D., Lugiez, D.: XML schema, tree logic and sheaves automata. Applicable Algebra in Engineering, Communication and Computing 17(5), 337–377 (2006)

Counting Multiplicity over Infinite Alphabets

Amaldev Manuel and R. Ramanujam

Institute of Mathematical Sciences
Chennai, India - 600113
{amal,jam}@imsc.res.in

Abstract. In the theory of automata over infinite alphabets, a central difficulty is that of finding a suitable compromise between expressiveness and algorithmic complexity. We propose an automaton model where we count the multiplicity of data values on an input word. This is particularly useful when such languages represent behaviour of systems with unboundedly many processes, where system states carry such counts as summaries. A typical recognizable language is: "every process does at most k actions labelled a". We show that emptiness is elementarily decidable, by reduction to the covering problem on Petri nets.

1 Summary

Consider a system of concurrently running sequential processes. When there is no a priori bound on the number of processes, though at any point of time only finitely many are active, the necessity of the system to distinguish one process from another involves potentially unbounded data. Typically, system states carry summary information about processes that are known to be active, and hence the set of system configurations is infinite. Such systems arise in the study of web services, communication protocols and software systems with recursive concurrent threads of execution.

Infinite state systems are not unfamiliar in theory of computation; a rich body of results exists on counter systems, pushdown systems and Petri nets. Most reachability properties of such infinite state systems are either undecidable or have such high complexity that algorithmic verification is impractical. On the other hand, if we restrict ourselves to only finite state systems, we can reason only about systems where the set of processes is fixed and known a priori, and we do not (as yet) have clear abstractions that allow us to transfer the results of such reasoning to systems of unbounded processes. Hence there is a clear need for formal models that work with unbounded systems but yet restrict expressiveness to allow decidable verification.

Notice that interesting properties of such systems do not involve process names (or identifiers) explicitly. A specification that restricts attention only to processes P_1, P_2 and P_3 can be implemented by a finite state system. On the other hand, consider a specification such as: "at least k processes get to perform an a action": this necessitates remembering potentially unboundedly many values, thus leading to infinite state systems.

O. Bournez and I. Potapov (Eds.): RP 2009, LNCS 5797, pp. 141–153, 2009.

This paper is situated in such a context and while we have no definitive answers, we consider "state summaries" that allow elementary decidability. The model we use is that of finite automata over infinite alphabets and we use counters to record the intended "summaries". The main result is that emptiness if Expspace-complete for such a class of automata. Unfortunately, the automata are not closed under complementation, and even the word problem is intractable, suggesting that we have more work to be done to further restrict expressiveness.

The study of automaton mechanisms over infinite alphabets has gained interest in recent years, especially from the viewpoint of database theory. In this approach, data values are modelled using a countably infinite domain, and structures are finite words labelled by this infinite alphabet. Typically the alphabet is presented as a product $(\Sigma \times \mathbb{D})$, where Σ is finite and $\mathbb{D}$ is countable. For our purposes, we can think of $\mathbb{D}$ as process names and Σ as the finite set of events they participate in, or conditions that hold.

The study of languages over infinite alphabets were initiated in [ABB80] and [Ott85], where the approach was to define the notion of regularity for languages over infinite alphabets in terms of morphisms to languages over finite alphabets. There are many automaton mechanisms for studying word languages over infinite alphabets: register automata ([KF94]), pebble automata ([NSV01], [NSV04]), data automata ([BMS$^+$06]), nested words ([AM06]), class memory automata ([BS07]) and automata on Gauss words ([LPS09]), with different expressive power and complexity. Logic based approaches include monadic second order logics ([Bou02], [Bac03]), two variable first order logics ([BMS$^+$06]) and temporal logics with special "freeze" quantifiers ([DL06]) or predicate abstraction ([LP05], [LP09]). Algebraic approaches involve quasi-regular expressions ([KT06]), or register monoid mechanisms ([BPT01]). All these involve interesting tradeoffs between expressiveness and complexity of decision procedures. A unifying framework placing all these models in perspective is as yet awaited (see [Seg06] for an excellent survey).

While register automata have polynomial complexity, they are effectively finite state; data automata are more expressive, but emptiness is not known to be elementary. What we present here is a restriction of class memory automata: these automata that can not only test for existence of data values, but can also count the multiplicity of occurrences of data values, subject to constraints on such counts. However, these counters are *monotone*, and hence the constraints are limited in expressive power: we can compare counts against constants, but not much more. We show that such a model of **Class counting automata (CCA)** is interesting, for several reasons; specifically, we get elementary decidability. We see this as "populating the landscape" of classes of data languages, in the sense of [BS07].

From the viewpoint of reasoning about unbounded systems of processes, it is unclear what exactly is the expressiveness needed. For instance, consider the specification: "No two successive positions carry the same data value"; this is naturally implemented using a register mechanism. But this is a "hard" global scheduling constraint: after any process event is scheduled, the succeeding event must necessarily be from a different process; it is hardly clear that such a constraint is important for loosely coupled systems of processes. This indicates that while we do want to specify combinations of global and local properties, we need to nonetheless allow for sufficient flexibility.

2   Class Counting Automata

Let $k > 0$; we use $[k]$ to denote the set $\{1, 2, \ldots k\}$. When we say $[k]_0$, we mean the set $\{0\} \cup [k]$. By $\mathbb{N}$ we mean the set of natural numbers $\{0, 1, \ldots\}$. When $f : A \to B$, $(a, b) \in (A \times B)$, by $f \oplus (a, b)$, we mean the function $f' : A \to B$, where $f'(a') = f(a')$ for all $a' \in A, a' \neq a$, and $f'(a) = b$.

Customarily, the infinite alphabet is split into two parts: it is of the form $\Sigma \times D$, where Σ is a finite set, and D is a countably infinite set. Usually, Σ is called the *letter alphabet* and D is called the *data alphabet*. Elements of D are referred to as *data values*. We use letters a, b etc to denote elements of Σ and use d, d' to denote elements of D.

A **data word** w is an element of $(\Sigma \times D)^*$. A collection of data words $L \subseteq (\Sigma \times D)^*$ is called a *data language*. In this article, by default, we refer to data words simply as words and data languages as languages. As usual, by $|w|$ we denote the length of w.

Let $w = (a_1, d_1)(a_2, d_2) \ldots (a_n, d_n)$ be a data word. The *string projection* of w, denoted as $str(w) = a_1 a_2 \ldots a_n$, the projection of w to its Σ components. Let $i \in [n] = |w|$. The **data class** of d_i in w is the set $\{j \in [n] \mid d_i = d_j\}$. A subset of $[n]$ is called a data class of w if it is the data class of some $d_i, i \in [n]$. Note that the set of data classes of w form a partition of $[|w|]$.

The automaton we present below includes a bag of infinitely many monotone counters, one for each possible data value. When it encounters a letter - data pair, say (a, d), the multiplicity of d is checked against a given constraint, and accordingly updated, the transition causing a change of state, as well as possible updates for other data as well. We can think of the bag as a hash table, with elements of D as keys, and counters as hash values. Transitions depend only on hash values (subject to constraints) and not keys.

A **constraint** is a pair $c = (op, e)$, where $op \in \{<, =, \neq, >\}$ and $e \in \mathbb{N}$. When $v \in \mathbb{N}$, we say $v \models c$ if $v \, op \, e$ holds. Let $\mathcal{C}$ denote the set of all constraints. Define a *bag* to be a finite set $h \subseteq (\mathbb{D} \times \mathbb{N})$ such that whenever $(d, n_1) \in h$ and $(d, n_2) \in h$, we have: $n_1 = n_2$. Thus h defines a partial function from $\mathbb{D}$ to $\mathbb{N}$ which is defined on a finite subset of $\mathbb{D}$. By convention, we implicitly extend it to a total function on D by considering h to represent the set $h' = h \cup \{(d, 0) \mid$ there is no $n \in \mathbb{N}$ such that $(d, n) \in h\}$. Hence we (ab)use the notation $h(d) = n$ for a bag h. Let $\mathbb{B}$ denote the set of bags. Note that the notation $h \oplus (d, n)$ now stands for the bag $h' = (h - (\{d\} \times \mathbb{N})) \cup \{(d, n)\}$.

Below, let $Inst = \{\uparrow^+, \downarrow\}$ stand for the set of *instructions*: $\uparrow^+$ tells the automaton to increment the counter, whereas $\downarrow$ asks for a reset. Note that the instruction $(\uparrow^+, 0)$ says that we do not wish to make any update, and $(\uparrow^+, 1)$ causes a unit increment; we use the notation $[0]$ and $[+1]$ for these instructions below.

Definition 1. *A class counting automaton, abbreviated as CCA, is a tuple* $CCA = (Q, \Delta, I, F)$, *where Q is a finite set of states, $I \subseteq Q$ is the set of initial states, $F \subseteq Q$ is the set of final states. The transition relation is given by:* $\Delta \subseteq (Q \times \Sigma \times C \times Inst \times U \times Q)$, *where C is a finite subset of $\mathcal{C}$ and U is a finite subset of* $\mathbb{N}$.

Let A be a CCA. A configuration of A is a pair (q, h), where $q \in Q$ and $h \in \mathbb{B}$. The initial configuration of A is given by (q_0, h_0), where h_0 is the empty bag; that is, $\forall d \in \mathbb{D}, h_0(d) = 0$ and $q_0 \in I$.

Given a data word $w = (a_1, d_1), \ldots (a_n, d_n)$, a run of A on w is a sequence $\gamma = (q_0, h_0)(q_1, h_1) \ldots (q_n, h_n)$ such that $q_0 \in I$ and for all $i, 0 \le i < n$, there exists a transition $t_i = (q, a, c, \pi, m, q') \in \Delta$ such that $q = q_i$, $q' = q_{i+1}$, $a = a_{i+1}$ and:

- $h_i(d_{i+1}) \models c$.
- h_{i+1} is given by:

$$h_{i+1} = \begin{cases} h_i \oplus (d_{i+1}, m') \text{ if } \pi = \uparrow^+, m' = h_i(d_{i+1}) + m \\ h_i \oplus (d_{i+1}, m) \text{ if } \pi = \downarrow \end{cases}$$

γ is an accepting run above if $q_n \in F$. The language accepted by A is given by $\mathcal{L}(A) = \{w \in (\Sigma \times \mathbb{D})^* \mid A \text{ has an accepting run on } w\}$. $\mathcal{L} \subseteq ((\Sigma \times \mathbb{D}))^*$ is said to be recognizable if there exists a CCA A such that $\mathcal{L} = \mathcal{L}(A)$. Note that the counters are either incremented or reset to fixed values.

We first observe that CCA runs have some useful properties. To see this, consider a bag h and $d_1, d_2 \in D$, $d_1 \ne d_2$ such that at a confguration (q, h), we have two transitions enabled on inputs (a_1, d_1) and (a_2, d_2) leading to configurations (q_1, h_1) and (q_2, h_2) respectively. Notice that for any condition c, if $h(d_2) \models c$ then so also $h_1(d_2) \models c$. Similarly, for any condition c', if $h(d_1) \models c'$ then so also $h_2(d_1) \models c'$. Thus when we have distinct data values, tests on them do not "interfere" with each other. We can extend this observation further: given data words u and v such that the data values in u are pairwise disjoint from those in v, if we have a run from (q, h) on u to (q, h_1) and on v from (q, h_1) to (q', h_2), then there is a configuration (q', h') and a run from (q, h) on v to (q', h'). This will be useful in the following.

Example 1. The language $L_{fd(a)} = $ *"Data values under a are all distinct"* is accepted by a CCA. The CCA accepting this language is the automaton $A = (Q, \Delta, q_0, F)$ where $Q = \{q_0, q_1\}$, q_0 is the only initial state and $F = \{q_0\}$. Δ consists of:

- $(q_0, a, (=, 0), q_0, [+1]); (q_0, a, (=, 1), q_1, [0]);$
- $(q_0, b, (\ge, 0), q_0, [0]); (q_1, \Sigma, (\ge, 0), q_1, [0]).$

Since the automaton above is deterministic, by complementing it, that is, setting $F = \{q_1\}$, we can accept the language $\overline{L_{fd(a)}} = $ *"There exists a data value appearing at least twice under a"*. On the other hand, since every data word can mention only finitely many data values, trivially every word has a value

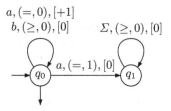

Fig. 1. Automaton in the Example 1

that appears less than twice under a (namely zero times). Hence the statement above can be strengthened to saying that the language $L_{\exists a,\neq n} =$ " *There exists a data value whose multiplicity under a is not 2*" is recognizable. But as we show below, its complement language, $L_{\forall,= n} =$ "*All data values under a occur exactly twice*" is not recognizable. Thus, CCA- recognizable data languages are not closed under complementation.

Proposition 1. *The language $L_{forall,= n} =$ "All data values under a occur exactly twice" is not recognizable.*

Proof. Suppose there is a CCA A with m states accepting this language. Consider the data word

$$w = (a,d_1)(a,d_2)!..(a,d_{m+1})(a,d_1)(a,d_2)..(a,d_{m+1})$$

Clearly, $win L_{\forall,= n}$. Therefore, there is a successful run of A on w. Then there is a state q repeating in the suffix of length $m + 1$. Let us say this splits w as $u \cdot v \cdot v'$, where the configurations at the repeating state after u with configuration (q, h) to (q, h_1) on v and to (q', h_2) on v'. Then by the remarks we made earlier, we can find a run from (q, h) to a configuration (q', h') on v' as well. Thus we have "chopped" off a part of the run so that we have an accepting run on a word $u \cdot v'$. But then $u \cdot v'$ is not in $L_{forall,= n}$. □

The following statement is easily proved:

Proposition 2. *CCA-recognizable data languages are closed under union and intersection but not under complementation.*

The following observation will be useful for decision questions that follow. Given a CCA $A = (Q, \Delta, q_0, F)$ let m be the maximum constant used in Δ. We define the following equivalence relation on $\mathbb{N}$, $c \simeq_{m+1} c'$ iff $c < (m+1) \vee c' < (m+1) \Rightarrow c = c'$. Note that if $c \simeq_{m+1} c'$ then a transition is enabled at c if and only if it is enabled at c'. We can extend this equivalence to configurations of the CCA as follows. Let $(q_1, h_1) \simeq_{m+1} (q_2, h_2)$ iff $q_1 = q_2$ and $\forall d \in \mathbb{D}, h_1(d) \simeq_{m+1} h_2(d)$.

Lemma 1. *If C_1, C_2 are two configurations of the CCA such that $C_1 \simeq_{m+1} C_2$, then $\forall w \in ((\Sigma \times \mathbb{D}))^*$, $C_1 \vdash^*_w C'_1 \implies \exists C'_2, C_2 \vdash^*_w C'_2$ and $C'_1 \simeq_{m+1} C'_2$.*

Proof. Proof by induction on the length of w. For the base case observe that any transition enabled at C_1 is enabled at C_2 and the counter updates respects the equivalence. For the inductive case consider the word $w.a$. By induction hypothesis $C_1 \vdash^*_w C_1' \implies \exists C_2', C_2 \vdash^*_w C_2'$ and $C_1' \simeq_{m+1} C_2'$. If $C_1' \vdash_a C_1''$ then using the above argument there exists C_2'' such that $C_2' \vdash_a C_2''$ and $C_1'' \simeq_{m+1} C_2''$.

In fact the lemma holds for any $N \geq m+1$, where m is the maximum constant used in Δ. This observation paves the way for proving the decidability of the emptiness problem (in the next section).

3   Decision Problems

Since the space of configurations of a CCA is infinite, reachability is in general non-trivial to decide. We now show that the emptiness problem is elementarily decidable.

Theorem 1. *The non-emptiness problem for CCA is* EXPSPACE-*complete.*

3.1   Upper Bound

We reduce the emptiness problem of CCA to the covering problem on Petri nets. For checking emptiness, we can omit the $\Sigma \times D$ labels from the configuration graph; we are then left with counter behaviour. However since we have unboundedly many counters, we are led to the realm of vector addition systems.

Definition 2. *An ω-counter machine B is a tuple (Q, Δ, q_0) where Q is a finite set of states, $q_0 \in Q$ is the initial state and $\Delta \subseteq (Q \times C \times Inst \times U \times Q)$, where C is a finite subset of $\mathcal{C}$ and U is a finite subset of $\mathbb{N}$.*

A configuration of B is a pair (q, h), where $q \in Q$ and $h : \mathbb{N} \to \mathbb{N}$. The initial configuration of B is (q_0, h_0) where $h_0(i) = 0$ for all i in $\mathbb{N}$. A run of B is a sequence $\gamma = (q_0, h_0)(q_1, h_1) \ldots (q_n, h_n)$ such that for all i such that $0 \leq i < n$, there exists a transition $t_i = (p, c, \pi, m, q) \in \Delta$ such that $p = q_i$, $q = q_{i+1}$ and there exists j such that $h(j) \models c$, and the counters are updated in a similar fashion to that of CCA.

The reachability problem for B asks, given $q \in Q$, whether there exists a run of B from (q_0, h_0) ending in (q, h) for some h ("Can B reach q?").

Lemma 2. *Checking emptiness for CCA can be reduced to checking reachability for ω-counter machines.*

Proof. It suffices to show, given a CCA, $A = (Q, \Delta, q_0, F)$, where $F = \{q\}$, that there exists a counter machine $B_A = (Q, \Delta', q_0)$ such that A has an accepting run on some data word exactly when B_A can reach q. (When F is not a singleton, we simply repeat the construction.) Δ' is obtained from Δ by converting every transition (p, a, c, π, m, q) to (p, c, π, m, q). Now, let $L(A) \neq \emptyset$. Then there exists a data word w and an accepting run $\gamma = (q_0, h_0)(q_1, h_1) \ldots (q_n, h_n)$ of A on w, with $q_n = q$. Let $g : \mathbb{N} \to D$ be an enumeration of data values. It is easy to see that $\gamma' = (q_1, h_0 \circ g)(q_1, h_1 \circ g) \dots (q_n, h_n \circ g)$ is a run of B_A reaching q.

($\Leftarrow$) Suppose that B_A has a run $\eta = (q_0, h_0)(q_1, h_1) \dots (q_n, h_n)$, $q_n = q$. It can be seen that $\eta' = (q_0, h_0 \circ g^{-1})(q_1, h_1 \circ g^{-1}) \dots (q_n, h_n \circ g^{-1})$ is an accepting run of A on $w = (a_1, d_1) \dots (a_n, d_n)$ where w satisfies the following. Let (p, c, π, m, q) be the transition of B_A taken in the configuration (q_i, h_i), and d_k such that $h_i(d_k) \models c$. Then by the definition of B_A there exists a transition (p, a, c, π, m, q) in Δ. Then it should be the case that $a_{i+1} = a$ and $d_{i+1} = g(d_k)$.

Proposition 3. *Checking non-emptiness of ω-counter machines is decidable.*

Let $s \subseteq \mathbb{N}$, and c a constraint. We say $s \models c$, if for all $n \in s$, $n \models c$.

We define the following partial function Bnd on all finite and cofinite subsets of $\mathbb{N}$. Given $s \subseteq_{fin} \mathbb{N}$, $Bnd(s)$ is defined to be the least number greater than all the elements in s. Given $s \subseteq_{cofinite} \mathbb{N}$, $Bnd(s)$ is defined to be $Bnd(\mathbb{N} \backslash s)$. Given an ω-counter machine $B = (Q, \Delta, q_0)$ let $m_B = max\{Bnd(s) \mid s \models c, c \text{ is used in } \Delta\}$.

We construct a Petri net $N_B = (S, T, F, M_0)$, where,

- $S = Q \cup \{i \mid i \in \mathbb{N}, 1 \leq i \leq m_B\}$.
- T is defined according to Δ as follows. Let $(p, c, \pi, n, q) \in \Delta$ and let i be such that $0 \leq i \leq m_B$ and $i \models c$. Then we add a transition t such that $\bullet t = \{p, i\}$ and $t^\bullet = \{q, i'\}$, where (i) if π is $\uparrow^+$ then $i' = min\{m_B, i + n\}$, and (ii) if π is $\downarrow$ then $i' = n$.
- The flow relation F is defined according to $\bullet t$ and $t^\bullet$ for each $t \in T$.
- The initial marking is defined as follows. $M_0(q_0) = 1$ and for all p in S, if $p \neq q_0$ then $M_0(p) = 0$.

The construction above glosses over some detail: Note that elements of these sets can be zero, in which case we add edges only for the places in $[m_B]$ and ignore the elements which are zero.

Let M be any marking of N_B. We say that M is a *state marking* if there exists $q \in Q$ such that $M(q) = 1$ and $\forall p \in Q$ such that $p \neq q$, $M(p) = 0$. When M is a state marking, and $M(q) = 1$, we speak of q as the state marked by M. For $q \in Q$, define $M_f(q)$ to be set of state markings that mark q. It can be shown, from the construction of N_B, that in any reachable marking M of N_B, if there exists $q \in Q$ such that $M(q) > 0$, then M is a state marking, and q is the state marked by M.

We now show that the counter machine B can reach a state q iff N_B has a reachable marking which covers a marking in $M_f(q)$. We define the following equivalence relation on $\mathbb{N}$, $m \simeq_{m_B} n$ iff $(m < m_B) \vee (n < m_B) \Rightarrow m = n$. We can lift this to the hash functions (in ω-counters) in the natural way: $h \simeq_{m_B} h'$ iff $\forall i \, (h(i) < m_B) \vee (h'(i) < m_B) \Rightarrow h(i) = h'(i)$. It can be easily shown that if $h \simeq_{m_B} h'$ then a transition is enabled at h if and only if it is enabled at h'.

Let μ be a mapping B-configurations to N_B-configurations as follows: given $\chi = (q, h)$, define $\mu(\chi) = M_\chi$, where

$$M_\chi(p) = \begin{cases} 1 & \text{iff } p = q \\ 0 & \text{iff } p \in Q \backslash \{q\} \\ |[p]| & \text{iff } p \in P \backslash Q, p \neq 0 \end{cases}$$

Above $[p]$ denotes the equivalence class of p under $\simeq_{m_B}$ on $\mathbb{N}$ in h. Now suppose that B reaches q. Let the resulting configuration be $\chi = (q, h)$. We claim that the marking $\mu(\chi)$ of N_B is reachable (from M_0) and covers $M_f(q)$. Conversely if a reachable marking M of N_B covers $M_f(q)$, for some $q \in Q$, then there exists a reachable configuration $\chi = (q, h)$ of B such that $\mu(\chi) = M$. This is proved by a simple induction on the length of the run.

Since the covering problem for Petri nets is decidable, so is reachability for ω-counter machines and hence emptiness checking for CCA is decidable.

3.2   Lower Bound

The decision procedure above runs in EXPSPACE, and thus we have elementary decidability. We now show that the emptiness problem is also EXPSPACE-hard. Effectively this is a reduction of the covering problem again, but for technical convenience, we use multi-counter automata.

A k multi-counter automaton with weak acceptance is a tuple $A = (Q, \Sigma, \Delta, q_0, F)$ where Q is a finite set of states, $q_0 \in Q$ is the initial state and $F \subseteq Q$ is a set of final states. The transition relation is of the form $\Delta \subseteq_{fin} (Q \times \Sigma \times \mathbb{N}^k \times \mathbb{N}^k \times Q)$. The two vectors in the transition specify decrements and increments of the counters.

The automaton works as follows: it has k-counters, denoted by $\bar{v} = (v_1, \ldots v_k)$ which hold non-negative counter values. A configuration of the machine is of the form $(q, \bar{v})$ where $q \in Q$ and $\bar{v} \in \mathbb{N}^k$. The initial configuration is $(q_0, \bar{0})$. Given a configuration $(q, \bar{v})$ the automaton can go to a configuration $(q', \bar{v}')$ on letter a if there is a transition $(q, a, v_{\bar{dec}}, v_{\bar{inc}}, q')$ such that $\bar{v} - v_{\bar{dec}} \geq \bar{0}$ (pointwise) and $\bar{v}' = \bar{v} - v_{\bar{dec}} + v_{\bar{inc}}$. A final configuration is one in which the state is final.

The problem of checking non-emptiness of a multicounter automaton with weak acceptance is known to be (at least) EXPSPACE-hard ([Lip76]).

Any multicounter automaton $M = (Q, \Sigma, \Delta, q_0, F)$ can be converted to another (in a "normal form"): $M' = (Q', \Sigma, \Delta', q_0, F)$ such that $L(M)$ is non-empty if and only if $L(M')$ is non-empty and M' uses only unit vectors or zero vectors in its transitions. A unit vector is of the form $(b_1, b_2, \ldots, b_k)$ where there is a unique $i \in [k]$ such that $b_i = 1$ and for $j \neq ik$, $b_j = 0$. That is M' decrements or increments at most one counter in each transition.

Δ' is obtained as follows. Let $t = (q, a, v_{\bar{dec}}, v_{\bar{inc}}, q')$. Let $\bar{u}_1, \bar{u}_2, \ldots, \bar{u}_n$ be a sequence of unit vectors such that $v_{\bar{dec}} = \Sigma_i \bar{u}_i$ and $\bar{u}_1', \bar{u}_2', \ldots, \bar{u}_m'$ be a sequence of unit vectors such that $v_{\bar{inc}} = \Sigma_i \bar{u}_i'$. We add intermediate states to rewrite t by the following sequence of transitions,

$$(q, a, \bar{u}_1, \bar{0}, q_{(t, \bar{u}_1)}), (q_{(t, \bar{u}_1)}, a, \bar{u}_2, \bar{0}, q_{(t, \bar{u}_2)}), \ldots,$$

$$(q_{(t, \bar{u}_n)}, a, \bar{0}, \bar{u}_1', q_{(t, \bar{u}_1')}), (q_{(t, \bar{u}_1')}, a, \bar{0}, \bar{u}_2', q_{(t, \bar{u}_2')}), \ldots,$$

$$(q_{(t, u_{\bar{m-1}}')}, a, \bar{0}, \bar{u}_m', q')$$

Lemma 3. $L(M)$ is non-empty if and only if $L(M')$ is non-empty.

Proof. By an easy induction on the length of the run. It is easy to see that for every accepting run ρ of M we have an accepting run ρ' of M', this is achieved by replacing every transition t in the run ρ by the corresponding sequence of transitions. For the reverse direction, we need to show that every run accepting run ρ' of M' can be translated to an accepting run ρ of M. This is possible since the intermediate states added to obtain the transitions in M' are unique for each transition t in M. Hence for every sequence of transitions taking M' from q_1 to q_2 where $q_1, q_2 \in Q$ there is a unique transition t which takes M from q_1 to q_2. By doing an induction on the number of states occuring in ρ' which are from Q we can show that there is a valid run ρ which is accepting.

Next we convert M' to a CCA thus establishing a lowerbound of EXPSPACE for the emptiness problem. Let $M' = (Q, \Sigma, \Delta, q_0, F)$ be a k-multicounter automaton in normal form. We construct the automaton $A = (Q, \Sigma, \Delta_A, q_0, F)$. Let $t = (q, a, \bar{u}, \bar{u}', q')$ where $\bar{u}, \bar{u}'$ are either unit or zero vectors. If $\bar{u}$ is a i-th unit vector and $\bar{u}'$ is a zero vector, we add a transition $t_A = (q, a, (x = i), (\downarrow, 0), q')$ to Δ_A. If $\bar{u}$ is a i-th unit vector and $\bar{u}'$ is j-th unit vector, we add a transition $t_A = (q, a, (x = i), (\downarrow, j), q')$ to Δ_A. If $\bar{u}$ is a zero vector and $\bar{u}'$ is a j-th unit vector, we add a transition $t_A = (q, a, (x = 0), (\downarrow, j), q')$ to Δ_A.

Lemma 4. *$L(M')$ is non-empty if and only if $L(A)$ is non-empty.*

Proof. The proof is by induction on the length of the run. First we define a mapping from configurations of A to configurations of M' in the following manner, $\mu((q, \bar{h})) = (q, \bar{v})$ where $v_i = |\{j \mid \bar{h}(j) = i\}|$. We show, by induction on the length of the run, that for every configuration χ reachable by A there is a configuration ψ of M' such that $\mu(\chi) = \psi$ and conversely for every configuration ψ reachable by M' there is a configuration χ reachable by A such that $\mu(\chi) = \psi$.

For the base case, it is evident that $\mu((q_0, \bar{h}_0)) = (q_0, \bar{0})$.

Suppose that $\chi = (q, \bar{h})$ is a configuration reachable in l steps, and that the transition $t = (q, a, x = j, (\downarrow, i), q')$ is enabled at χ. Therefore there is a counter holding the value j. By induction hypothesis there exists a configuration ψ such that $\mu(\chi) = \psi = (q, \bar{v})$ such that $v_j > 0$. After the transition t, the number of counters holding the value j decreases by one and the number of counters holding the value i increases by one(if $i \neq 0$). This is achieved by the transition $(q, a, \bar{u}_j, \bar{u}_i, q')$ in Δ', preserving the map μ.

Conversely, suppose a configuration $\psi = (q, \bar{v})$ is reachable by M' in l steps. Then by induction hypothesis we have a configuration χ reachable by the automaton A such that $\mu(\chi) = \psi$. Suppose a transition $t' = (q, a, \bar{u}_i, \bar{u}_j, q')$ is enabled in ψ resulting in ψ'.

Consider the case where $\bar{u}_i \neq \bar{0}$ and $\bar{u}_j \neq \bar{0}$. By construction t' is obtained from a transition $t = (q, a, (x = i), \downarrow, j, q')$. We choose the smallest counter holding the value zero and apply the transition t, resulting in ξ' such that $\mu(\xi') = \psi'$. The remaining cases are similar.

3.3   Inclusion and Word Problem

The next interesting algorithmic question is that of checking inclusion among accepted languages. It turns out that this problem is undecidable, which can be shown by reduction from the Post Correspondence Problem. We postpone the discussion on this until we discuss alternation later.

Since emptiness checking is of such high complexity, one may wonder whether the model is complex enough to render even the word problem to be hard: the simplest algorithmic question of how one can check whether a given word is accepted or not. The important thing to note is that during a run, the size of the configuration is bounded by the length of the input data word. Therefore a nondeterministic Turing machine can easily guess a path in polynomial time and check for acceptance. Hence the word problem is easily seen to be in NP. Interestingly, it turns out to be NP-hard as well.

Theorem 2. *The word problem for CCA is NP-complete.*

The proof is by reduction of the satisfiability problem for 3-CNF formulas to the word problem for CCAs. Given the 3-CNF formula, we code it up as a data word, where data values are used to remember the identity of literals in clauses. We use a two letter alphabet with $+, -$ indicating whether a propositional variable occurs positively or negatively. Data values stand for the propositional variables themselves. Thus a pair $((+, d_1)$ asserts that the first boolean variable occurs positively.

We show the coding by an example, let $\varphi \equiv (p_1 \vee \neg p_3 \vee p_4) \wedge (\neg p_2 \vee p_5 \vee p_1) \wedge (\neg p_3 \vee \neg p_4 \vee p_5)$, we construct the corresponding word $w = (+, d_1)(-, d_3)(+, d_4) (\#, d) (-, d_2)(+, d_5)(+, d_1) (\#, d) (-, d_3)(-, d_4)(+, d_5)(\#, d) \in (\{+, -, \#\} \times \mathbb{D})^*$.

The nondeterministic automaton checks satisfiability in the following way. Every time the automaton encounters a new data value (representing a propositional variable), the automaton nondeterministically assigns a boolean value and stores it in the counter (1 for $\bot$ and 2 for $\top$) corresponding to the data value, in the future whenever the same data value occurs the counter is consulted to obtain the assigned value to the propositional variable. The automaton evaluates each clause and carries the partial evaluation in its state. Finally the automaton accepts the word if the formula evaluates to $\top$.

4   Discussion

We first observe that the model admits many extensions, without substantially affecting the main decidability result.

4.1   Extensions

1. Instead of working with one bag of counters, the automaton can use several bags of counters, much as multiple registers are used in the register automaton. It is easy to formally define CCA with k-bags, using k-tuples of constraints on guards.

2. Another strengthening involves checking for the presence of *any* counter satisfying a given constraint and updating it.
3. The language of constraints can be strengthened: any syntax that can specify a finite or co-finite subset of $\mathbb{N}$ will do. Indeed, we can work with constraints specifying *semilinear sets* without affecting the technical results, and the syntax can be any formula in Presburger arithmetic.

On the other hand, some natural extensions of the model do affect the decidability of non-emptiness problem. One such is *alternation*. However, we then find that *the non-emptiness problem for the class of alternating class counting automata is undecidable*. The proof of this proceeds by reduction of the Post Correspondence Problem to this one in a manner similar to the one in [BMS$^+$06]. From this, we can show that the inclusion problem for CCAs is undecidable as well.

Other interesting extensions relate to the kind of updates allowed and to acceptance conditions. While adding decrements to counters in CCA leads to undecidability of the emptiness problem, we can add resets to counters preserving decidability. A reset operation sets the corresponding counter value to zero. The acceptance condition we have in CCA is *global* in the sense that it relates only to the global control state rather than multiplicities encountered. We can strengthen the acceptance condition as follows: $A = (Q, \Delta, q_0, F, C)$ where $(Q, q_0, \Delta, F$ are as before, and $C \subseteq_{fin} N$. We say a final configuration (q, h) is accepting if $q \in F$ and $\forall d \in \mathbb{D}, h(d) \in C$ or $h(d) = 0$.

We then find that the non-emptiness problem (for CCAs with reset and counter conditions) continues to be decidable but becomes as hard as Petri net reachability, which is not even known to be elementarily decidable.

4.2   Other Automata Models

CCA are situated among a family of automata models that have been proposed for data languages. The simplest form of memory is a finite random access read-write storage device, traditionally called *register*. In *finite memory* automata [KF94], the machine is equipped with finitely many registers, each of which can be used to store one data value. Every automaton transition includes access to the registers, reading them before the transition and writing to them after the transition. The new state after the transition depends on the current state, the input letter and whether or not the input data value is already stored in any of the registers. If the data value is not stored in any of the registers, the automaton can choose to write it in a register. The transition may also depend on which register contains the encountered data value. Because of finiteness of the number of registers, in a sufficiently long word the automaton cannot distinguish between all data values. On the other hand, register automata have the capability of keeping the "latest information", a capability that deterministic CCA do not have.

In class memory automata (CMA, [BS07]), a function assigns to every data value d the state of the automaton that was assumed after reading the previous position with value d. We can think of this as using hash tables, with values coming from a finite set. On reading a (a, d), the automaton reads the table entry corresponding to d and makes a transition dependent on the table entry, the input letter a and the current state. The transition causes a change of state as well as updating of the table entry.

We can show that the class of CCA-recognizable languages is strictly contained in the class of CMA-recognizable languages, but when we add resets and counter acceptance conditions as above, the class becomes exactly as expressive as CMAs. Indeed, we see CCA as a natural restriction of CMAs yielding elementary decidability of the non-emptiness problem.

Another simple computational model, based on transducers is the data automaton model introduced in [BMS+06], and [BS07] shows that this model is exactly as expressive as CMA.

4.3 Restrictions

With an NP-hard word problem, Expspace-hard non-emptiness question and undecidable language inclusion, working with data languages does seem daunting. However, given the need for verifying properties of systems with unboundedly many processes, the abstraction of infinite alphabets is yet worth preserving. What we need to look at are restrictions that are meaningful for systems of unbounded processes, and we are studying some proposals in this regard.

References

[ABB80]   Autebert, J.-M., Beauquier, J., Boasson, L.: Langages sur des alphabets infinis. Discrete Applied Mathematics 2, 1–20 (1980)

[AM06]   Alur, R., Madhusudan, P.: Adding nesting structure to words. In: Ibarra, O.H., Dang, Z. (eds.) DLT 2006. LNCS, vol. 4036, pp. 1–13. Springer, Heidelberg (2006)

[Bac03]   Baclet, M.: Logical characterization of aperiodic data languages. Research Report LSV-03-12, Laboratoire Spécification et Vérification, ENS Cachan, France, 16 p. (September 2003)

[BMS+06] Bojanczyk, M., Muscholl, A., Schwentick, T., Segoufin, L., David, C.: Two-variable logic on words with data. In: LICS, pp. 7–16. IEEE Computer Society, Los Alamitos (2006)

[Bou02]   Bouyer, P.: A logical characterization of data languages. Inf. Process. Lett. 84(2), 75–85 (2002)

[BPT01]   Bouyer, P., Petit, A., Thérien, D.: An algebraic characterization of data and timed languages. In: Larsen, K.G., Nielsen, M. (eds.) CONCUR 2001. LNCS, vol. 2154, pp. 248–261. Springer, Heidelberg (2001)

[BS07]   Björklund, H., Schwentick, T.: On notions of regularity for data languages. In: Csuhaj-Varjú, E., Ésik, Z. (eds.) FCT 2007. LNCS, vol. 4639, pp. 88–99. Springer, Heidelberg (2007)

[DL06]   Demri, S., Lazic, R.: Ltl with the freeze quantifier and register automata. In: LICS 2006: Proceedings of the 21st Annual IEEE Symposium on Logic in Computer Science, pp. 17–26. IEEE Computer Society, Los Alamitos (2006)

[KF94]   Kaminski, M., Francez, N.: Finite-memory automata. Theor. Comput. Sci. 134(2), 329–363 (1994)

[KT06]   Kaminski, M., Tan, T.: Regular expressions for languages over infinite alphabets. Fundam. Inform. 69(3), 301–318 (2006)

[Lip76]   Lipton, R.: The reachability problem requires exponential space. Research Report 62, Yale University (1976)

[LP05]   Lisitsa, A., Potapov, I.: Temporal logic with predicate lambda-abstraction. In: TIME, pp. 147–155 (2005)

[LP09]   Lisitsa, A., Potapov, I.: On the computational power of querying the history. Fundam. Inform. 91(2), 395–409 (2009)

[LPS09]   Lisitsa, A., Potapov, I., Saleh, R.: Automata on gauss words. In: LATA, pp. 505–517 (2009)

[NSV01]   Neven, F., Schwentick, T., Vianu, V.: Towards regular languages over infinite alphabets. In: Sgall, J., Pultr, A., Kolman, P. (eds.) MFCS 2001. LNCS, vol. 2136, pp. 560–572. Springer, Heidelberg (2001)

[NSV04]   Neven, F., Schwentick, T., Vianu, V.: Finite state machines for strings over infinite alphabets. ACM Trans. Comput. Log. 5(3), 403–435 (2004)

[Ott85]   Otto, F.: Classes of regular and context-free languages over countably infinite alphabets. Discrete Applied Mathematics 12, 41–56 (1985)

[Seg06]   Segoufin, L.: Automata and logics for words and trees over an infinite alphabet. In: Ésik, Z. (ed.) CSL 2006. LNCS, vol. 4207, pp. 41–57. Springer, Heidelberg (2006)

The Periodic Domino Problem Is Undecidable in the Hyperbolic Plane

Maurice Margenstern

Laboratoire d'Informatique Théorique et Appliquée, EA 3097,
Université de Metz, I.U.T. de Metz,
Département d'Informatique,
Île du Saulcy,
57045 Metz Cedex, France
margens@univ-metz.fr

Abstract. In this paper, we consider the periodic tiling problem which was proved undecidable in the Euclidean plane by Yu. Gurevich and I. Koriakov, see [3]. Here, we prove that the same problem for the hyperbolic plane is also undecidable.

Keywords: undecidability, hyperbolic plane, tilings, periodic tiling problem.

1 Introduction

A lot of problems deal with tilings. Most of them are considered in the setting of the Euclidean plane. A certain number of these problems turn out to be undecidable in this frame, thanks to the facility to simulate the computation of a Turing machine in this setting. The most famous case of such a problem is the **general tiling problem** proved to be undecidable by Berger in 1966, see [1]. In 1971, R. Robinson gave a simplified proof of the same result, see [12]. Sometimes, the general problem is simply called the **tiling problem**. The reason of these different names lies in the fact that several conditions were put on the problem, leading to different settings, and a dedicated proof was required each time when the problem turned out to be undecidable. Among these variations, the most well-known is the **origin-constrained** problem, proved to be undecidable by Wang in 1958, see [14].

The general tiling problem consists in the following. Given a finite set of tiles T, is there an algorithm which says whether it is possible to tile the plane with copies of the tiles of T or not? The **origin-constrained** problem consists in the same question to which a condition is appended: given a finite set of tiles T and a tile $T_0 \in T$, is there an algorithm which says whether i is possible or not to tile the plane with copies of the tiles of T or not, the first tile being T_0? In the general problem there is no condition on the first tile: it can be a copy of any tile of T.

There are a lot of variants of these problems and the reader is referred to [12], where an account is given on several such conditions.

O. Bournez and I. Potapov (Eds.): RP 2009, LNCS 5797, pp. 154–165, 2009.

The **periodic tiling problem** is a bit different question. Given a finite set of tiles T, is there a way to tile the plane with copies of T in a **periodic** way? The problem was proved undecidable for the Euclidean plane by Yu. Gurevich and I. Koriakov in 1972. Now, it turns out that the notion of period is well defined in the Euclidean plane, but it is not clear how to define it in the hyperbolic plane. As many authors do, we shall consider that a tiling of the hyperbolic plane is periodic if it is unchanged under a non trivial shift.

The general tiling problem for the hyperbolic plane was raised by R. Robinson in his 1971 paper, see [12]. In 1978, R. Robinson proved that the origin-constrained problem is undecidable in the hyperbolic plane, see [13]. The general tiling problem for the hyperbolic plane remained pending a long time. The problem was solved in 2007 by the present author, see [10]. At the same time, J. Kari established the same result, using a completely different approach, see [4]. Note that the proof of [10] is fully constructive as it has the following additional property: the proof constructs a family $\{T_M\}$ of tilings which are indexed by the members M of a set of Turing machines; now, if we are given a 1-bit oracle that there is a solution to tile the hyperbolic plane with copies of T_M for a given M, then the algorithm of the proof of [10] constructs such a solution, of course in infinite time.

In this paper, we prove that:

Theorem 1. *The periodic domino problem is undecidable in the hyperbolic plane.*

The solution combines the construction given in [9,10] with an argument of [8,11] and a construction given in [6].

In the next section, section 2, we very sketchily remind the solution to the tiling problem of [10] in a simplified setting. In section 3, we prove the theorem.

2   The Interwoven Triangles

The solution of the domino problem which we now consider takes place in the tiling $\{7,3\}$ of the hyperbolic plane. It consists in delimiting infinitely many regions of infinitely many sizes in which the simulation of the same Turing machine is performed.

The construction takes place in the tiling $\{7,3\}$ of the hyperbolic plane, in which we construct a grid thanks to a particular tiling, the **background**. In this tiling, we implement a construction which is based on what we call the **abstract brackets**, which is a construction on the line. We lift the intervals which are defined by the construction up to triangles in the Euclidean plane, with parallel legs whose heights lie on the same line. In this way, we can see the previous intervals as a projection of the triangles on the line of their heights. Then, we implement these triangles in the background.

First, we sketchily describe the tiling $\{7,3\}$ of the hyperbolic plane and the background.

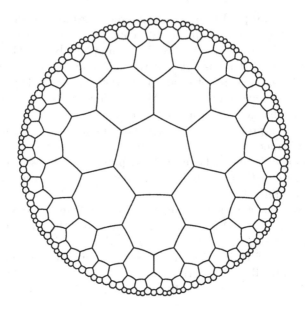

Fig. 1. The heptagrid: the tiling $\{7,3\}$ of the hyperbolic plane in the Poincaré's disc model

2.1   The Tiling $\{7,3\}$ of the Hyperbolic Plane

The tiling $\{7,3\}$ is obtained from the regular heptagon with an interior angle of $\dfrac{2\pi}{3}$ by reflection in its edges and, recursively, by reflection of the images in their edges. The existence of the tiling is a corollary of Poincaré's theorem on a sufficient condition for tiling the hyperbolic plane by triangles. It is enough to consider the rectangular triangle of the hyperbolic plane with the acute angles $\dfrac{\pi}{7}$ and $\dfrac{\pi}{3}$. Below, figure 1 illustrates the tiling $\{7,3\}$ which we later call the **heptagrid**.

In [2], we introduced a way to exhibit a generating tree of the tiling which is basically the same as the generating tree of the pentagrid, the tiling $\{5,4\}$ of the hyperbolic plane. This tiling is constructed by a process, similar to the one used for constructing the heptagrid, but it is used with the regular rectangular pentagon. This tree is called the **standard Fibonacci tree**, simply **Fibonacci tree** in the sequel, see [5] for more details on this tree.

The way to exhibit the Fibonacci tree is based on the **mid-point lines**, which we introduced in [2]. As suggested by their name, these lines join the mid-points of two consecutive edges of a heptagon, see figure 2. It turns out that the angular sector determined by two rays obtained by two mid-point lines meeting at a mid-point C, joining the two mid-points of the two other edges which meet at C, exactly contains a set of tiles spanned by a Fibonacci tree. This structure will play an important rôle in what follows.

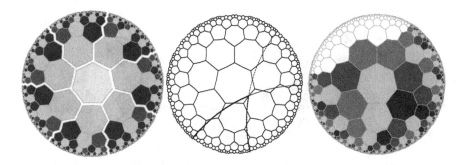

Fig. 2. Left-hand side: the standard Fibonacci trees which span the heptagrid. Middle: the mid-point lines. Right-hand side: the background of the constructions.

The correspondence of the black and white hue of the tiles with the letters is given as follows:

   G: ▨ , Y: ▦ , B: ▢ , O: ■

2.2   The Background: Constructing a Grid in the Heptagrid

In the Euclidean plane the square grid is defined by horizontals and verticals which are regularly spaced, with the same distance between horizontals and between verticals. If we look at the simulation which is performed in [3] as well as the construction in [1] and [12], we can notice that the metric constraints of the square grid are not relevant to the problem. As an example, it is possible to define the distance of a tile A to a tile B by the smallest number of tiles which have to be crossed to go from A to B. Reflection in a horizontal and in a vertical can be defined in this terms and not in those of the usual Euclidean metric. Now, in this case, the metric distance between the horizontals and the verticals can change from one line to another. Also, horizontals and verticals need not be straight lines.

We shall see that even verticals are not fully needed: it is enough to have **half-verticals** at our disposal.

This can rather easily be implemented as follows in the heptagrid, see [7].

We define four colours, green, G, blue, B, orange, O and yellow, Y and we assign to each tile of the heptagrid one of these colours in such a way that the following rules are observed:

$$G \to YBG, \qquad Y \to YBG, \qquad O \to YBO, \qquad B \to BO$$

There are infinitely many such tilings and in fact there are uncountably many of them, see the right-hand side picture of Figure 2.

In such tilings, we can implement two different kinds of Fibonacci trees at the same time. If we consider B as a black node and G, Y and O as white ones, we get a **central Fibonacci tree**, see [5]. Now, if we consider B and O as white nodes and G with Y as black nodes, we get standard Fibonacci trees with the rules

$$G \to YB, \qquad Y \to YB, \qquad O \to YBO, \qquad B \to GBO$$

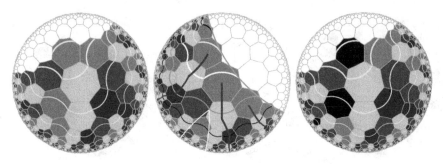

Fig. 3. The construction of the grid in the heptagrid. From left to right: definition of the horizontals, of the half-verticals and of the seeds. The seeds are black on the right-hand side picture.

The structure of the standard Fibonacci trees of this new setting allows us to define the horizontals as follows: say that G- and Y-tiles have a convex arc joining the mid-points of the sides shared with their O-neighbour and with their B-one while B- and O-tiles have a concave arc joining the mid-pints of the sides shared with their Y-neighbour and with their G-one or to their G-neighbour with their O-one. A tile has exactly two neighbours which abut one of its sides: the sides to which abut its arc. So that the arc of a tile can be continued by those of its two neighbours abutting its arc. This defines curves which we call **isoclines**. They are the **horizontals** of our construction, see the left-hand side picture of Figure 3.

The **half-verticals** start from a G or a Y-tile, go to the B-son of the tile and next, endlessly repeat the same pattern: go to the O-son of the B-tile and then to the B-son of the O-tile.

We number the isoclines from 0 to 7, repeating this numbering periodically, downwards and upwards. By definition, the sons of a tile are on the next isocline of the tile, where 0 is the successor of 7 in this order.

Now, we can define an important notion: say that a G-node is a **seed** if it is on an even isocline, if its father is a Y-node and if its grand-father is a G-node. Now, consider a seed S. Let A be the mid-point of the edge shared by the Y-father of S and the B-neighbour of this Y-node. Then, the rays issued from S which are supported by the mid-point lines passing through S which also intersect S define a sector spanned by a Fibonacci tree. By definition, this tree is a **tree of the grid** which we constructed. The set of tiles which are included in the sector determined by the rays is called the **area** of the tree, and the rays are called its **borders**. We shall also call border the set of tiles which are in contact with the rays. The trees of the grid have a very important property, see [10]:

Lemma 1. *Consider two trees of the grid. Their borders never meet. Either their areas are disjoint or the area of one contains the area of the other.*

Now, half-verticals also have an important property, see [10]:

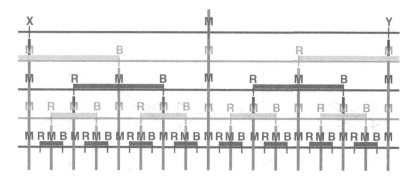

Fig. 4. The silent and active intervals with respect to mid-point lines. The light green vertical signals send the mid-point of the concerned interval to the next generation. The colours are chosen to be easily replaced by red or blue in an opposite way. The ends X and Y indicate that the figure can be used to study both active and silent intervals.

Lemma 2. *Consider a tree of the grid T. Let v be a vertical which starts from a G-node of its right-hand side border or a Y-node of its left-hand side border or from its root. Then, for any tree of the grid whose area A is contained in that of T, v does not meet A.*

2.3 The Abstract Brackets

By this name, we call the following process of construction of intervals on the line, illustrated by figure 4 which is performed by successive **generations**.

Generation 0 consists of points on a line which are regularly spaced. The points are labelled R, M, B, M, in this order, and the labelling is periodically repeated. An interval defined by an R and the next B, on its right-hand side, is called **active** and an interval defined by a B and the next R on its right-hand side is called **silent**. Generation 0 is said to be **blue**.

Blue and red are said **opposite**. Assume that the generation n is defined. For the generation $n+1$, the points which we take into consideration are the points which are still labelled M when the generation n is completed. Then, we take at random an M which is the mid-point of an active interval of the generation n, and we label it, either R or B. Next, we define the active and silent intervals in the same way as for generation 0. The active and silent intervals of the generation $n+1$ have a colour, opposite to that of the generation n.

When the process is achieved, we get an **infinite model**. The model has interesting properties, see [10].

In an interval of the generation n, consider that a letter of a generation m, $m \leq n$, which is inside an active interval is hidden for the generations k, $k \geq n+1$. Also, a letter has the colour of its generation. Now, we can prove that in the blue active intervals, we can see only one red letter, which is the mid-point of the interval. However, in a red active interval of the generation $2n+1$, we can see $2^{n+1}+1$ blue letters.

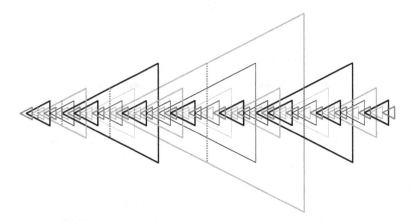

Fig. 5. An illustration for the interwoven triangles

Cut an infinite model at some letter and remove all active intervals which contain this letter. What remains on the right-hand side of the letter is called a **semi-infinite model** which is also called a **cut** of an infinite model.

It can be proved that in a semi-infinite model, any letter y is contained in at most finitely many active intervals, see [10].

2.4   The Interwoven Triangles in the Euclidean Plane

As indicated in the introduction, we lift up the active intervals as **triangles** in the Euclidean plane. The triangles are isosceles and their heights are supported by the same line, called the **axis**, see figure 5.

We also lift up silent intervals of the infinite model up to again isosceles triangles with their heights on the axis. To distinguish them from the others, we call them **phantoms**. We shall speak of **trilaterals** for properties shared by both triangles and phantoms.

We have very interesting properties for our purpose.

Lemma 3. *Triangles of the same colour do not meet nor overlap: they are disjoint or embedded. Phantoms can be split into* **towers** *of embedded phantoms with the same mid-point and alternating colours. Trilaterals can meet by a basis cutting the half of leg which contains the vertex.*

From these properties, we prove in [10] that:

Lemma 4. *There is a set of* 190 *tiles which force the construction of a tiling of the Euclidean plane which implements the interwoven trilaterals.*

3   The Tiling Problem in the Hyperbolic Plane

We sketchily remember the implementation of the interwoven triangles in the hyperbolic plane.

To this purpose, note that the inclusion between areas of trees of the grid defines a partial order. We call **threads** the maximal sequences of totally ordered trees of the grid. Threads are indexed by $I\!N$ or by $Z\!\!\!Z$. When it is indexed by $Z\!\!\!Z$, the thread is called an **ultra-thread**. From [10], we know that two ultra-thread coincide, starting from a certain index. We also noticed in [10] that there can be realizations of the grid with ultra-threads as well as without them.

Now, in the hyperbolic plane, we implement different **cuts** of the **same** infinite model of the abstract brackets, taking a thread as the guideline for the realization of interwoven triangles.

The idea is that the triangles will be defined by the borders of trees of the grid as legs and by a piece of an isocline as the basis. Now, the isoclines will play the rôles of the letters in the abstract brackets.

From Section 2, we know that if the isocline which passes through a seed S is numbered 0, the closest seed inside the area of the tree defined by S is on isocline 0. We select isoclines 0, 2, 4 and 6 to play the rôles of the letters. These isocline are needed for the construction of the generations. Once generation 0 is installed, the further generations are obtained by the algorithm suggested by the figures 4 and 5, see [10] for a precise description. Sketchily, the seeds grow legs until a green line is met. The green line is stopped by a triangle but not by a phantom. Afterwards, the leg go on growing until they meet the basis of their colour. For red triangles, we proceed, at the same time, to the detection of the **free rows** which are the isoclines corresponding to a free letter in a red active interval. For this process, the legs of the triangle diffuse a horizontal red signal outside the triangle and this, already from generation 1. Accordingly, a row is free in a given red triangle T if and only if there is no horizontal red signal on the considered isocline inside T.

Due to the presence of several trilaterals on the same interval of isoclines inside a given trilateral in the hyperbolic plane, it is needed to **synchronize** the processes which occur along different threads. In particular, when the threads merge, the processes may also merge, as they implement the same cut of the same infinite model.

To do this, we decide that triangles and phantoms of a given generation will have their basis and their roots on the same isoclines. Note that this is already the case of generation 0. We simply extend this property to all the generations. To obtain this result, we decide that all bases of a given isocline merge into a unique basis signal which runs over the whole isocline. Now, the difference will be made by the presence of a horizontal upper signal of the colour of the trilateral, above the basis signal, on the same isocline. It may be realized as another channel on the same tiles, at a higher level than the channel used by the basis signal. Now, for coherence with the detection of the free rows in a red triangle, the vertices of the trilaterals also emit a horizontal signal of their colour but, this time, in a lower position: below the basis signal, again on the same isocline. The signals emitted by the legs of triangles at points which are neither the vertex nor the corner, are also of the colour of the triangle, but they are upper horizontal signals.

Moreover, the signals have a laterality given by the leg which they cross when they are of the same colour. Also, we allow signals of opposite lateralities to meet when they come from directions which are opposite to their lateralities. The other case of meeting is ruled out, except for the lower signals, where such a meeting is realized by the vertex itself. With these indications, a leg always know to which kind of basis it has to deal when it meets one of them.

Now, we select the red triangles and forget the others as well as the phantoms. The free rows and the verticals inside the triangle will be used to implement the space-time diagram of a Turing machine. As in Berger's and Robinson's proof, we consider the same Turing machine starting from an empty tape, whose simulation is performed in each computing area.

When the Turing machine does not stop, the computation is stopped by the basis of the concerned triangle, as its area is finite. As we have areas of infinitely many sizes, this allows to tile the plane. If the Turing machine halts, in one of the areas, the halting state will be called by the Turing machine. It is easy to associate to this state a tile which blocks the continuation of the tiling.

4   The Periodic Tiling Problem

In [8,11], we proved that the finite tiling problem is undecidable. In fact, we shall use similar tiles to construct a periodic tiling of the hyperbolic plane when the simulated Turing machine halts. What is performed in [11] is that we have tiles which adapts to the halting tiles in order to encapsulate the computing area in a closed signal which runs along the legs and the basis of the concerned triangle.

However, this is not enough to prove the undecidability of the periodic tiling problem. We need another important ingredient coming from [7].

Before going on, we shall remark an important property. Consider again a seed. Then, replace the right-hand side ray by another one which delimits what we shall call a **black** tree. Such a tree is obtained by the same rules as for a standard Fibonacci tree, but its root is a black node. Technically, the root of a tree of the grid is also a black node, but everything happens as if we apply a white-node rule to the root and we change all white nodes which are on the right-hand side border into black nodes. In fact, a tree of the grid can be realized as a stack of black trees and standard Fibonacci trees which we shall call **white** trees to simplify the denotation. Now, with black trees we have the same properties as for the white ones: for any black trees of the grid, either their areas are disjoint or the area of one contains the area of the other. Accordingly, we can repeat the same construction process by replacing all trees of the grid by black trees. We refer the reader to [8,11] for a figure representing the tiles of such a border.

Accordingly, when the simulated Turing machine halts, we may construct two encapsulated areas B and W where the whole computation of the machine takes place. The area W will be called white and the area B will be called black.

Moreover, from what we have noticed, both areas have the same height. Now, we shall make four super-prototiles with such computing areas. The border of an area has an interior side and an exterior one. It is not difficult to see that the difference can be noticed by the orientation of the tiles with respect to their father: number the edges of a tile counter-clockwise from 1 to 7, giving the number 1 to the edge which is shared by the father. Then the left-hand side border always go from the edge 1 to the edge 4 and the right-hand side border always go from the edge 1 to the edge 5. The difference between a black and a white tree is performed by the root. If the left-hand side border is also defined from the edge 1 to the edge 4 in the root, then the right-hand side border goes from the edge 2 to the edge 6 for a white tree and it goes from the edge 7 to the edge 4 for a black tree.

This distinction will allow us to look at the border as two-sided: one side is interior and the other is exterior. Now we shall consider that the left- and right-hand side borders of B are black inside and white outside. Now, from W, we shall make three copies, W_1, W_2 and W_3. For W_1, W_2 and W_3, the inside of the borders is always white. For W_1, the outside of the left-hand and right-hand side borders is black. For W_2, the outside of the left-hand side border is black and that of the right-hand side border is white. For W_3, the outside of the left-hand side border is white and that of the right-hand side border is black.

Now, we consider B, W_1, W_2 and W_3 as super-prototiles with which we shall construct a periodic tiling. In [6], we have considered sets of tiles which we called **quarters** and **bars**. A quarter of size n, denoted by Q_n, is the set of tiles spanned by a standard Fibonacci tree restricted to its first n levels. A bar of size n, denoted by R_n, is defined in the same way as a quarter but with a black Fibonacci tree. In [6], we proved that Q_{n+m} can be split into Q_n and f_{2n} copies of Q_m and f_{2n-1} copies of R_m. Note that f_{2n} is the number of white nodes on the level n of a standard Fibonacci tree and that f_{2n-1} is the number of black nodes. In fact, the property comes from the fact that the trees rooted at two consecutive nodes of the same level of a Fibonacci trees are disjoint and that there is no node in between. Now, we can take $m = n$ and repeat the process as long as we wish. Using the property also proved in [6] that the hyperbolic plane can be viewed as the union of a growing sequence of quarters, we can construct the periodic tiling as follows.

Initial step:
Take a tile of the tiling $\{7, 3\}$ which will be called the **origin**. Take a copy of W_2, and place it in such a way that the root of the copy of W_2 coincides with the origin. Call this just defined region of the tiling $\mathcal{T}_0$. The root of this copy of W_2 is also called the top of $\mathcal{T}_0$ and its bottom border is called the bottom border of $\mathcal{T}_0$. Now, we define copies of B and W_i, with $i \in \{1, 2, 3\}$ which we call $\mathcal{R}_0$ and $\mathcal{Q}_0^i$ respectively. We define the roots and the bottom borders of these regions as we did for $\mathcal{T}_0$.

Now, in what follows, a particular role will be played by the node of the last level of W_2 whose coordinate is a term of the Fibonacci sequence. Call it the **junction point**. Now, on the tiling $\{7, 3\}$, draw the line which passes through the mid point of the origin and through the junction point of T_0. Call it the **axis**. This line is simply an auxiliary tool in our construction.

Induction step:

Assume that T_n is constructed, as well as regions R_n and Q_n^i, which have the following particularity: the left- and right-hand side borders of Q_n^i also have an outside and an inside parts and the outside and inside part of the left- and right-hand side borders of Q_n^i have the same colour as the corresponding elements of W_i, for each $i \in \{1, 2, 3\}$.

Then, take a copy of W_2 and put it above T_n in such a way that the top of T_n is the middle son of the junction point of W_2. Now, on the bottom border of W_2, proceed as follows, starting from the leftmost node: if we have a black node, place a copy of R_n, then a copy of Q_n^1; if we have a white node, place copies of R_n, Q_n^2 and Q_n^3, in this order. This is the first step: we get a set of tiles which is alike a copy of Q_m for an appropriate m. Now, consider the bottom border of this region, and proceed as follows, starting from the leftmost node: if we have a black node, place a copy of B, then a copy of W_1; if we have a white node, place copies of B, W_2 and W_3, in this order. Now, we get a new region T_{n+1} which strictly contains T_n: the tiles which belong to the border of T_n do not meet the tiles which belong to the border of T_{n+1}. Now, we construct R_{n+1} and Q_{n+1}^i in a similar way, starting from, respectively R_n and Q_n^i and completing them by two rows of copies of B and W_i's which are placed as above indicated.

Now, it is plain that $\underset{n \in N}{\cup} T_n$ is the hyperbolic plane. Also, from the construction, it is plain that this tiling is invariant under the shift along the axis which transforms the origin in the middle son of the junction point of T_0.

To complete the proof, assume that the simulated Turing machine does not halt.

If there is a solution, the tiles for the border of the super-tiles always admit near them a seed S which is of generation 0: hence, it does not bear the root of a copy of a super-tile. Next, we know that the computation goes on endlessly, thanks to the construction of [10]: as the halting state cannot be met, the occurrence of tiles of the border of super-tiles cannot be triggered inside the tree of the grid rooted at S. If the tiling would be periodic, then from the invariance under a shift, we would easily get that as S does not contain tiles of the border of a super-tile, this is also the case for the whole plane. Now, this means that we have a solution of the tiling constructed in [10] and, as we know, such a solution cannot be periodic. Indeed, a shift should keep the isoclines globally invariant and then, there is no shift which would match the triangles of a certain generation and those which are of a higher generation. But this contradicts the assumption of a periodic solution. Accordingly, if the Turing machine does not halt, there is no periodic solution. The just produced argument indicates that in this case there are solutions, but they are not periodic.

This completes the proof of theorem 1.

References

1. Berger, R.: The undecidability of the domino problem. Memoirs of the American Mathematical Society 66, 1–72 (1966)
2. Chelghoum, K., Margenstern, M., Martin, B., Pecci, I.: Cellular automata in the hyperbolic plane: proposal for a new environment. In: Sloot, P.M.A., Chopard, B., Hoekstra, A.G. (eds.) ACRI 2004. LNCS, vol. 3305, pp. 678–687. Springer, Heidelberg (2004)
3. Yu, G., Koriakov, I.: A remark on Berger's paper on the domino problem. Siberian Mathematical Journal 13, 459–463 (1972)
4. Kari, J.: The Tiling Problem Revisited. In: Durand-Lose, J., Margenstern, M. (eds.) MCU 2007. LNCS, vol. 4664, pp. 72–79. Springer, Heidelberg (2007)
5. Margenstern, M.: New Tools for Cellular Automata of the Hyperbolic Plane. Journal of Universal Computer Science 6(12), 1226–1252 (2000)
6. Margenstern, M.: Theory. In: Cellular Automata in Hyperbolic Spaces, OCP, Philadelphia, vol. 1, 422 p. (2007)
7. Margenstern, M.: Implementation and computations. In: Cellular Automata in Hyperbolic Spaces, OCP, Philadelphia, vol. 2, 360 p. (2008)
8. Margenstern, M.: The finite tiling problem is undecidable in the hyperbolic plane, *arxiv:cs.CG*/0703147, 8 p. (March 2007)
9. Margenstern, M.: The Domino Problem of the Hyperbolic Plane is Undecidable. Bulletin of the EATCS 93, 220–237 (2007)
10. Margenstern, M.: The domino problem of the hyperbolic plane is undecidable. Theoretical Computer Science 407, 29–84 (2008)
11. Margenstern, M.: The Finite Tiling Problem Is Undecidable in the Hyperbolic Plane. International Journal of Foundations of Computer Science 19(4), 971–982 (2008)
12. Robinson, R.M.: Undecidability and nonperiodicity for tilings of the plane. Inventiones Mathematicae 12, 177–209 (1971)
13. Robinson, R.M.: Undecidable tiling problems in the hyperbolic plane. Inventiones Mathematicae 44, 259–264 (1978)
14. Wang, H.: Proving theorems by pattern recognition. Bell System Tech. J. 40, 1–41 (1961)

Games with Opacity Condition

Bastien Maubert and Sophie Pinchinat

IRISA, France

Abstract. We describe the class of games with opacity condition, as an adequate model for security aspects of computing systems. We study their theoretical properties, relate them to reachability perfect information games and exploit this relation to discuss a search approach with heuristics, based on the directing-word problem in automata theory.

1 Introduction

We describe a class of two-player imperfect information games that we call *games with opacity condition*. In these games, the players are Robert (for "robber") and Gerald (for "guardian"). Imperfect information is asymmetric between the players: Robert has imperfect information as opposed to Gerald who has perfect information. The model we used for games with opacity condition uses the classic imperfect information arenas, as defined in [12,4,1], but it differs in the nature of the winning objectives: in games with opacity, Gerald aims at maintaining the uncertainty of Robert regarding the actual position in the game along the play.

Games with opacity conditions easily relate to computer systems security issues, since in practice interactive systems are expected to have a policy against intruders that attempt to reach a secret, modelled e.g as perfect information in the model.

Our claim that games with opacity condition are natural and adequate models for practical applications is all the more sustained by very recent contributions of the literature [13,5]. These results mainly arise from the analysis of discrete-event systems and their theory of control. We believe that the abstract setting provided by the game-theoretical paradigm enables to focus on essential aspects such as circumventing the complexity of the problems and synthesizing strategies.

In this contribution, we first establish that deciding the opacity-guarantee problem translates into the problem of solving a perfect information safety game – which, according to determinacy in the perfect information setting, is dual to a perfect information reachability game. This is a key point of our approach: although standard bottom-up techniques to solve safety perfect information games are intractable in this case, due to a blow-up in the translation, top-down methods may be worth considering. Moreover, these methods may be enriched with heuristics, preventing the search from a useless exhaustive exploration of the entire state space.

We therefore discuss a search-based approach in an AND/OR graph (the perfect information arena of a reachability game). The search is sustained by

O. Bournez and I. Potapov (Eds.): RP 2009, LNCS 5797, pp. 166–175, 2009.

heuristics arising from a standard problem in automata theory: the *directing-word problem* [3,10], which addresses the existence of a finite word that leads every state of a non-deterministic automaton to a unique single state; the literature also refers to the *synchronizing word* or the *reset* problem.

The paper is organized as follows. In Section 2 we introduce the model and the notion of opacity, and we define the opacity–guarantee and opacity–violate problems. Theoretical analysis of games with opacity condition is done in Section 3, where their non–determinacy is proved, and the equivalence of the opacity–guarantee and opacity–violate problems with a safety, respectively reachability perfect information game is established as well as their connection with the directing-word problem. Finally, we end by Section 4 where we discuss a search approach with heuristics based on directing-word techniques.

2   Games with Opacity Condition

2.1   Arena, Strategies

An *imperfect-information arena* over the alphabet Σ and the set of observations Γ is a structure $A = (V, \Delta, \mathrm{obs}, \mathrm{act})$ where V is a finite set of *positions*, $\Delta : V \times \Sigma \to 2^V$ is a transition function, $\mathrm{obs} : V \to \Gamma$ is an observation function and $\mathrm{act} : \Gamma \to 2^\Sigma \backslash \emptyset$ assigns to each observation the non–empty set of available actions. The fact that act is defined on Γ reflects the fact that available actions must be identical for observationally equivalent positions.

We sometimes write γ instead of $\mathrm{obs}^{-1}(\gamma)$ to denote the set of positions $v \in V$ whose observation is γ.

In an arena $A = (V, \Delta, \mathrm{obs}, \mathrm{act})$, the players Robert and Gerald play as follows.

First, before the game starts, Gerald chooses an initial position v_0. We refer to the game A just after v_0 has been chosen in the first round by A_{v_0}. Then Robert chooses an action $a_1 \in act(v_0)$, and Gerald chooses a position $v_1 \in \Delta(v_0, a_1)$. In the next round, we process similarly but from position v_1 where Robert is given the information $\mathrm{obs}(v_1)$ to choose a suitable action $a_2 \in \Sigma$. A *concrete play* in A_{v_0} is an infinite sequence $\rho = v_0 a_1 v_1 a_2 v_2 a_3... \in v_0(\Sigma V)^\omega$ that results from an interaction of Robert and Gerald in this game.

We now extend obs as a morphism $\mathrm{obs} : (V \cup \Sigma)^* \to (\Gamma \cup \Sigma)^*$, by letting $\mathrm{obs}(a) = a$, for all $a \in \Sigma$. The imperfect information setting leads Robert to partially observe a concrete play ρ as the *abstract play* $\mathrm{obs}(\rho) \in \gamma_0(\Sigma\Gamma)^\omega$, where $\gamma_0 := \mathrm{obs}(v_0)$.

Since Gerald has perfect information on how the play progresses, a strategy of Gerald in A_{v_0} is a mapping of the form

$$\beta : v_0(\Sigma V)^* \Sigma \to V$$

On the contrary, because the information revealed to Robert is based on observations, a strategy of Robert in A_{v_0} is a mapping of the form

$$\alpha : \gamma_0(\Sigma\Gamma)^* \to \Sigma$$

For every natural number $k \in \mathbb{N}$, we denote by $\pi^k \in \gamma_0(\Sigma\Gamma)^k$ the k-th prefix of π, defined by $\pi^k := \gamma_0 a_1 \gamma_1 a_2 \gamma_2 \ldots a_k \gamma_k$, with the convention that $\pi^0 = \gamma_0$. We denote by π^+ an arbitrary prefix of π, and we may use analogous notations for concrete plays.

Given strategies α and β of Robert and of Gerald respectively, we say that a play $\rho = v_0 a_1 v_1 \ldots$ is *induced by* α if $\forall i \geq 1$, $a_i = \alpha(\text{obs}(\rho^{i-1}))$, and ρ is *induced by* β if $\forall i \geq 1$, $v_i = \beta(\rho^{i-1} a_i)$.

2.2   Opacity Condition

Let us fix an abstract play $\pi = \gamma_0 a_1 \gamma_1 a_2 \gamma_2 \ldots$. Note that every k-th prefix of π characterizes a unique *information set* $I(\pi^k) \subseteq V$ consisting of the set of plausible actual concrete positions of Robert in the game after k rounds. Formally, $I(\pi^0) := \gamma_0$, and $I(\pi^{k+1}) := \Delta(I(\pi^k), a_{k+1}) \cap \gamma_{k+1}$, for $k \in \mathbb{N}$. For a concrete play ρ we define $I(\rho^k) := I(\text{obs}(\rho^k))$.

A (concrete) play ρ satisfies *the opacity property*, or is *opaque*, if for every natural number k, $I(\rho^k)$ is not a singleton, that is $|I(\rho^k)|$[1] is strictly greater than 1.

Informally, the opacity condition means that the actual position along the play is never revealed to Robert.

We investigate effective methods to solve *games with opacity condition*, that is to answer the following *opacity-guarantee problem*: *Given an imperfect-information arena* $A = (V, \Delta, \text{obs}, \text{act})$ *and an initial position* v_0, *does Gerald have a strategy* β *in* A_{v_0} *such that any play induced by* β *is opaque?*

Actually, driven by the natural application domains underlying this game-theoretic problem, we also expect to compute a winning strategy for Gerald, when it exists. We also define the *opacity–violate problem*, dual to the opacity–guarantee problem, that consists in deciding the existence of a strategy α for Robert such that no play induced by α is opaque. If the answer to the opacity-guarantee problem is positive, v_0 is a *winning position* for Gerald. Similarly, if the answer to opacity–violate problem is positive, then v_0 is a winning position for Robert.

3   Results on Games with Opacity Condition

We first establish the non–determinacy of games with opacity condition. We next show how the opacity-guarantee and the opacity–violate problems can be rephrased in terms of solving a safety perfect information game and a reachability perfect information game respectively. Finally we introduce the *directing-word problem* and show a polynomial time reduction to the opacity–violate problem. From the above, we end the section by inferring complexity results.

3.1   Non–determinacy

We recall that a game is *determined* if each position is winning for one player or the other. It is well known that perfect–information games are determined [9],

---

[1] the cardinal of $I(\rho^k)$.

and that imperfect–information games are not determined in general. We prove the following:

Theorem 1. *Games with opacity condition are not determined in general.*

Proof. Consider the game on Figure 1. Note that the dashed sets represent observation classes. We first prove that Robert does not have a winning strategy in the initial position v_0.

Robert has information set I, and he must play a. Next Gerald chooses one of the two reachable positions v and v' and Robert now knows the information set I'. There are two possibilities: Robert can either play a or b. If he plays a, then if the actual position is v, Robert wins (he reaches v'' that is alone in its observation class). But if the actual position is v', then Gerald can whether choose to loop, whether move to v. Notice that in both cases, Robert still knows information set I': he never gains information, thus can never know if he should play a or b. Then the strategy of playing a at the second round is not winning. Reversing the roles of a and b in this reasoning yields the result that playing b at the second round is not winning neither. Robert does not have a winning strategy.

We now prove that Gerald does not have a winning strategy either. As we said, at first Robert can only choose a. If Gerald chooses v, then Robert can win by playing a, and if he chooses v', Robert can win by playing b. So there is no winning strategy for Gerald neither. □

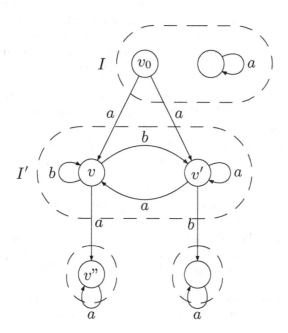

Fig. 1. A game with opacity condition

3.2   Reductions to Perfect Information Games

We informally describe a powerset construction that leads to solve an alternating reachability problem in a perfect information game. This construction is strongly inspired from the one of [12].

Let $A = (V, \Delta, \text{obs}, \text{act})$ be an imperfect-information arena, and v_0 be the initial position chosen by Gerald. We define a two-player perfect information arena $\widetilde{A}_{v_0}$, where the players are Roberta and SuperGeraldine[2].

A position of $\widetilde{A}_{v_0}$ is either I where I is a reachable information set in the game A_{v_0} – it is a position of Roberta –, or (I, a) where I is a reachable information set in A_{v_0}, and $a \in \text{act}(I)$ – it is a position of SuperGeraldine.

The game is played as follows. It starts in the initial position $I_0 := \text{obs}(v_0)$ of Roberta. In a position I, Roberta chooses $a \in \text{act}(I)$ and moves to position (I, a). Next, define O the set of reachable observations from I by a: let Π_I denote the set of prefix plays ρ^+ in A_{v_0} such that $I(\rho^+) = I$. Now pose $O := \{\text{obs}(v') \mid v' \in \Delta(v, a), \ v = \text{last}(\rho^+), \ \rho^+ \in \Pi_I\}$. SuperGeraldine chooses a non empty information set $\Delta(I, a) \cap \gamma$, where γ ranges over O. In $\widetilde{A}_{v_0}$, a play $I_0(I_0, a_1)I_1(I_1, a_2)\ldots$ is winning for Roberta if it reaches a position of the form $\{v\}$, otherwise it is winning for SuperGeraldine.

Theorem 2. *Robert has a winning strategy in A_{v_0}, if and only if, Roberta has a winning strategy in the perfect information game $\widetilde{A}_{v_0}$.*

Theorem 2 has been proved by Reif in [12]. He establishes a 1–1 correspondence between winning strategies in A_{v_0} and winning memoryless strategies in $\widetilde{A}_{v_0}$. However since our model, though equivalent to his, looks different, we explicate the correspondence between strategies in our model, but do not provide its proof of correctness as it exactly matches the one in [12, page 288]:

– Let α be a winning strategy of Robert. Define the memoryless strategy $\widetilde{\alpha}$ of Roberta by $\widetilde{\alpha}(I) := (I, \alpha(\text{obs}(\rho^+))$, for some prefix concrete play ρ^+ in the game A_{v_0} such that $I(\rho^+) = I$.
– Let $\widetilde{\alpha}$ be a memoryless winning strategy of Roberta in $\widetilde{A}_{v_0}$. Define the strategy α of Robert in A_{v_0} by: for any prefix abstract play π^+, $\alpha(\pi^+) := a$, with $(I(\pi^+), a) = \widetilde{\alpha}(I(\pi^+))$.

We now establish Theorem 3 demonstrating a powerset construction for Gerald, leading to a safety perfect information game $\widehat{A}_{v_0}$. In this game, we maintain an extra information on how Gerald is playing in A_{v_0}. The players in $\widehat{A}_{v_0}$ are SuperRoberta[3] and Geraldine. A position in $\widehat{A}_{v_0}$ is either of the form (I, v) where I is a reachable information set in A_{v_0}, and $v \in I$ – it is a position of

---

[2] We use the superlative "Super" here because in general the winning strategies of SuperGeraldine do not reflect any winning strategy of Gerald in A_{v_0}. She has "more power" than Gerald.

[3] We use the superlative "Super" as, contrary to what Roberta could do in the game $\widetilde{A}_{v_0}$, SuperRoberta can take advantage of the extra information.

SuperRoberta –, or of the form (I, v, a) where I is a reachable information set in A_{v_0}, $v \in I$, and $a \in \text{act}(v)$ – it is a position of Geraldine. The initial position is $(\text{obs}(v_0), v_0)$. In position (I, v), SuperRoberta chooses $a \in \text{act}(v)$, and moves to (I, v, a). In position (I, v, a), Geraldine chooses $v' \in \Delta(v, a)$ and moves to (I', v') where $I' = \Delta(I, a) \cap \text{obs}(v')$. In $\widehat{A}_{v_0}$, a play $(I_0, v_0)(I_0, v_0, a_1)(I_1, v_1) \ldots$ is winning for SuperRoberta if it reaches a position (I, v) or (I, v, a) where $|I| = 1$, otherwise it is winning for Geraldine.

Theorem 3. *Gerald has a winning strategy in A_{v_0}, if and only if, Geraldine has a winning strategy in the perfect information game $\widehat{A}_{v_0}$.*

Proof. We establish a 1–1 correspondence between winning strategies in A_{v_0} and winning memoryless strategies in $\widehat{A}_{v_0}$.

- Let β be a winning strategy of Gerald. Define the strategy $\widehat{\beta}$ of Geraldine by
$$\widehat{\beta}((I_0, v_0)(I_0, v_0, a_1)(I_1, v_1) \ldots (I_n, v_n, a_{n+1})) := (I_{n+1}, v_{n+1})$$
  with $v_{n+1} = \beta(v_0 a_1 v_1 \ldots v_n a_{n+1})$ and $I_{n+1} = \Delta(I_n, a_{n+1}) \cap \text{obs}(v_{n+1})$.
  We prove by contradiction that $\widehat{\beta}$ is winning for Geraldine in $\widehat{A}_{v_0}$. Assume $\widehat{\beta}$ is not winning, we show that β is not winning for Gerald in A_{v_0}. There exists $\widehat{\rho}^n = (I_0, v_0)(I_0, v_0, a_1) \ldots (I_n, v_n)$ a prefix of a play $\widehat{\rho}$ in $\widehat{A}_{v_0}$ induced by $\widehat{\beta}$ such that $|I_n| = 1$. From the definition of $\widehat{\beta}$ we have that $\rho^n = v_0 a_1 v_1 \ldots v_n$ is a prefix of a play in A_{v_0} induced by β. We show that this prefix is losing for Gerald by proving that $\forall i \leq n, I(\rho^i) = I_i$. We proceed by induction over i: clearly $I(\rho^0) = \text{obs}(v_0) = I_0$. Suppose $I(\rho^i) = I_i$, for $0 \leq i < n$.
$$\begin{aligned} I(\rho^{i+1}) &= \Delta(I(\rho^i), a_{i+1}) \cap \text{obs}(v_{i+1}) \\ &= \Delta(I_i, a_{i+1}) \cap \text{obs}(v_{i+1}) \\ &= I_{i+1} \end{aligned}$$

  So $|I(\rho^n)| = |I_n| = 1$, and β is not winning. By contradiction, $\widehat{\beta}$ is winning.
- Let $\widehat{\beta}$ be a winning strategy of Geraldine.
  For a prefix $\rho^n = v_0 a_1 v_1 \ldots v_n$ and an action $a_{n+1} \in \text{act}(v_n)$, we define the strategy β of Gerald by $\beta(\rho^n a_{n+1}) := v_{n+1}$ with $(I_{n+1}, v_{n+1}) = \widehat{\beta}((I(\rho^0), v_0)(I(\rho^0), v_0, a_1) \ldots (I(\rho^n), v_n, a_{n+1}))$. We prove again by contradiction that β is winning for Gerald in $\widehat{A}_{v_0}$. Assume β is not winning. There exists a prefix $\rho^n = v_0 a_1 v_1 \ldots v_n$ of a play ρ induced by β such that $|I(\rho^n)| = 1$.
  Let $\widehat{\rho} = (I(\rho^0), v_0)(I(\rho^0), v_0, a_1) \ldots (I(\rho^n), v_n)$. It is a prefix of a play in $\widehat{A}_{v_0}$ that is losing for Gerald. We need to prove that it is induced by $\widehat{\beta}$. For $i < n$, let I_{i+1} be the information set such that $\widehat{\beta}((I(\rho^0), v_0)(I(\rho^0), v_0, a_1) \ldots (I(\rho^i), v_i)) = (I_{i+1}, v_{i+1})$.
$$\begin{aligned} I_{i+1} &= \Delta(I(\rho^i), a_{i+1}) \cap \text{obs}(v_{i+1}) \text{ from the construction of } \widehat{A}_{v_0} \\ &= I(\rho^{i+1}) \text{ by definition of } I \end{aligned}$$

  $\widehat{\rho}^n$ is induced by $\widehat{\beta}$ and is losing for Gerald, so $\widehat{\beta}$ is losing. Contradiction. $\square$

3.3   The Directing-Word Problem

We define the directing-word problem, a classic problem in automata theory originally considered in [11,3].

Given a non-deterministic complete finite-state automaton $\mathcal{A} = (Q, X, \delta)$ over alphabet X, a *directing word* in $\mathcal{A}$ is some $w \in X^*$ such that $|\delta(Q, w)| = 1$.

The *directing-word problem* is a decision problem: *Given a non-deterministic complete finite-state automaton $\mathcal{A}$, does there exist a directing-word in $\mathcal{A}$?*

Proposition 1. *The directing-word problem is in* PSPACE.

Proof. Not surprisingly, a powerset construction and a guess on how a subset of the form $\{q\}$ is reachable from the full subset Q, shows a solution of the problem in NPSPACE, which equals PSPACE by the Theorem of Savitch [14].   □

However, we are not aware whether the directing-word problem is PSPACE-hard or not. Under the hypothesis that the automata are deterministic, the problem, known as the *synchronizing word problem* [2] has been extensively studied. It particular, it is NP-complete to decide whether there exists a synchronizing word of length $\leq k$, and the Cerny conjecture states that if a synchronizing word exists, then so does a synchronizing word of length at most $(n-1)^2$ [10,2]. In the general case, the powerset construction in Proposition 1 shows an exponential bound on the length of a minimal directing word [6].

We establish a polynomial reduction of the directing-word problem into the opacity-violate problem. Let $\mathcal{A} = (Q, X, \delta)$ be a non-deterministic complete finite-state automaton. We construct the arena $A^{\mathcal{A}} = (Q, \delta, \text{obs}, \text{act})$ over X and $\{\gamma\}$ (a fresh symbol), such that Proposition 2 holds. Let $\text{act}(v) = X$, for every v, since $\mathcal{A}$ is complete, and obs be the constant mapping sending any position to the unique observation γ; notice that Robert is consequently blindfold – in the sense of [12]. Let v_0 be any position in Q.

Proposition 2. *Robert wins the game $A^{\mathcal{A}}_{v_0}$ if, and only if, there exists a directing word in $\mathcal{A}$.*

Proof. Assume there exists a directing word $w = x_1 x_2 \ldots x_\ell$ in $\mathcal{A}$ of length ℓ, which leads any state of $\mathcal{A}$ to the state q_w. We use w to define the winning strategy α_w of Robert in the game $A^{\mathcal{A}}_{v_0}$ as:

$$\begin{cases} \alpha_w(\gamma x_1 \gamma x_2 \ldots \gamma x_i \gamma) := x_{i+1}, & \text{for all } 0 \leq i < \ell, \\ \alpha_w(\gamma x_1 \gamma x_2 \ldots (x_\ell \gamma)^k) := x_\ell, & \text{for all } k > 0. \end{cases}$$

Reciprocally, assume there exists a winning strategy α for Robert in $A^{\mathcal{A}}_{v_0}$. Since there is only one observation, the only possible abstract play induced by this strategy is $\pi = \gamma \alpha(\gamma) \gamma \alpha(\gamma \alpha(\gamma) \gamma) \ldots$ Projecting the least prefix π^+ of π such that $|I(\pi^+)| = 1$ on X gives a directing word for $\mathcal{A}$.   □

3.4   On the Complexity of Opacity Problems

We let the size of a game be the size of its arena, that is the number of positions. We study the complexity of the opacity problems.

First, note that Theorem 3 gives an EXPTIME upper bound to the opacity–guarantee problem: For an instance $A = (V, \Delta, \mathrm{obs}, \mathrm{act})$ and initial position v_0 of this problem, the safety game $\widehat{A}_{v_0}$ of Theorem 3 can be solved in polynomial time. Indeed, as $\widehat{A}_{v_0}$ is a perfect information game, it is determined, and the existence of a winning strategy for Geraldine can be decided by verifying whether her opponent, SuperRoberta, has a winning strategy. This amounts to solving a perfect information reachability game, and can be done in polynomial time [12], for example by a backward iteration from the target positions. Now, because the game $\widehat{A}_{v_0}$ arises from a powerset construction, its size is exponential in the size of A. For the same reasons, thanks to Theorem 2, the opacity–violate problem also has an EXPTIME upper bound.

Still considering the opacity–violate problem, Proposition 2 provides a polynomial reduction of the D1–directing word problem, but cannot bring any tight lower bound, even if the D1–directing word problem would be proved PSPACE-complete.

To our knowledge, the exact complexity of the opacity-guarantee and opacity–violate problems are an open question.

However, in our attempt to develop efficient algorithms for the opacity-guarantee problem, we somehow rely on Theorem 3 and promote a top-down approach in the graph $\widehat{A}_{v_0}$. This approach should compete with the straightforward intractable bottom-up method to solve alternating reachability in $\widehat{A}_{v_0}$, that leads to the EXPTIME algorithm.

4   Towards a Search-Based Algorithm

In this section we present the idea of an algorithm that, given a game with opacity-condition $A = (V, \Delta, \mathrm{obs}, \mathrm{act})$ over Σ and Γ, with v_0 as initial position, decides the existence of a winning strategy for Gerald and returns one if it exists.

The algorithm is based on a search approach in the graph of the perfect-information game $\widehat{A}_{v_0}$ from Theorem 3. We distinguish between nodes in which it is SuperRoberta's turn to play and those in which it is Geraldine's. The first ones correspond to positions of the form (I, v) in $\widehat{A}_{v_0}$, the second ones to positions of the form (I, v, a). Since we want the computed strategy to be winning whatever SuperRoberta does, we have to provide a solution in all sons of SuperRoberta's nodes, entailing an AND-node interpretation of SuperRoberta's nodes. Dually in a Geraldine's node, it is sufficient to provide a solution for one of its sons to have a winning strategy, hence the OR-node interpretation of Geraldine's nodes. General search algorithms with heuristics on AND-OR graphs have already been studied [7,8], but our setting is more involved. The halting condition of the search is subtle because we consider safety conditions in graphs that may contain cycles.

Halting conditions: There are only three ways to stop the exploration of a branch. The current node is:

- A losing position, thus this branch is cut.
- An OR-node (a Geraldine position) for which a safe strategy has already been found.
- An OR-node whose associated position is also associated to an ancestor.

The third point needs some justification. Assume we find an OR-node n' with an ancestor n both associated to position (I, v, a). Two cases can be distinguished.

- The choice made at node n is not part of a winning strategy. If we expand the node n', we have to be coherent with the strategy currently being constructed, thus the subtree rooted at n' is the same as the one rooted at n. It implies that the choice made at n can be proved wrong without expanding n'.
- The choice made at node n is part of a winning strategy. In this case n' doesn't need to be expanded neither since a solution has already been defined for the corresponding position, and exploring the rest of the subtree rooted at n will prove this choice correct.

Pruning: In this section we describe how we prune some branches during the search.

In an OR–node n, before expanding a son n' associated to position (I', v'), we check a sufficient condition for n' to be a position from which there is no winning strategy for Geraldine. This condition is that there exists a sequence of actions $a_1 \ldots a_n$ that, if played by SuperRoberta from n', will lead to a losing position whatever Geraldine does. This can be rephrased as a generalized D1–directing word problem in the non-deterministic automaton $\mathcal{A}_A = (V, \Sigma, \Delta')$, where transitions are added in order to obtain a complete automaton:

$$\Delta'(v, a) = \begin{cases} \Delta(v, a) & \text{if non–empty,} \\ \{\bot\} & \text{else.} \end{cases}$$

The problem becomes: does there exist a directing word w to a singleton different from $\{\bot\}$? Depth-first search techniques seem appropriate, and due to efficiency purposes, we may limit the length of the directing word by some parameter k_1.

Heuristics: In OR-nodes, we use heuristics to order the expansion of unpruned sons. To compute the values assigned to these sons, we seek synchronizing words of minimal length in a deterministic automaton that, unlike $\mathcal{A}_A$, does not abstract Geraldine's moves. A synchronizing word w of length at most k_2 (a parameter) in this automaton reveals a winning play for SuperRoberta. The heuristics is that the longer the minimal synchronizing word, the more chances to avoid the singleton position. We can use breadth-first search techniques to compute minimal length directing words, no longer than k_2.

5 Conclusion and Perspectives

We have defined and studied in detail games with opacity condition, which address theoretical questions related to security aspects of computer systems. In order to bypass the intractable powerset–based procedure, we have proposed to exploit synchronizing words techniques from automata theory as heuristics for a top–down search algorithm.

We are currently developing this algorithm, with the pruning condition. Also, the proposed heuristics arises from an intuitive argument that deserves being validated in practice (by tuning parameters k_1 and k_2), and next theoretically justified.

Acknowledgement. We are very grateful to Dietmar Berwanger for initial discussions on this topic.

References

1. Berwanger, D., Doyen, L.: On the power of imperfect information. In: Hariharan, R., Mukund, M., Vinay, V. (eds.) IARCS Annual Conference on Foundations of Software Technology and Theoretical Computer Science, Dagstuhl, Germany (2008), Schloss Dagstuhl - Leibniz-Zentrum fuer Informatik
2. Černý, J.: Poznámka k. homogénnym experimentom s konecnými automatmi. Mat. fyz. čas SAV 14, 208–215 (1964)
3. Černý, J., Pirická, A., Rosenauerova, B.: On directable automata. Kybernetica 7, 289–298 (1971)
4. Chatterjee, K., Henzinger, T.A.: Semiperfect-information games. In: Sarukkai, S., Sen, S. (eds.) FSTTCS 2005. LNCS, vol. 3821, pp. 1–18. Springer, Heidelberg (2005)
5. Dubreil, J., Darondeau, P., Marchand, H.: Opacity enforcing control synthesis. In: Workshop on Discrete Event Systems, Gothenburg, Sweden (March 2008)
6. Imreh, Steinby: Directable nondeterministic automata. ACTACYB: Acta Cybernetica 14 (1999)
7. Kumar, V., Nau, D.S.: A general branch-and-bound formulation for and/or graph and game tree search. In: Search in Artificial Intelligence, pp. 91–130 (1988)
8. Mahanti, A., Bagchi, A.: AND/OR graph heuristic search methods. Journal of the ACM (JACM) 32(1), 28–51 (1985)
9. Martin, D.: Borel determinacy. Annales of Mathematics 102, 363–371 (1975)
10. Pin, J.-E.: Le problème de la synchronisation et la conjecture de černý. In: De luca, A. (ed.) Non-commutative structures in algebra and geometric combinatorics, CNR, Roma. Quaderni de la Ricerca Scientifica, vol. 109, pp. 37–48 (1981)
11. Pin, J.-E.: On two combinatorial problems arising from automata theory. Annals of Discrete Mathematics 17, 535–548 (1983)
12. Reif: The complexity of two-player games of incomplete information. JCSS: Journal of Computer and System Sciences 29 (1984)
13. Saboori, A., Hadjicostis, C.N.: Opacity-enforcing supervisory strategies for secure discrete event systems. In: IEEE Conference on Decision and Control (CDC), Cancun Mexico (December 2008)
14. Savitch, W.J.: Relationships between nondeterministic and deterministic tape complexities. J. Comput. System. Sci. 4, 177–192 (1970)

Abstract Counterexamples for Non-disjunctive Abstractions

Kenneth L. McMillan[1] and Lenore D. Zuck[2,*]

[1] Cadence Research Labs
[2] University of Illinois at Chicago

Abstract. Counterexample-guided abstraction refinement (CEGAR) is an important method for tuning abstractions to properties to be verified. The method is commonly used, for example in selecting predicates for predicate abstraction. To date, however, it has been applied primarily to powerset abstractions, which allow one to speak of an abstract transition system and abstract states. Here, we describe a general framework for CEGAR in non-disjunctive abstractions by introducing a generalized notion of abstract counterexample, and methods for computing such counterexamples. We apply this framework to Indexed Predicate Abstraction (IPA), a promising technique for synthesizing quantified inductive invariants of infinite-state systems. In principle, it can be applied to other non-disjunctive abstractions occurring in program analysis.

1 Introduction

Effective application of abstract interpretation depends on choosing the right abstract domain. This domain must be rich enough to contain an inductive invariant that proves a given property, but not so rich as to make analysis intractable. One very fruitful approach to choosing abstractions has been abstraction refinement. That is, when our abstract domain fails to prove a given property, we analyze this failure, producing a refined abstract domain that rules out some class of failures. This process repeats until either the property is proved, or analysis reveals that the property is false, or computational resources are exhausted. A particularly successful form of abstraction refinement is *counterexample-guided abstraction refinement*, or CEGAR [10,2]. In this approach, when the abstract domain fails to prove the property, we produce an *abstract counterexample*. This is a sequence of abstract states in which every transition is allowed by the abstract transformer, and the property is violated. An abstract state is, in effect, an atom of the abstract lattice. We refine the abstract domain so as to rule out the abstract counterexample. Abstract counterexamples both focus and simplify the refinement process, since they allow us to consider a limited class of behaviors.

---

* This material was based on work supported by the National Science Foundation, while Lenore Zuck was working at the Foundation. Any opinion, finding, and conclusions or recommendations expressed in this article are those of the author and do not necessarily reflect the views of the National Science Foundation.

O. Bournez and I. Potapov (Eds.): RP 2009, LNCS 5797, pp. 176–188, 2009.

CEGAR has been applied effectively to a variety of domains, including localization abstractions [10] and predicate abstraction [18]. Its use is limited, however, by the fact that it applies only to abstract domains that are *disjunctive* (*i.e.*, closed under union). It is this condition that allows us to construct abstract counterexamples. For example, CEGAR cannot be applied directly to indexed predicate abstraction [11] (IPA) because this abstraction is not disjunctive.

In this work, we generalize the notion of abstract counterexample to a construct we call a *minimal sufficient explanation* (MSE). An MSE is a sequence of elements of the abstract domain that may not be atoms. In the case of an abstract lattice that is atomistic and disjunctive, however, it reduces to the standard notion of abstract counterexample. An MSE may be used to focus abstraction refinement in much the same way as an abstract counterexample, for example, using the interpolation approach [7]. Our primary motivation in this work is to be able to effectively refine indexed predicate abstractions, and we will use this method as an example application.

Related Work. Existing work on refinement of non-disjunctive abstractions is not based on abstract counterexamples. Typically, the weakest liberal precondition operator is iterated. This allows us to find the first point in the abstract fixed point series in which lost information resulted in inclusion of a bad concrete state (one reaching a state violating the property). The abstraction is refined at this point. For example, Gulavani and Rajamani do this by eliminating widenings at specific points in the fixed point series [6].

In the terminology of this paper, the sequence of (negations of) the weakest preconditions of the property is a *sufficient explanation* for the failure to prove the property. However, it is not *minimal* with respect to the given abstract domain, nor is it generally even expressible in that domain. Using a *minimal* sufficient explanation allows us to focus on a restricted set of concrete behaviors. In this way, we hope to gain both efficiency and better focus on relevant refinements, as in CEGAR. Moreover, this avoids having to deal with the series of weakest preconditions, which may have deeply nested quantifiers in the case of programs with input or non-deterministic choice.

Since one of the goals of this work is to produce quantified inductive invariants, we mention some other work in this area. Lahiri presents a collection of heuristics based on the weakest precondition operator for guessing indexed predicates [11], but leaves open the question of how to apply CEGAR. Henzinger, *et al.*, use interpolants for predicate refinement [7], but without index variables. The method of *invisible invariants* [16] can effectively synthesize quantified invariants, but only for families of finite-state systems. IPA can also handle infinite-state systems.

Outline. The paper is organized as follows. Section 2 introduces the notion of MSE, or generalized abstract counterexample. Section 3 then reviews the method of indexed predicate abstraction, and shows how to compute MSE's for this application. In section 4, we show how indexed predicates can be derived from interpolants, and how, in principle, MSE's can be used to drive this process in a CEGAR loop.

2   Generalized Abstract Counterexamples

In this section, we generalize the concept of abstract counterexample. The idea is to view an abstract counterexample not as a run of an abstract transition system, but rather as a minimal sufficient explanation of the failure of the abstraction to prove a given property. We will see that in the case of powerset abstractions, these two notions coincide.

Abstract Interpretation. First we review some concepts from abstract interpretation [3]. Consider a concrete transition system with set of states S, initial states $I \subseteq S$, and transition relation $T \subseteq S \times S$. We can define a concrete transformer $\tau(s) = I \cup T(s)$, where $T(s)$ is the image of s with respect to T. The least fixed point of τ is the set of concrete reachable states of the system.

An *abstraction* of the system is defined by an abstract lattice L and a monotone concretization function $\gamma : L \rightarrow S$. The abstract lattice is ordered by $\sqsubseteq$, with least upper bound operator $\sqcup$ and greatest lower bound operator $\sqcap$, usually referred to as "join" and "meet" respectively. If we think of L as a logical language, then γ defines the semantics of the language, with $\gamma(p)$ giving the extension of predicate $p \in L$.

We will assume that L is finite and intersection-closed, that is, for any $p, q \in L$, there exists $r \in L$ such that $\gamma(r) = \gamma(p) \cap \gamma(q)$. In this case, γ is the upper adjoint of a Galois connection, whose lower adjoint is:

$$\alpha(s) = \sqcap\{p \mid s \subseteq \gamma(p)\}$$

The abstraction function α gives the best abstract approximation of a set of states s, which can be thought of as the conjunction of all the predicates in L that are valid over s.

This in turn gives us a *best abstract transformer*, $\tau^\sharp = \alpha \circ \tau \circ \gamma$. For any predicate $p \in L$, this function yields the best abstract approximation of the set of successors of states in p. The fixed points of $\tau^\sharp$ are all the inductive invariants in L, and the least fixed-point is the strongest of these. Thus, to prove that a given set $F \subseteq S$ is unreachable, we compute the least fixed point of $\tau^\sharp$, as the stable limit of the series $(\tau^\sharp)^i(\bot)$. Then, if $\mathrm{lfp}(\tau^\sharp) \sqcap \alpha(F) = \bot$ we say the abstraction proves F unreachable. On the other hand, if $(\tau^\sharp)^i(\bot) \sqcap \alpha(F) \neq \bot$ for any $i > 0$, then the abstraction fails to prove unreachability of F.

Explanation of Failures. What then would constitute a minimal sufficient explanation for such a failure? Consider first the case of a single transition. Given two predicates $p, q \in L$, we will say that p is a *minimal sufficient precondition* (MSP) of q when $\tau^\sharp(p) \sqsupseteq q$ and there is no $\dot{p} \sqsubset p$ such that $\tau^\sharp(\dot{p}) \sqsupseteq q$. That is, p is a minimal element of the abstract lattice sufficient to guarantee at least q at the next time. Put another way, p is an explanation of why the abstraction produced q at the next time.

Now we extend this notion to a reachability computation. We will say that a sequence $x_0, \ldots x_k \in L^*$ is a *minimal sufficient explanation* (MSE) for failure to prove unreachability of F, when it is pointwise minimal such that:

- $x_0 = \bot$ and
- for all $0 \leq i < k$, $\tau^\sharp(x_i) \sqsupseteq x_{i+1}$, and
- $x_k \sqcap \alpha(F) \neq \bot$

That is, each element of the sequence is a MSP of its successor, and the last element fails to rule out F.

The notion of MSE corresponds precisely to the notion of "abstract counterexample" in the traditional CEGAR framework. This framework applies only to *powerset abstractions*. This means that γ is disjunctive (join-preserving) in the sense that $\gamma(p \sqcup q) = \gamma(p) \cup \gamma(q)$. Moreover, it requires that L be *atomistic*, in that every element is the join of some set of atoms (elements that cover $\bot$). For example, in predicate abstraction, the join operation is logical disjunction (*i.e.*, union over sets of states) and the atoms are the minterms over the abstraction predicates P (a minterm over P is a conjunction of literals over P in which each predicate in P occurs once).

In this case, we can think of the atoms of L as "states" of an abstract transition system. That is, because of the disjunctive join and atomicity, $\tau^\sharp$ is point-wise over atoms:

$$\tau^\sharp(p) = \sqcup\{\tau^\sharp(a) \mid a \in \text{atoms}(p)\} \sqcup \tau^\sharp(\bot)$$

It follows that an MSP of any atom is an atom or $\bot$. That is, if $\tau^\sharp(p) \sqsupseteq q$ and $p \neq \bot$, then p contains some atom a such that $\tau^\sharp(a) \sqsupseteq q$. Thus, in a disjunctive, atomistic abstraction such as predicate abstraction, MSE's contain only atoms, and we can think of them as sequences of abstract "states".

This is not true in the general case, however. For example, indexed predicate abstraction is atomistic but not disjunctive. As a result, an MSE is a sequence of *sets* of atoms. One way to view the occurrence of multiple atoms at some point in the MSE is that the abstract interpretation has lost information in merging multiple execution paths. Thus, no one path is sufficient to "explain" the successor state.

Computing Generalized Counterexamples. Now we consider the problem of computing an MSE for failure to prove unreachability of F. Suppose that we have computed a sequence of fixed point approximations $x_i = (\tau^\sharp)^i(\bot)$ for $i = 0 \ldots k$ such that $x_k \sqcap \alpha(F) \neq \bot$. There is a simple but inefficient backward approach to computing an MSE. We start by setting x_k to any atom in $x_k \sqcap \alpha(F)$. Then for $i = k - 1$ down to 1, we greedily reduce x_i in the lattice order so long as $\tau^\sharp(x_i) \sqsupseteq x_{i+1}$. If L is atomistic, this means greedily removing atoms from x_i. When x_i cannot be further reduced, it is an MSP for x_{i+1}, and we move on to x_{i-1}. At the end of this process, we have an MSE for the failure.

This simple approach could be computationally costly, because the height of the abstract lattice is typically exponential in some parameter of the abstraction (for example, in predicate abstraction it is exponential in the number of abstraction predicates). Thus, the number of reduction steps can also be exponential. To avoid this, we need some way of putting an upper on the MSP so that the number of reduction steps necessary to reach the MSP is also bounded. We will show how to do this in some special cases of practical interest.

Our basic problem is to compute a MSP for predicate q that is dominated by some predicate p. For any monotone transformer τ, we will write the set of sufficient preconditions of q dominated by p as:

$$\mathrm{SP}(p,\tau,q) = \{\hat{p} \mid \hat{p} \sqsubseteq p \text{ and } \tau(\hat{p}) \sqsupseteq q\}$$

The set of minima of this set will be denoted $\mathrm{MSP}(p,\tau,q)$. Our general approach will be to compute a SP, then iteratively remove atoms until it becomes a MSP.

To do this, we can rely on several useful properties of SP. First, SP is join-preserving, that is, if $\hat{p}_i \in \mathrm{SP}(p,\tau,q_i)$ then $\sqcup_i \hat{p}_i \in SP(p,\tau,\sqcup_i q_i)$. This is due simply to monotonicity of τ. It means that to compute a SP for q, we can simply take the join of SP's for the individual atoms of q. As we observed above, when τ is join-preserving, we need only consider SP's for atoms that are atoms. In addition, we can make use of the following results:

Theorem 1. *If τ is meet-preserving, then $SP(p,\tau,q)$ is closed under meets. Moreover, if $SP(p,\tau,q)$ is non-empty, the unique element of $MSP(p,\tau,q)$ is $\sqcap SP(p,\tau,q)$.*

Theorem 2 (Meet rule). *Let $\tau(s) = \sqcap_i \tau_i(s)$, and suppose $\hat{p}_i \in MSP(p,\tau_i,q)$. Then $\sqcup \hat{p}_i \in MSP(p,\tau,q)$.*

Theorem 3 (Chain rule). *If $\tau = \tau_2 \circ \tau_1$ and $\hat{p} \in MSP(p,\tau,r)$, then there exists $\hat{q}$ such that $\hat{p} \in MSP(p,\tau_1,\hat{q})$ and $\hat{q} \in MSP(\tau_1(p),\tau_2,r)$.*

The chain rule allows us to compute MSP's for a composition of transformers by working backward. We can think of the MSE computation as being one long application of this rule.

3   Indexed Predicate Abstraction

We now apply the notion of MSE to the problem of abstraction refinement for indexed predicate abstraction. We will apply IPA to transition systems represented symbolically using first-order logic. We break the abstract transformer for IPA into a composition of a join-preserving and a meet-preserving transformer. Then we use the chain rule and the meet rule to compute MSP's. By this means, we use a number of decision procedure calls which is quadratic in the final number of atoms in the MSE.

Symbolic Transition Systems. Let Σ be a first-order signature consisting of individual variables and uninterpreted n-ary functional and propositional constants. A *state formula* is a first-order formula over Σ, (which may include various interpreted symbols, such as $=$ and $+$). We can think of a state formula ϕ as representing a set of states, namely, the set of first-order models of ϕ. We will express the proposition that an interpretation σ over Σ models ϕ by $\phi[\sigma]$, or $\sigma \models \phi$. If s is a set of interpretations, we will write $s \models \phi$ to mean that every element of s models ϕ.

We also assume a first-order signature Σ', disjoint from Σ, and containing for every symbol $v \in \Sigma$, a unique symbol v' of the same type. For any formula or term ϕ over Σ, we write ϕ' for the result of replacing every occurrence of a symbol v in ϕ with v'. Similarly, for any interpretation σ over Σ, we will denote by σ' the interpretation over Σ' such that $\sigma'v' = \sigma v$. A *transition formula* is a first-order formula over $\Sigma \cup \Sigma'$. We think of a transition formula T as representing a set of state pairs, namely the set of pairs (σ_1, σ_2), such that $\sigma_1 \cup \sigma_2'$ models T. We will express the proposition that $\sigma_1 \cup \sigma_2'$ models T by $T[\sigma_1, \sigma_2]$.

A *symbolic transition system* is a pair (I, T), where I is a state formula and T is a transition formula. We interpret this as a transition system whose initial states are represented by I and whose transition relation is represented by T.

Indexed Predicate Abstraction. Indexed predicate abstraction [11] is similar to predicate abstraction, except that the predicates contain free variables that are implicitly universally quantified. We start with a distinguished set $J = \{i, j, k, \ldots\}$ of individual variables called the *index variables*, not occurring in I or T, and a finite set P of atomic formulas (possibly containing index variables). Our abstract lattice is the lattice L_P of Boolean combinations over P. In this lattice, the atoms are the minterms over P, which we can think of as either truth assignments to P, or conjunctions of literals over P. To avoid confusion between an *atom* of the abstract lattice and an *atomic* predicate, from here on we will refer to the lattice atoms as *minterms*.

Each element in L_P is a set of minterms. The lattice order $\sqsubseteq$ is set inclusion, and the meet and join are intersection and union, respectively. Alternately, we can think of meet and join as propositional conjunction and disjunction, and the lattice order as propositional implication (*i.e.*, where the propositions in P are uninterpreted). The concretization function is defined by:

$$\gamma(p) = \{\sigma \mid \sigma \models \forall J.\ p\}$$

That is, p represents the set of concrete states that model p for all valuations of the index variables. Because universal quantification distributes over conjunction, γ is meet-preserving. Thus the corresponding abstraction function is, by definition:

$$\alpha(s) = \sqcap\{p \mid s \models \forall J.\ p\}$$

which is equivalent to:

$$\alpha(s) = \{m \in \mathrm{minterms}(P) \mid \sigma \models \exists J.\ m \text{ for some } \sigma \in s\}$$

This identity allows us to write the best abstract transformer as

$$\tau^\sharp(p) = \{m \in \mathrm{minterms}(P) \mid \sigma \models (\exists J.\ m) \text{ for some } \sigma \in \tau(\gamma(p))\}$$
$$= \{m \in \mathrm{minterms}(P) \mid (\forall J.\ p) \wedge T \wedge m' \text{ is sat.}\} \sqcup \alpha(I)$$

That is, computing $\tau^\sharp(p)$ amounts to deciding $2^{|P|}$ satisfiability problems in first-order logic, one for each minterm over P. However, since first-order logic is undecidable, we make a further over-approximation by heuristically choosing a finite set of instantiations for the quantifiers.[1]

A *substitution* ρ for a set W of individual variables is a function that maps each variable in W to a first-order term. We will write $\rho(\phi)$ for the application of substitution ρ to formula ϕ (*i.e.*, the simultaneous replacement of each occurrence of variable $w \in W$ by $\rho(w)$). Given a set of substitutions $\mathcal{I}$, we write $\mathcal{I}(\phi)$ to denote $\bigwedge\{\rho(\phi) \mid \rho \in \mathcal{I}\}$. Note that if $\mathcal{I}$ is a set of substitutions for J, then $\forall J : \phi$ implies $\mathcal{I}(\phi)$.

Now we choose a finite set of substitutions $\mathcal{I}$ for the index variables and a suitable instantiation $\dot{T}$ of transition formula T. We might use, for example, the quantifier instantiation heuristics used in provers such as Simplify [5] for this purpose. The problem of instantiation is inherent in IPA, and not specifically related to abstraction refinement. The incompleteness of instantiation heuristics results in over-approximation. We can express the over-approximation of the best transformer as:

$$\dot{\tau}^\sharp(p) = \{m \in \mathrm{minterms}(P) \mid \mathcal{I}(p) \wedge \dot{T} \wedge m' \text{ is sat.}\} \sqcup \alpha(I)$$

After instantiation, the satisfiability problems are quantifier-free, which means we can use an appropriate decision procedure, or reduce the problem to Boolean satisfiability. In [11], ALL-SAT methods are used to efficiently compute all the satisfying minterms. Even with these methods, $\dot{\tau}^\sharp$ may still be costly to compute in practice, so a weaker approximation may be called for.

Abstract Counterexamples for IPA. Notice that indexed predicate abstraction is not disjunctive. This is because universal quantification does not distribute over disjunction. In general, if p and q are two predicates in the abstract lattice, the concretization of their disjunction $\forall J.(p \vee q)$ is not equivalent to the disjunction of their concretizations $(\forall J.\ p) \vee (\forall J.\ q)$. As an example of this, suppose that a system chooses arbitrarily two process indices x and y, and transitions to a particular control state s when process x is in state p, but process y is not in state p. If we start from the minterm $p(i) \wedge \neg s$, then clearly we cannot transition to state s, since this represents $\forall i :\ p(i) \wedge \neg s$, states in which all processes are in state p. Similarly, if we start from the minterm $\neg p(i) \wedge \neg s$, we also cannot transition to s. However, if we start from disjunction $(p(i) \wedge \neg s) \vee (\neg p(i) \wedge \neg s)$, then we *can* transition to s. The abstract transformer is not point-wise over minterms, in the sense that a pair of minterms can have a successor that neither individual minterm has. For this reason, IPA does not yield abstract counterexamples.

However, we *can* compute MSE's and use these as an aid in refining the abstraction. We first observe that the computation of $\dot{\tau}^\sharp(p)$ can be broken into a sequence of two transformers: the instantiation of the index variables, followed by a forward image operation. The first transformer is function $\eta : L_P \to L_{\dot{P}}$, where $\dot{P} = \{\rho(\phi) \mid \phi \in P, \rho \in \mathcal{I}\}$, such that $\eta(p) = \mathcal{I}(p)$.

---

[1] We could, of course, restrict ourselves to a decidable fragment, such as Presburger arithmetic, but this would not allow us to model, for example, parametrized protocols, or programs with unbounded arrays.

As an example, suppose that $J = \{i\}$, $P = \{s, p(i)\}$ and $\mathcal{I} = \{\rho_1, \rho_2\}$, where $\rho_1(i) = x$ and $\rho_2(i) = y$. Then $\dot{P} = \{s, p(x), p(y)\}$, and $\eta(p(i) \wedge \neg s) = p(x) \wedge p(y) \wedge \neg s$.

The second transformer is function $\delta : L_{\dot{P}} \to L_P$ that computes the predicate image with respect to $\dot{T}$. That is, let

$$\delta(p) = \{q \in \text{minterms}(P) \mid p \wedge \dot{T} \wedge q' \text{ is sat.}\}$$

This gives us $\dot{\tau}^{\sharp}(p) = \delta(\eta(p)) \sqcup \alpha(I)$. We show how to compute MSP's for δ and η, then combine these steps using the chain rule to compute MSP's for $\dot{\tau}^{\sharp}$.

Since δ is pointwise, the MSP's for a minterm q with respect to δ are also minterms. We can write the set of MSP's of minterm q as:

$$\text{MSP}(p, \delta, q) = \{m \in \text{minterms}(\dot{P}) \mid m \sqsubseteq p \text{ and } \delta(m) \sqsupseteq q\}$$
$$= \{m \in \text{minterms}(\dot{P}) \mid m \wedge p \wedge \dot{T} \wedge q' \text{ is sat.}\}$$

Thus, finding one MSP for a minterm is a satisfiability problem. Because SP is join-preserving, we can compute a SP for any predicate q as the join of MSP's for the minterms of q. This SP can then be reduced to a MSP. Testing whether one minterm can be removed from the SP requires $|q|$ satisfiability tests (the number of minterms in q). Thus, in the worst case, the number of tests needed to compute an MSP is quadratic in $|q|$.

The transformer η is a meet over a finite set of transformers, that is, the individual substitutions:

$$\eta(p) = \sqcap\{\rho(p) \mid \rho \in \mathcal{I}\}$$

Thus, we can apply the meet rule, computing a MSP of η as a join over MPS's of the individual substitutions. Moreover, since substitutions are meet preserving, their MSP's are unique. Given a substitution ρ and a minterm $m \in L_{\dot{P}}$, there is a unique minterm $n \in L_P$ such that $\rho(n) \sqsupseteq m$. This is the one minterm such that for every literal l occurring in n, $\rho(l)$ occurs in m. Put another way, if we think of a minterm over P as a truth assignment to the predicates in P, then for all predicates $\phi \in P$, $n(\phi) = m(\rho(\phi))$.

Continuing the previous example, suppose that m is the minterm $s \wedge p(x) \wedge \neg p(y)$. Then the unique MSP of m with respect to ρ_1 is the minterm $s \wedge p(i)$. This is because m contains $\rho_1(s) = s$ and $\rho_1(p(i)) = p(x)$. In general, if m is a minterm in $L_{\dot{P}}$, we have:

$$\text{MSP}(p, \eta, m) = \{\lambda\phi.\ m(\rho(\phi))\}$$

By the meet rule and the join-preserving property, the unique MSP for a predicate q with respect to η is:

$$\text{MSP}(p, \eta, q) = \{\sqcup \cup \{\text{MSP}(p, \rho, m) \mid \rho \in \mathcal{I},\ m \in \text{minterms}(q)\}\}$$

Continuing our previous example, if $m = s \wedge p(x) \wedge \neg p(y)$ and $p = \top$, then $\text{MSP}(p, \eta, m) = (s \wedge p(i)) \vee (s \wedge \neg p(i)) = s$. The first disjunct derives from ρ_1 and

Algorithm 1
Input: A pair $p, q \in L_P$
Output: An MSP $\hat{p}$ of q, such that $\hat{p} \sqsubseteq p$
   1) Let $M = \emptyset$
   2) For each minterm m in $q \setminus \alpha(I)$:
   3)   let $\hat{m}$ be a minterm over $\dot{P}$ s.t. $\mathcal{I}(p) \wedge \hat{m} \wedge \dot{T} \wedge m'$ is sat.
   4)   add $\hat{m}$ to M
   5) Greedily remove minterms from M, while $\delta(M) \sqsupseteq q$
   6) For each minterm $m_i \in M$:
   7)   let $n_i = \sqcup\{\{\lambda\phi.\ m_i(\rho(\phi)\} \mid \rho \in \mathcal{I}\}$
   8) Return $\sqcup_i n_i$

Fig. 1. MSP computation for IPA

the second from ρ_2. This is a case where no single minterm serves as an MSP for a minterm.

With this result, we can now compute an MSP for the composition $\delta \circ \eta$ using the chain rule. To find an element of $\text{MSP}(p, \delta \circ \eta, r)$, we first compute $q = \eta(p)$. Then let $\hat{q}$ be an element of $\text{MSP}(q, \delta, r)$, and finally find an element of $\text{MSP}(p, \eta, \hat{q})$. The resulting algorithm for computing a MSP in indexed predicate abstraction is shown in Figure 1.

In lines 1–4, we compute a SP of q with respect to δ, restricted to $\eta(p)$. At line 5, this is reduced to an MSP. In lines 6–8, we then compute the unique MSP with respect to η.

Notice that the number of satisfiability tests at line 3 is just $|q|$, the number of minterms in q. Each test of $\delta(M) \sqsupseteq q$ at line 5 could cost $|q|$ satisfiability tests, making the total number of tests quadratic in $|q|$. Alternatively, the test of $\delta(M) \sqsupseteq q$ might be done with a BDD image computation. The total number of decision problems we encounter is quadratic in the largest element of the MSE and linear in its length. Of course, this does not mean the number of minterms in the MSE cannot be exponential in $|P|$. For example, the number of minterms might double at each backward step.

4   Indexed Predicates from Interpolants

Interpolation has been used to derive relevant predicates for ordinary predicate abstraction from the refutation of counterexamples [7,8]. This is one possible approach to counterexample-guided abstraction refinement. In this section, we extend this CEGAR technique to indexed predicate abstraction, using the notion of MSE introduced above. In effect, we make use of MSE's as constraints to simplify and focus the interpolant computation process.

Bounded Model Checking. For any symbol s, and natural number i, we will use the notation $s^{\langle i \rangle}$ to represent the symbol s with i primes added. Thus, $s^{\langle 3 \rangle}$ is s'''. A symbol with i primes will be used to represent the value of that symbol at time i. We also extend this notation to formulas. Thus, the formula $\phi^{\langle i \rangle}$ is the result of adding i primes to every uninterpreted symbol in ϕ.

Now, given a system (I, T), we define a symbolic transformer formula $\mathcal{T} = T \vee I'$. This is defined so that the image of a set of states s with respect to $\mathcal{T}$ is exactly $\tau(s)$. The following formula is satisfiable exactly when $\tau^k(\bot) \cap F \neq \emptyset$, that is, when a state in F is reachable in $k - 1$ steps or fewer from a state in I:

$$I^{\langle 1 \rangle} \wedge \mathcal{T}^{\langle 1 \rangle} \wedge \cdots \mathcal{T}^{\langle k-1 \rangle} \wedge F^{\langle k \rangle}$$

We will refer to this as a *bounded model checking* formula [1], since by testing satisfiability of such formulas, we can determine the reachability of a given condition within a bounded number of steps.

Interpolants from Proofs. Given a pair of formulas (A, B), such that $A \wedge B$ is inconsistent, an *interpolant* for (A, B) is a formula $\hat{A}$ with the following properties:

- A implies $\hat{A}$,
- $\hat{A} \wedge B$ is unsatisfiable, and
- $\hat{A}$ refers only to the common symbols of A and B.

Here, "symbols" excludes symbols such as $\wedge$ and $=$ that are part of the logic itself. Craig showed that for first-order formulas, an interpolant always exists for inconsistent formulas [4]. Of more practical interest is that, for certain proof systems, an interpolant can be derived from a refutation of $A \wedge B$ in linear time. For example, a purely propositional refutation of $A \wedge B$ using the resolution rule can be translated to an interpolant in the form of a Boolean circuit having the same structure as the proof [9,17].

In [13] it is shown that linear-size interpolants can be derived from refutations in a first-order theory with uninterpreted function symbols and linear arithmetic. This translation has the property that whenever A and B are quantifier-free, the derived interpolant $\hat{A}$ is also quantifier-free. In [14], a method is described for computing universally quantified interpolants in first-order logic with equality, when such interpolants exist. In the sequel, we will assume that interpolants are universally quantified, and that the quantified variables are always drawn from J, the index set.

Heuristically, the chief advantage of interpolants derived from refutations is that they capture the facts that the prover derived about A in showing that A is inconsistent with B. Thus, if the prover tends to ignore irrelevant facts and focus on relevant ones, we can think of interpolation as a way of filtering out irrelevant information from A. Thus, atomic predicates occurring in the interpolant may be considered "relevant" to the refutation.

We can generalize the notion of interpolant to sequences of formulas. That is, given a sequence of formulas $\Gamma = \Gamma_1, \ldots, \Gamma_n$, we say that $A_0, \ldots A_n$ is an *interpolant* for Γ when

- $A_0 = \text{TRUE}$ and $A_n = \text{FALSE}$ and,
- for all $1 \leq i \leq n$, $A_{i-1} \wedge \Gamma_i$ implies A_i and

– for all $1 \leq i < n$, A_i refers only to common symbols between the prefix $\Gamma_1 \ldots \Gamma_i$ and the suffix $\Gamma_{i+1} \ldots \Gamma_n$.

An interpolant for a sequence can also be derived from a refutation of its constituent formulas.

We can use this concept to derive new indexed predicates sufficient to rule out a given abstract counterexample. In what follows, we will use a subscript to indicate the predicate set used to obtain a given quantity. Thus $\tau_P^\#$ is the abstract transformer obtained with predicate set P and so on. Now, suppose that using indexed predicate abstraction with predicates P, we obtain an MSE $X_P = x_0, \ldots, x_n$. The concretizations of X_P are all the sequences of concrete states $s_1, \ldots, s_n$ such that each s_i models $\gamma(x_i)$. If any such concretization is actually a failing run of (I, T), then the property is false. Otherwise, we would like to refine P by adding a set of predicates β sufficient to rule out X_P.

To this end, let Γ, the concretization sequence, be the following sequence of formulas:

$$I^{<1>}, (\mathcal{T} \wedge \gamma(x_1))^{<1>}, (\mathcal{T} \wedge \gamma(x_2))^{<2>}, \ldots, (\mathcal{T} \wedge \gamma(x_{n-1}))^{<n-1>}, (F \wedge x_n)^{<n>}$$

The conjunction of these formulas, $\wedge\Gamma$ is just a bounded model checking formula for (I, T) with the added constraint $\gamma(x_i)$ at each time i. The models of this conjunction are precisely the concretizations of X_P that are failing runs of (I, T). If we can prove $\wedge\Gamma$ unsatisfiable, then X_P is a false counterexample. We can then extract from the proof an interpolant as a sequence of quantified formulas for the form $\text{TRUE}, A_1^{<1>}, \ldots, A_n^{<n>}, \text{FALSE}$. We assume these are universal formulas, such that each $A_i = \forall J.\phi_{A_i}$. We now let β be the set of atomic predicates occurring in ϕ_{A_i} for any i. These are the predicates we will add to P in the next iteration of the refinement loop.

We can show that these predicates rule out future abstract counterexamples consistent with X_P. This is because the interpolant properties guarantee that $\tau_\beta^\#(A_i \wedge x_i) \Rightarrow A_{i+1}$. Thus, if X_P were a sufficient explanation for $\tau_\beta^\#$, then by induction we could show $x_i \Rightarrow A_i$, which implies that the interpolant sequence cannot end in FALSE. This also implies that we cannot have $\beta \subseteq P$, since in this case X_p must also be a sufficient explanation for $\tau_\beta^\#$. Thus, the refinement step is guaranteed to add at least one new predicate. The practical function of the abstract counterexample is to act as a constraint on the bounded model checking problem, thus helping to make the refutation tractable.

We should note that using the method of [14] to generate interpolants has some limitations. The logic supported in that work is first order logic with equality. If other theories are needed, such as the theory of arrays or arithmetic, these must be axiomatized. This is necessarily incomplete for theories that have no finite axiomatization and can also be inefficient. In [14], however, it is shown that interpolation can successfully find invariants for simple heap manipulating programs, using simple arithmetic and an array theory with reachability predicates. Thus, there is some reason to think it may be effective for the more restricted task of abstraction refinement. The method described here can benefit from any future advances in interpolation methods. Moreover, the use of MSE's in abstraction refinement is not limited to interpolation methods, or for that matter to IPA. It might be applied to a variety of program analyses with non-disjunctive joins.

5   Conclusion

We have defined a notion of abstract counterexample for non-disjunctive abstract domains, called a minimal sufficient explanation. The notion of MSE reduces to the traditional notion of abstract counterexample for powerset abstractions. We showed how to compute MSE's for a particular example of a non-disjunctive abstraction, indexed predicate abstraction. The purpose of an abstract counterexample in the CEGAR framework is to simplify and focus the abstraction refinement process. We showed how this could be done with MSE's, using an interpolant-based refinement approach. In particular, we saw that universally quantified interpolants can provide the indexed predicates needed to rule out a given MSE, thus guaranteeing refinement progress.

The hope is that, by restricting the analysis to a smaller set of concrete behaviors, the MSE approach may significantly lessen the burden on the first-order prover used for interpolant generation. Of course, it remains to be seen whether in practice the use of abstract counterexamples in IPA is more efficient than the alternatives, such as proof-based abstraction [15] or similar approaches based on weakest preconditions [6].

We also think it is possible that MSE's can be applied to refinement of other non-disjunctive abstractions, such as the partially-disjunctive shape abstractions of [12].

References

1. Biere, A., Cimatti, A., Clarke, E.M., Zhu, Y.: Symbolic model checking without BDDs. In: Cleaveland, W.R. (ed.) TACAS 1999. LNCS, vol. 1579, pp. 193–207. Springer, Heidelberg (1999)
2. Clarke, E.M., Grumberg, O., Jha, S., Lu, Y., Veith, H.: Counterexample-guided abstraction refinement. In: CAV, pp. 154–169 (2000)
3. Cousot, P., Cousot, R.: Abstract interpretation: A unified lattice model for static analysis of programs by construction of approximation of fixed points. In: POPL, pp. 238–252 (1977)
4. Craig, W.: Three uses of the Herbrand-Gentzen theorem in relating model theory and proof theory. J. Symbolic Logic 22(3), 269–285 (1957)
5. Detlefs, D., Nelson, G., Saxe, J.B.: Simplify: A theorem prover for program checking. Technical Report HPL-2003-148, HP Labs (2003)
6. Gulavani, B.S., Rajamani, S.K.: Counterexample driven refinement for abstract interpretation. In: Hermanns, H., Palsberg, J. (eds.) TACAS 2006. LNCS, vol. 3920, pp. 474–488. Springer, Heidelberg (2006)
7. Henzinger, T.A., Jhala, R., Majumdar, R., McMillan, K.L.: Abstractions from proofs. In: POPL, pp. 232–244 (2004)
8. Jhala, R., McMillan, K.L.: A practical and complete approach to predicate refinement. In: Hermanns, H., Palsberg, J. (eds.) TACAS 2006. LNCS, vol. 3920, pp. 459–473. Springer, Heidelberg (2006)

9. Krajíček, J.: Interpolation theorems, lower bounds for proof systems, and independence results for bounded arithmetic. J. Symbolic Logic 62(2), 457–486 (1997)

10. Kurshan, R.P.: Computer-Aided-Verification of Coordinating Processes. Princeton University Press, Princeton (1994)

11. Lahiri, S.K., Bryant, R.E.: Predicate abstraction with indexed predicates. ACM Trans. Comput. Log. 9(1) (2007)

12. Manevich, R., Sagiv, S., Ramalingam, G., Field, J.: Partially disjunctive heap abstraction. In: Giacobazzi, R. (ed.) SAS 2004. LNCS, vol. 3148, pp. 265–279. Springer, Heidelberg (2004)

13. McMillan, K.L.: An interpolating theorem prover. Theor. Comput. Sci. 345(1), 101–121 (2005)

14. McMillan, K.L.: Quantified invariant generation using an interpolating saturation prover. In: Ramakrishnan, C.R., Rehof, J. (eds.) TACAS 2008. LNCS, vol. 4963, pp. 413–427. Springer, Heidelberg (2008)

15. McMillan, K.L., Amla, N.: Automatic abstraction without counterexamples. In: Garavel, H., Hatcliff, J. (eds.) TACAS 2003. LNCS, vol. 2619, pp. 2–17. Springer, Heidelberg (2003)

16. Pnueli, A., Ruah, S., Zuck, L.D.: Automatic deductive verification with invisible invariants. In: Margaria, T., Yi, W. (eds.) TACAS 2001. LNCS, vol. 2031, pp. 82–97. Springer, Heidelberg (2001)

17. Pudlák, P.: Lower bounds for resolution and cutting plane proofs and monotone computations. J. Symbolic Logic 62(2), 981–998 (1997)

18. Saïdi, H., Graf, S.: Construction of abstract state graphs with PVS. In: CAV, pp. 72–83 (1997)

Cross-Checking - Enhanced Over-Approximation of the Reachable Global State Space of Component-Based Systems

Mila Majster-Cederbaum and Christoph Minnameier*

Institut für Informatik
Universität Mannheim, Germany
cmm@informatik.uni-mannheim.de

Abstract. State space explosion causes most relevant behavioral questions for component-based systems to be PSPACE-hard. Here, we exploit the structure of component-based systems to obtain a first approximation of the reachable global state space. In order to improve this approximation we introduce a new technique we call cross-checking. The resulting approximation can be used to study global properties of component-based systems, which we demonstrate here for local deadlock-freedom.

1 Introduction

In this paper we deal with reachability in the global state space of a component-based system with n components. We present a technique that builds on the analyses of certain subsystems generated by $d << n$ components, where d is fixed. We explain our approach using the model of *interaction systems* introduced in [GS03] by Gössler and Sifakis as a model for component-based systems. As typical for component-based systems, the description of interaction systems strictly separates the description of the components from the way they are put together, i.e. the *glue code*. I/O-Automata [LT89] and interface automata [dAH01], e.g. can be considered as a subclass of interaction systems, for the latter feature a more general notion of communication. More details about interaction systems and their properties can be found in [Sif04, Sif05, GGM+07b, GGM+07a, GS05, BBS06, MMM07a]. A framework for component-based modeling using interaction systems has been implemented in [BBS06, Goe06]. Please note that the ideas presented in this paper do not rely heavily on the model but can be transferred to other models as long as cooperation of systems forms the top level of system description.

For interaction systems the size of the global state space of a component based system may be exponential in the number n of components and it has been shown that deciding most important behavioral properties is PSPACE-complete [MM08b]. There are various ways to deal with this problem, e.g. partial order reduction or abstraction. Another approach is to establish conditions

---

* Corresponding author.

O. Bournez and I. Potapov (Eds.): RP 2009, LNCS 5797, pp. 189–202, 2009.
© Springer-Verlag Berlin Heidelberg 2009

that use compositionality and can be tested in polynomial time to ensure the desired properties, see e.g. [AC05, MMM07a, MMM07b]. Moreover one may impose architectural constraints concerning the communication structure of the component system [AG97, BCD02, MM08a].

In this paper we first exploit the structure of the component system to obtain (in polynomial time) a first over-approximation of the global state space. For this, we consider subsystems built from a fixed number d of components (which can be considered a parameter) and perform reachability analyses there. Restricting our view to sets of subsystems can be considered a form of locality-based abstraction. Different related techniques have been studied in [BPR01, ASCN99, Kov]. A general and abstract treatment of locality-based abstraction can be found in [EGS05].

The contribution of this paper basically consists of the following two steps. First the straight-forwardly computed subsystem approximations are enhanced by a technique called *cross-checking*. Second, the resulting approximations can be used to derive conditions on the subsystems that guarantee global properties.

We demonstrate this for local deadlock-freedom. Deadlock-freedom is an important property in itself and moreover, establishing safety properties can be reduced to establishing deadlock-freedom.

The paper is structured as follows. Section 2 presents the model and an example that will be used throughout the paper. In Section 3 we explain the general approach of investigating subsytems in order to prove properties on the reachable global state space. Section 4 introduces and analyzes cross-checking. As an application we establish in Section 5 a polynomial time checkable condition for deadlock-freedom that is tested in subsystems and makes use of our approximation. Section 6 discusses related work. Section 7 depicts our (partially still in progress) implementations. Finally, we give a short conclusion in Section 8.

2   Interaction Systems

We consider here interaction systems, a model for component-based systems that was proposed and discussed in [GS03].

2.1   Syntax and Semantics

Definition 1. *Interaction Systems*
*An **interaction system** is a tuple $Sys = (K, \{A_i\}_{i \in K}, Int, \{T_i\}_{i \in K})$, where K is the set of **components**. W.l.o.g. we assume $K = \{1, \dots, n\}$. Each component $i \in K$ offers a finite set of **ports** (resp. **actions**) A_i for cooperation with other components. The **port sets** A_i are pairwise disjoint. Cooperation is described by the **interaction set** Int. Each component i is provided with a **local behavior** T_i.*

*An **interaction** is a finite, non-empty set of actions $\alpha \subseteq \bigcup_{i \in K} A_i$. An interaction $\alpha = \{a_{i_1}, \dots, a_{i_k}\}$ with $a_{i_j} \in A_{i_j}$ describes that the components $i_1, \dots, i_j$ cooperate via these ports. Interactions α are subject to the constraint that for each component i at most one action $a_i \in A_i$ is in α. An **interaction set** Int* *is a finite set of interactions, s.t. every action of every component occurs in at least one interaction of Int.*

The **local behavior** of each component i is described by a labeled transition system $T_i = (Q_i, A_i, \rightarrow_i, q_i^0)$, where Q_i is the finite set of local states, $\rightarrow_i \subseteq Q_i \times A_i \times Q_i$ is the local transition relation and $q_i^0 \in Q_i$ is the local starting state.

Definition 2. *Participation and Enabled Actions*
Given an interaction $\alpha \in Int$ and a component $i \in K$ we denote by $i(\alpha) := A_i \cap \alpha$ the **participation** of i in α.
For $q_i \in Q_i$ we define the set of **enabled actions** $ea(q_i) := \{a_i \in A_i \mid \exists q_i' \in Q_i,$ s.t. $q_i \xrightarrow{a_i}_i q_i'\}$. We assume that the T_i's are non-terminating, i.e. $\forall i \in K \; \forall q_i \in Q_i \; ea(q_i) \neq \emptyset$.

Definition 3. *Semantics*
The **global behavior** $T_{Sys} = (Q, Int, \rightarrow_{Sys}, q^0)$ of Sys (henceforth also referred to as global transition system) is obtained from the behaviors of the individual components, given by the transition systems T_i, and the interaction set Int in a straightforward manner:

- $Q = \prod_{i \in K} Q_i$, the Cartesian product of the Q_i, which we consider to be order independent. We denote states by tuples $(q_1, \ldots, q_n)$ and call them **global states**.
- The relation $\rightarrow_{Sys} \subseteq Q \times Int \times Q$, defined by
  $\forall \alpha \in Int \; \forall q, q' \in Q \quad q = (q_1, \ldots, q_n) \xrightarrow{\alpha}_{Sys} q' = (q_1', \ldots, q_n')$ iff
  $\forall i \in K \quad (q_i \xrightarrow{i(\alpha)}_i q_i'$ if $i(\alpha) \neq \emptyset$ and $q_i' = q_i$ otherwise$)$.
- $q^0 = (q_1^0, \ldots, q_n^0)$ is the global starting state for Sys.

Less formally, a transition labeled by α may take place in the global transition system when each component i participating in α is ready to perform its part $i(\alpha)$.

Example 1. In the following we consider an interaction system that models Tanenbaum's solution [Tan08] to Dijkstra's Dining Philosophers problem. Tanenbaum suggests that each of the philosophers is provided with a separate semaphore that she has to set in order to leave her thinking state. A semaphore however can only be set if its "neighbour" semaphores are unset. Once a philosopher has eaten, she puts back the forks and resets her semaphore. This can be considered an elegant solution as it is symmetric and allows for maximum efficiency (meaning that it still allows for a global state where every second philosopher is in her eating state). On the other hand this is a deadlock-free system with a natural interaction structure whose reachable global state space is exponential in p. This solution can be modeled as an interaction system as follows, where p is the number of philosophers:
$DP(p) = (K(p), \{A_i\}_{i \in K(p)}, Int(p), \{T_i\}_{i \in K(p)})$, where
$K(p) = \{Phil_0, \ldots, Phil_{p-1}, Fork_0, \ldots, Fork_{p-1}, Sem_0, \ldots, Sem_{p-1},\}$,
$Int(p) = \bigcup_{0 \leq i \leq p-1} \{\{pickleft_{Phil_i}, occupy_{Fork_i}\}, \{pickright_{Phil_i}, occupy_{Fork_{i-1}}\},$
$\{priority_{Phil_i}, down_{Sem_i}, allow_{Sem_{i-1}}, allow_{Sem_{i+1}}\},$
$\{drop_{Phil_i}, up_{Sem_i}, vacate_{Fork_{i-1}}, vacate_{Fork_i}\}\},$

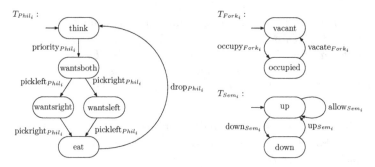

Fig. 1. Tanenbaum's Dining Philosophers - Local Transition Systems

where calculation is modulo p, and the local behaviors T_i and (implicitly the) port sets A_i are given in Figure 1.

2.2   Reachability

For most properties of interaction systems we must determine which states in Q are reachable from the global starting state. Here we propose to first investigate reachability in subsystems which is defined as follows.

Definition 4. $Reach(Sys) := \{q \in Q \mid q^0 \rightarrow^*_{Sys} q\}$, where $\rightarrow^*_{Sys}$ denotes the reflexive and transitive closure of $\rightarrow_{Sys}$.

Definition 5. *Substates*
Let $K' \subseteq K$ and q be a global state. Then $q \downarrow K'$ denotes the projection of q to the components in K' and we call $q' = q \downarrow K'$ a **substate**. We refer to the components K' that occur in q' by $K(q')$. We also use the $\downarrow$-operator to denote projections of substates. Finally, let $Q_{K'} = \prod_{i \in K'} Q_i$ and $Subs(K) = \bigcup_{K' \subseteq K} Q_{K'}$.

Definition 6. *Subsystems*
Let $K' \subseteq K$. The **subsystem** $Sys_{K'}$ is given by $(K', \{A_i\}_{i \in K'}, Int_{K'}, \{T_i\}_{i \in K'})$, where $Int_{K'} := \{\alpha_{K'} = \alpha \cap (\bigcup_{i \in K'} A_i) \mid \alpha \in Int\} \backslash \{\emptyset\}$. Note that $Sys_{K'}$ accords to our definition of an interaction system, so all definitions for interaction systems apply.

Definition 7. *Extensions*
Let q' be a substate. Then $Ext(q', K')$ for $K' \subseteq K$ denotes the set of **extensions** of q' in K' and is defined by $Ext(q', K') = q' \times \prod_{i \in K' \backslash K(q')} Q_i$. If $K' = K(q')$ let $Ext(q', K') = \{q'\}$. We say that a substate $\hat{q}'$ is an extension of a substate q' if $K(q') \subseteq K(\hat{q}')$ and $\hat{q}' \downarrow K(q') = q'$.

The definition of a subsystem implies that if a state q is reachable in the global transition system, then for every $K' \subseteq K$ the state $q \downarrow K'$ is reachable in the corresponding subsystem. We formalize this observation in the following lemma.
Lemma 1. Let $Sys = (K, \{A_i\}_{i \in K}, Int, \{T_i\}_{i \in K})$
$q \in Reach(Sys) \Rightarrow \forall K' \subseteq K, (q \downarrow K') \in Reach(Sys_{K'})$.

Corollary 1. *Let* $f(Reach(Sys_{K'})) := \bigcup_{q' \in Reach(Sys_{K'})} Ext(q', K)$. *Then*
$Reach(Sys) \subseteq \bigcap_{K' \subseteq K, \text{ with } |K'|=d} f(Reach(Sys_{K'}))$

Remark 1. Each $Reach(Sys_{K'})$ is a compact representation of $f(Reach(Sys_{K'}))$ which is in turn a very coarse over-approxmation of $Reach(Sys)$.

3 Proving Properties on Overapproximations

Let us assume we want to check a property P on each state of the reachable global state space of Sys. In a first approach we might proceed as follows.

- We choose a parameter $d << n$ and calculate the reachable states for each subsystem with d components. Each reachable substate $q' = (q_{i_1}, \ldots, q_{i_d})$ is a compact representation of $Ext(q', K)$.
- We formulate a predicate P' such that the validity of P on a global state q is implied by the validity of P' on the projections of q,
  $[\forall K' \subseteq K, |K'| = d \ P'(q \downarrow K')] \Rightarrow P(q)$, hence
  $[\forall K' \subseteq K, |K'|=d \ \forall q \in Reach(Sys) \ P'(q \downarrow K')] \Rightarrow P(Reach(Sys))$

Clearly, we do not want to handle explicitly global states at all. Instead we propose to use the implication
$[\forall K' \subseteq K, |K'|=d \forall q' \in Reach(Sys_{K'}) \ P'(q')] \Rightarrow P(Reach(Sys))$

The advantage of this approach is immense: Instead of a complexity that is exponential in $|K| = n$, we have a complexity that is polynomial (with degree d) in n and m. This is because for $K' \subseteq K$ with $|K'| = d$, $Reach(Sys_{K'})$ can be computed in $O(m^d)$, where $m = \max_{i \in K'} |Q_i|$ thus we may compute resp. store the reachable state spaces of all subsystems with d components in time resp. space $O(\binom{n}{d} \cdot m^d)$.

Example 2. For the dining philosophers example with $p = 6$ (i.e. $|K| = 18$) and $d = 4$ the sum of the sizes of the investigated substate spaces is 229.095 compared to 64.000.000 global states in the original system. Obviously the advantage is much greater for a larger parameter p.

However, there is an obvious drawback to our present approach:

Considering subsystems with $d << n$ components neglects much information. Indeed there will be many reachable substates that do not originate from a projection of a global reachable state but are "artefacts". If we check condition P' on many such artefacts we run the risk that P' is violated and we can not conclude P.

Example 3. For the dining philosophers example with $p = 6$ (i.e. $|K| = 18$) and $d = 4$, only 43212 of the 229.095 states in the state spaces of the subsystems are unreachable. This corresponds to 18,85%.

4   Cross-Checking

In this section, we introduce cross-checking as a technique to eliminate artefacts. In a first step, we consider the unreachable states.

Lemma 2. *Let* $\overline{Reach}(Sys_{K'}) = Q_{K'} \setminus Reach(Sys_{K'})$ *be the set of states that can not be reached in* $Sys_{K'}$, $Refuted = \bigcup_{K'' \subseteq K, |K''|=d} f(\overline{Reach}(Sys_{K''}))$ *and* $X_{K'} := f(Reach(Sys_{K'})) \setminus Refuted$.
Then i) $Refuted \subseteq \overline{Reach}(Sys)$ *and*
ii) $Reach(Sys) \subseteq X_{K'}$

As we noted before, the sets $Reach(Sys_{K'})$ may contain many artefacts. We want to use $Refuted$ to reduce the number of artefacts. However we can not use Lemma 2 directly as it involves the evaluation of the function f. Therefore we define $Ref(Sys_{K'}) := \{q' \in Subs(Q_{K'}) \mid Ext(q', K') \cap Reach(Sys_{K'}) = \emptyset\}$ and compute (in polynomial time) a $Reach'(Sys_{K'})$ with $f(Reach(Sys_{K'})) \supseteq f(Reach'(Sys_{K'})) \supseteq X_{K'}$ as follows.

Theorem 1. *Let*
$Reach'(Sys_{K'}) := Reach(Sys_{K'}) \setminus Ext(\bigcup_{K'' \subseteq K, |K''|=d} Ref(Sys_{K''}) \cap Subs(K'), K')$.
Then $X_{K'} \subseteq f(Reach'(Sys_{K'}))$.

By *cross-checking* we refer to the computation of the various sets $Reach'(Sys_{K'})$ by removing the elements in $Ext(\bigcup_{K'' \subseteq K, |K''|=d} Ref(Sys_{K''}) \cap Subs(K'), K')$ from $Reach(Sys_{K'})$ as described in Algorithm 1. For reasons of efficiency Algorithm 1 represents each set $Reach(Sys_{K'})$ by an array $reach(Sys_{K'})$ (that we refer to as "table") of booleans for all states in $Q_{K'}$. In such a table we have "$reach(Sys_{K'})[q'] = true$" iff $q' \in Reach(Sys_{K'})$. Also for reasons of efficieny Algorithm 1 does not loop over the various sets $Reach(Sys_{K'})$ and the therein reachable substates but rather over all states in $\{q'' \in Subs(K) \mid |q''| < d\}$. For a state q'' we decide (by looking up the reachability flags of its extensions in the various tables $reach(Sys_{K'})$) whether it belongs to $\bigcup_{K' \subseteq K, |K'|=d} Ref(Sys_{K'})$. If this is the case, we set all reachability flags of all extensions of q'' to false. If this is not the case, we add q'' to an initially empty list "list-of-possible-substates" that will be needed in Section 5.1.

Example 4. Let $K_1' = \{Phil_1, Phil_2, Fork_1, Fork_2\}$. In $Sys_{K_1'}$ we are able (by performing the connectors $\{priority_1\}$ and $\{priority_2\}$) to reach the substate $q' = (priority_{Phil_1}, priority_{Phil_2}, vacant_{Fork_1}, vacant_{Fork_2})$. However if we consider the substate $q'' = (priority_{Phil_1}, priority_{Phil_2})$ of q' and its occurrence in the subsystem that is implied by $K_2' = \{Phil_1, Phil_2, Sem_1, Sem_2\}$ we learn that no extension of q'' is in $Reach(Sys_{K''})$. Thus $q'' \in Ref(Sys_K'') \cap Subs(K')$, so we have $q' \notin Reach'(Sys_{K'})$.

After the first application of cross-checking for the subsystem reachabilites, we will have marked 147561 of the 229095 substates unreachable. This corresponds to 64,41%.

Lemma 3. *The sets* $Reach'(Sys_{K'})$ *for all subsystems* $Sys_{K'}$ *with* d *components can be computed in an overall amount of time that is in* $O(d \cdot n^d \cdot m^d)$.

Proof: In the following we present Algorithm 1 *Cross-Checking* that computes the sets $Reach'(Sys_{K'})$ within the specified time bounds.

Remark 2. Apart from the factor d (which can be considered a constant), our cross-checking algorithm remains within the asymptotic time bounds already given by the first step of performing the reachability analyses of the subsystems. We consider this to be an **important** property, as any refinement approach that attempts to increase the number d of considered components would instead result in a complexity in $\Omega(n^{d+1})$.

Remark 3. Note that Algorithm 1 *may* be applied iteratively to the result of the previous application thus further reducing the number of states that are marked reachable until we reach a fixpoint. It is an open question how many iterations will be needed.

Algorithm 1 Cross-Checking

1: PROCEDURE Cross-Checking
2: **for** $x := 1$ to $(d-1)$ **do**
3:       **for all** subsets $K'' = \{i_1, \ldots, i_x\}$ of K **do**
4:             **for all** $q'' = (q_{i_1}, \ldots, q_{i_x}) \in Q_{K''}$ **do**
5:                   reachable := true;
6:                   **for all** subsystems $Sys_{K'}$ with $K'' \subseteq K'$ (and $|K'| = d$) **do**
7:                         occurrence := false;
8:                         **for all** $q' \in Ext(q'', K')$ **do**
9:                               occurrence := occurrence OR $reach(Sys_{K'})[q']$;
10:                        **end for**
11:                        reachable := reachable AND occurrence;
12:                  **end for**
13:                  **if** reachable = false **then**
14:                        **for all** subsystems $Sys_{K'}$ with $K'' \subseteq K'$ (and $|K'| = d$) **do**
15:                              **for all** $q' \in Ext(q'', K')$ **do**
16:                                    $reach(Sys_{K'})[q')] :=$ false;
17:                              **end for**
18:                        **end for**
19:                  **else** add q'' to *list-of-possible-substates*;
20:                  **end if**
21:            **end for**
22:      **end for**
23: **end for**
24: END Cross-Checking

---

5   Detecting Deadlocks

Deadlock-freedom is an important property in itself and in addition establishing safety properties can be reduced to establishing deadlock-freedom. In this section, we present a definition of some locally checkable predicate P' that implies (in the sense that was described in Section 3) local deadlock-freedom for interaction systems.

Definition 8. *(Minimal) local Deadlock*
Given an interaction system Sys and a global state q we say that a set of com-
*ponents $D \subseteq K$ is a **local deadlock** in q if every interaction in which any of*
the components in D could (in its present local state) participate is blocked by
another component in D. More formally:
$$\forall i \in D \, \forall \alpha \in Int : (ea(q_i) \cap \alpha \neq \emptyset) \Rightarrow (\exists j \in D \, j(\alpha) \nsubseteq ea(q_j)).$$
Obviously if we reach a global state q such that some set $D \subseteq K$ is a local dead-
lock in q no component in D can ever again participate in any interaction.
$D \subseteq K$ is a minimal local deadlock in q if no proper subset of D is a local dead-
*lock in q. A system Sys is **locally deadlock-free** if no state q is reachable such*
that there is a local deadlock in q.

Example 5. Let us consider a global state q, where
$q \downarrow \{Phil_1, Fork_1, Phil_2, Fork_2, Phil_3\} =$
$(wantsleft_{Phil_1}, occupied_{Fork_1}, wantsboth_{Phil_2}, occupied_{Fork_2}, wantsright_{Phil_3})$. In
this case, $D = \{Phil_1, Fork_1, Phil_2, Fork_2, Phil_3\}$ is a minimal local deadlock
in q. (However, no such q is reachable in any of the systems $DP(p)$.)

Definition 9. *Small vs. Large Deadlocks*
When we compute the subsystem reachabilities as described in Section 3 we
choose a value for the parameter d. Henceforth we will call local deadlocks D
*with $|D| \leq d$ **small** local deadlocks and local deadlocks D with $|D| > d$ **large***
local deadlocks.

In order to prove for a system *Sys* the predicate $P = $ "Local Deadlock-Freedom"
it is sufficient to prove for some fixed $d << n$ that there are neither small nor
minimal large local deadlocks reachable in *Sys*. When we traverse the reachable
substates in the various $Reach'(Sys_{K'})$ we will be able to identify deadlocks of
size $|D| \leq d$ directly, whereas the existence of deadlocks of size $|D| > d$ will have
to be excluded by a sufficient condition. In the following subsections, we describe
a locally checkable (i.e. checkable in the subsystems) P' that - when true on all
substates in all $Reach'(Sys_{K'})$ - ensures the validity of P.

5.1   Defining and Checking a Condition for Small Deadlocks

To deal with the question of small local deadlocks it is sufficient to prove that
there are no deadlocks of size $\leq d$ in the substates that are marked reachable
in the reachability tables of the investigated subsystems. Again, it is infeasible
to check all 2^d subsets of every substate of every subsystem, because this would
yield up to $2^d \cdot \binom{n}{d} \cdot m^d$ loop cycles in the first place. Instead, we will directly check
the substates of size 1 to $d-1$ that have been added to *list-of-possible-substates*
in Algorithm 1.

5.2   Defining a Condition for Large Deadlocks

Obviously, we can not directly identify a large local deadlock in a subsystem
with d components. Instead we are going to check a condition which is sufficient

for deadlock-freedom. In order to formulate this condition we first introduce the following relation between the local states of the components' transition systems.

Definition 10. "q_i waits for q_j", $(q_i \in Q_i, q_j \in Q_j)$
We say q_i waits for q_j if $\exists \alpha \in Int$, s.t. $\alpha \cap ea(q_i) \neq \emptyset \wedge j(\alpha) \not\subseteq ea(q_j)$.

I.e. q_i waits for q_j if j might prevent i from participating in an interaction in a corresponding global state q. If components prevent each other from participating in interactions then they might be in deadlock. The following definition assigns to a global state q (resp. a substate q') a directed graph based on the relation introduced in Definition 10.

Definition 11. *Wait-for-graph*
For a system Sys and a global state q we define the wait-for-graph $WFG(q) = (V, E)$ by:
$V = \{q_i \mid 1 \leq i \leq n\}$ and $E = \{(q_i, q_j) \in V \times V \mid q_i$ waits for $q_j\}$.
For $K' \subseteq K$ and a corresponding substate $q' = q \downarrow K'$ we denote by $WFG(q')$ the subgraph of $WFG(q)$ generated by $V' = \{q_i \in V \mid i \in K'\}$.

Given a large deadlock $D \subseteq K$ in a reachable global state q we will be able to detect (in at least one subsystem) the following pattern.

Theorem 2. *If Sys has a large minimal deadlock D in a global state q, then there is a subset $K' \subseteq D$ with $|K'| = d$ and a linear order $(i_1, \ldots, i_d)$ of the components in K' such that*
$k < l \Rightarrow q_{i_l}$ *is reachable from* q_{i_k} *in* $WFG(q \downarrow K')$.

Example 6. When we apply P' (for $DP(6)$ with $d = 4$) to the reachable state spaces $Reach(Sys_{K'})$ that we computed in the first place, we will detect 1584 (of reachable 185883) substates for which P' is not valid.

Applying P' (for $DP(6)$ with $d = 4$) to the reachable state spaces $Reach'(Sys_{K'})$ that we gain after applying cross-checking, the number of substates for which P' is not valid decreases to 432 (of reachable 81534).

These numbers induce that among the substates whose reachability was refuted via cross-checking there are indeed critical ones. Even more the percentage of reachable substates that are critical has decreased. This is due to a tendency in our approach to leave uncritical substates marked reachable.

5.3   Complexity of Checking Our Condition for Large Deadlocks

According to Section 3 and Theorem 2 we may prove that a system Sys does not contain any reachable minimal large deadlocks by proving that for neither of the subsystems $Sys_{K'}$ (with $K' \subseteq K$ and $|K'| = d$) and their substates q' in $Reach'(Sys_{K'})$ there is a linear order as described above for the nodes in $WFG(q')$.

To do so, we first construct, for every subsystem with d components and every therein reachable substate q', the graph $WFG(q')$. Then we could apply the following procedure **Order** which finds a linear order as described in Theorem 2, if there is any.

Preprocessing: Constructing the wait-for-relations
As a preprocessing we can (in $O((n \cdot m)^2)$) compute a $(n \cdot m \times n \cdot m)$-matrix W with $W(q_i, q_j) = 1$ if q_i waits for q_j, $W(q_i, q_j) = 0$ otherwise. This matrix includes for every substate q' the information about $WFG(q')$. Thus for a substate q', we simply create the $d \times d$-matrix for $WFG(q')$ and fill it by copying the relevant information from our matrix W. This can be done in $O(d^2)$.

Procedure Order:
Perform breadth-first search for every local state in $WFG(q')$. If there is one state q_j from which all other states can be reached make $i_1 := j$. Now find a state from which all remaining (not yet ordered) states are reachable and so on. Whenever such a state cannot be found, abort. Return the order when all components are ordered.

Proof of Correctness:
It is obvious that if the Procedure Order is not aborted then a returned order suffices our requirements. We show that if there is a linear order as described in Theorem 2, then Procedure Order will find one.

Note that if there is a linear order of the components in K' as described in Theorem 2 then this also holds for every subset of K' (w.r.t. the graph $WFG(q')$). This means in every step of Procedure Order we can choose the next component for the linear order and it is always guaranteed that the linear order so far can be enhanced (by a linear order of the remaining components) to a correct linear order for all d components.

Actual Implementation and Complexity:
The description of Procedure Order above will not be implemented directly but rather acts as a makeshift for ease of understanding and for our proof of correctness. For our implementation, we first compute the transitive closure of $WFG(q')$. This is possible in $O(d^3)$. Thus, instead of performing up to d breadth-first-searches we can simply determine the next component in our linear order (d times) by examining for each of the d components, if the remaining (not yet ordered) components are reachable from it (a comparison which can be done in $O(d)$ using the transitive closure).

So the Procedure Order can be performed in an overall time in $O(d^3)$.

The overall complexity of our check for large deadlocks is thus bounded by $O(m^d \cdot n^d \cdot d^3)$.

5.4   Connected Subsystems

Definition 12. *Interaction Graph & Connected Systems*
*For an interaction system Sys we define the **interaction graph** $IG = (V, E)$ by: $V = K$ and $E = \{\{i, j\} \mid \exists \alpha \in Int \, (\alpha \cap A_i \neq \emptyset \wedge \alpha \cap A_j \neq \emptyset)\}$*
*We call an interaction system **connected** if its interaction graph is connected.*

Note that for the purpose of deadlock detection we may restrict our attention to connected systems *Sys* (as for an unconnected system it is equivalent to prove its interaction graph's connected components deadlock-free). However if the original interaction system is connected, we may restrict all our observations (i.e. reachability analyses, cross-checking and the checks for small resp. large deadlocks) - without loss of correctness or even loss of information - to connected subsystems.

Example 7. For Tanenbaum's dining philosphers as modelled here, the maximum degree of a node in the interaction graph is 9. This makes it easy to derive that the maximum number of *connected* subsystems is bounded by $n \cdot 9^{d-1} = O(n)$ for a fixed choice of d.

6   Related Work

Many important approaches have been developed in the past to tackle the problem of state space explosion. A wide spectrum of methods for approximation and/or reduction of the state space, ranging from partial order reduction, exploiting equivalences to abstraction/refinement techniques have been investigated and e.g. incorporated in model checking tools and abstract interpretation approaches. Our approach to establish properties of component based systems is in a certain sense complementary but can nevertheless be put into comparison with some existing techniques. The basic principles of our approach can be summarized as follows.

1) We exploit the knowledge about the interaction structure of the system, i.e. the interaction graph. For this we determine the connected subgraphs with d nodes in this graph ($d << n$ a constant).
2) Then we calculate the reachable state spaces for the subsystems belonging to these subgraphs.
3) We apply cross-checking to delete "artefacts" within these subsystems.
4) We establish a condition that is to be checked on the subsystems and when satisfied guarantees a global property.
5) All steps can be performed in polynomial time (bounded by a polynomial with degree d).

Step 2) can be seen as a locality based abstraction in the sense of [EGS05] which is the most general paper on locality-based abstraction we know of. When we consider a subsystem with d components then this corresponds to an observer that has access to these d components of the system. (If each observer has access to exactly one component, the special case of Cartesian abstraction [BPR01, Arn94] arises.) However a closer look to the notion of partial transition relation in locality based abstractions of [EGS05] and hence the notion of local reachability, shows that our approach has to be distinguished from theirs. In [EGS05] one condition for a partial state p_1 to evolve to state p_2, i.e. $\underline{t}(p_1, p_2)$, is that p_1 matches some state p with respect to the kernel of the transition t.(Please note that each transition relation t of [EGS05] corresponds to an interaction α of our model.)

In one of our local systems given by the components $\{i_1, ...i_d\}$ a local transition can take place in a sub-state if all partners of an interaction α that are part of the subsystem offer their part of the interaction, i.e. we then perform $\alpha \cap \bigcup_{1 \leq i \leq d} A_{i_d}$.

As we may obtain artefacts by proceeding in such a way, called loss of information in [EGS05], we then apply cross-checking in step 3) to eliminate as many artefacts as possible by comparing the subsystems. To the best of our knowledge this technique has not been applied in the context of state space investigations. Cartesian abstraction [BPR01, Arn94] which involves finding the smallest Cartesian product that contains a given set has a similar purpose for the case $d = 1$. Cross-checking is however closely related to techniques employed in the relational model of data bases. Speaking in data base terminology we decompose an initial data base scheme into sub-schemes (corresponding to our subsystems). Applying the join operation $\bowtie$ to all these subsystems, i.e. $Reach(Sys_{K'_1}) \bowtie Reach(Sys_{K'_2}) \bowtie \ldots \bowtie Reach(Sys_{K'_k})$ (where $k = \binom{n}{d}$) would yield the best over-approximation one can get when using locality based abstraction if no further knowledge is available. However calculating these joins leaves us with the same complexity issue as calculating the reachable global state space. So we avoid evaluating this sequence of joins and perform instead the comparison of pairs of subsystems.

Concerning step 4), the conditions on subsystems that guarantee global properties, the closest work to ours is by [AC05] on deadlock-freedom who base their work on concurrent programs but employ a similar notion of subsystems and local transition relation. They consider subsystems of size 3 but apply no technique comparable no cross checking.

7 Implementation

All presented techniques have been implemented in our tool "PrInSESSA" [MS08] where in addition we also apply a variant of cross-checking to the detection of minimal large deadlocks.

In order to allow for a quantitative comparison with other tools, a BDD-based framework is presently being implemented. First benchmarks show that the BDD-based variant can prove instances as large as $DP(500)$ deadlock-free within minutes.

8 Conclusion and Further Work

We presented a method to obtain an enhanced over-approximation of the global state space of a component-based system with n components in polynomial time. The method consists of choosing a value for d, investigating subsystems consisting of d components in a first step (in $O(n^d \cdot m^d)$) and then improving this approximation by cross-checking (in $O(d \cdot n^d \cdot m^d)$). The computation of the first step can be improved in various respects. Firstly, for the purpose of proving deadlock-freedom we do not have to consider all $\binom{n}{d}$ subsystems but only such that are connected. Secondly, the computation of the various sets $Reach(Sys_{K'})$ can be performed in parallel. Our approximation can be used to investigate global properties by considering subsystems and checking conditions on them which requires only polynomial cost. We showed how this can be achieved for the property of local deadlock-freedom. Interesting open theoretical questions are e.g. how many iterations are needed at most until the iterative application of cross-checking reaches a fixpoint and in which complexity class the exact computation of the set $X_{K'}$ defined in Section 4 lies.

References

[AC05]   Attie, P., Chockler, H.: Efficiently Verifiable Conditions for Deadlock-Freedom of Large Concurrent Programs. In: Cousot, R. (ed.) VMCAI 2005. LNCS, vol. 3385, pp. 465–481. Springer, Heidelberg (2005)

[AG97]   Allen, R., Garlan, D.: A Formal Basis for Architectural Connection. ACM Trans. Softw. Eng. Methodol. 6(3), 213–249 (1997)

[Arn94]   Arnold, A.: Finite transition systems: semantics of communicating systems (Translator-John Plaice). Prentice Hall International (UK) Ltd., Hertfordshire (1994), Translator-John Plaice

[ASCN99]   George, A.S., Clarke, L.A., Naumovich, G.: Data flow analysis for checking properties of concurrent java programs. In: ICSE 1999: Proceedings of the 21st international conference on Software engineering, pp. 399–410. ACM, New York (1999)

[BBS06]   Basu, A., Bozga, M., Sifakis, J.: Modeling Heterogeneous Real-time Components in BIP. In: Proceedings of SEFM 2006, pp. 3–12. IEEE Computer Society, Los Alamitos (2006)

[BCD02]   Bernardo, M., Ciancarini, P., Donatiello, L.: Architecting Families of Software Systems with Process Algebras. ACM Trans. on Software Engineering and Methodology 11, 386–426 (2002)

[BPR01]   Ball, T., Podelski, A., Rajamani, S.K.: Boolean and cartesian abstraction for model checking c programs. In: Margaria, T., Yi, W. (eds.) TACAS 2001. LNCS, vol. 2031, pp. 268–283. Springer, Heidelberg (2001)

[dAH01]   de Alfaro, L., Henzinger, T.: Interface Automata. In: FSE 2001, pp. 109–120 (2001)

[EGS05]   Esparza, J., Ganty, P., Schwoon, S.: Locality-based abstractions. In: Hankin, C., Siveroni, I. (eds.) SAS 2005. LNCS, vol. 3672, pp. 118–134. Springer, Heidelberg (2005)

[GGM+07a] Goessler, G., Graf, S., Majster-Cederbaum, M., Martens, M., Sifakis, J.: An Approach to Modelling and Verification of Component Based Systems. In: van Leeuwen, J., Italiano, G.F., van der Hoek, W., Meinel, C., Sack, H., Plášil, F. (eds.) SOFSEM 2007. LNCS, vol. 4362, pp. 295–308. Springer, Heidelberg (2007)

[GGM+07b] Goessler, G., Graf, S., Majster-Cederbaum, M., Martens, M., Sifakis, J.: Ensuring Properties of Interaction Systems by Construction. In: Reps, T., Sagiv, M., Bauer, J. (eds.) Wilhelm Festschrift. LNCS, vol. 4444, pp. 201–224. Springer, Heidelberg (2007)

[Goe06]   Goessler, G.: Component-based Design of Heterogeneous Reactive Systems in Prometheus. Technical Report 6057 INRIA (2006)

[GS03]   Goessler, G., Sifakis, J.: Component-based Construction of Deadlock-free Systems. In: Pandya, P.K., Radhakrishnan, J. (eds.) FSTTCS 2003. LNCS, vol. 2914, pp. 420–433. Springer, Heidelberg (2003)

[GS05]   Goessler, G., Sifakis, J.: Composition for Component-based Modeling. Sci. Comput. Program. 55(1-3), 161–183 (2005)

[Kov]   Kovalyov, A.V.:

[LT89]   Lynch, N.A., Tuttle, M.R.: An Introduction to Input/Output Automata. In: CWI-Quarterly, pp. 219–246 (1989)

[MM08a]   Majster-Cederbaum, M., Martens, M.: Compositional Analysis of Tree-Like Component Architectures. In: Proceedings of EMSOFT 2008 (2008)

[MM08b]   Majster-Cederbaum, M., Minnameier, C.: Everything is PSPACE-Complete in Interaction Systems. In: Fitzgerald, J.S., Haxthausen, A.E., Yenigun, H. (eds.) ICTAC 2008. LNCS, vol. 5160, pp. 216–227. Springer, Heidelberg (2008)

[MMM07a]   Majster-Cederbaum, M., Martens, M., Minnameier, C.: A Polynomial-time Checkable Sufficient Condition for Deadlock-Freedom of Component-based Systems. In: van Leeuwen, J., Italiano, G.F., van der Hoek, W., Meinel, C., Sack, H., Plášil, F. (eds.) SOFSEM 2007. LNCS, vol. 4362, pp. 888–899. Springer, Heidelberg (2007)

[MMM07b]   Majster-Cederbaum, M., Martens, M., Minnameier, C.: Liveness in Interaction Systems. In: Proceedings of FACS 2007. ENTCS (2007)

[MS08]   Minnameier, C., Schaube, R.: PrInSESSA - Proving Properties of Interaction Systems by Enhanced State Space Approximation (2008), http://134.155.88.3/main/chair_de/03/cmm_cross_checking/index_de.html

[Sif04]   Sifakis, J.: Modeling Real-time Systems. In: Keynote talk RTSS 2004 (2004)

[Sif05]   Sifakis, J.: A Framework for Component-based Construction. In: Proceedings of the Third IEEE International Conference on Software Engineering and Formal Methods, pp. 293–300. IEEE Computer Society, Los Alamitos (2005)

[Tan08]   Tanenbaum, A.: Modern Operating Systems, 3rd edn (2008)

Games on Higher Order Multi-stack Pushdown Systems

Anil Seth

Department of Computer Science and Engg.
I.I.T. Kanpur, Kanpur 208016, India
seth@cse.iitk.ac.in

Abstract. In this paper we define higher order multi-stack pushdown systems. We show that parity games over bounded phase higher order multi-stack pushdown systems are effectively solvable and winning strategy in these games can be effectively synthesized.

1 Introduction

Higher order pushdown systems (*hpds*) are a generalization of pushdown systems (*pds*) in that *hpds* can have nested stacks, such as stack of stacks. The order of an *hpds* depends on the depth of nested stacks allowed by it. Higher order *push* and *pop* operations are provided to push a copy of the topmost stack of any order and to pop it. These models, in their automata form, were introduced in [17] and were further studied in [16,15]. The *hpds* may be used to model higher order recursion, [16,8,4]. In recent years there has been considerable interest in model checking these systems and their variants [10,8,3,9].

Another generalization of pushdown systems is multi-stack pushdown systems (*mpds*). An *mpds* has a finite set of control states and a fixed number, l ($l > 1$), of independent stacks. The transition function of an *mpds* allows for a (nondeterministically chosen) push or a pop operation on any of its stack along with a change in its control state. Multi-stack pushdown systems can be used to model a class of programs with (order-1) recursion and threads. Each thread has its own stack for its procedures calls and communication among threads is through the common finite states of *mpds*. There has been quite some work in model checking *mpds* and its variants in recent years, see [7,5,6,2], as part of model checking of concurrent recursive programs. For effective model checking of *mpds* some restrictions however need to be imposed on *mpds* as even simple properties such as reachability from one configuration to another are undecidable for unrestricted *mpds*. One such restriction, called bounded context switching, was studied in [7]. This was generalized to bounded phases in [5]. The class of bounded phase *mpds* strictly includes the class of bounded context switching *mpds*.

In this paper we define higher order multi-stack pushdown systems (*hmpds*). An order-n *hmpds* (*n-hmpds*) has a fixed number (say, l) of order-n stacks. The transition function of an *n-hmpds* takes as input its control state and topmost symbols of a stack and may (nondeterministically) do a higher order push or

O. Bournez and I. Potapov (Eds.): RP 2009, LNCS 5797, pp. 203–216, 2009.

a pop operation on the stack along with a change in its control state. These systems can model a class of programs with *higher order* recursion and threads. Such programs may naturally arise while considering functional programs with threads as functional programs typically have some higher order operations. The notion of *bounded phase* defined in [5] for multi-stack pushdown automata can be lifted to *hmpds* as follows. In a k phase bounded *hmpds* only those runs of *hmpds* are considered which can be divided into k parts where each part is a consecutive sequence of moves from the run and is called a phase. In a single phase, *pop* operations (of any order) are performed only in one stack while *push* operations (of any order) can be performed on any stack.

Model checking of *hmpds* against a specification can also be formulated as solving a game over configuration graph of the *hmpds*. We show that parity games over bounded phase *hmpds* can be effectively solved. This implies decidability of properties expressed in a rich specification logic, μ-calculus, over configuration graphs of *hmpds*. Parity games over *hpds* were shown to be effectively solvable in [10] and parity games over *mpds* were shown to be effectively solvable in [2].

We solve parity games over *hmpds* by extending the technique of [2]. In order to explain our extension to the technique of [2], we briefly recall this technique. The solution in [2] is based on a fundamental technique of Walukiewicz [12] which shows how to reduce a parity game on a pushdown system to a parity game over a finite state space. In [12] each time a symbol is pushed in the stack, a set of states (along with priorities) is guessed by player 0, the game now divides into two independent parts. In the first sub game player 1 verifies that if the symbol is popped then it is in one of the guessed states, in the second sub game it is verified that if the pushed symbol is popped satisfying the guessed conditions then the game is winning for player 0. The key step in extending the technique of Walukiewicz to the *mpds* case in [2] is to define finite sets $N_{i,h}$ whose elements code relevant information summarizing the play between a push in stack-i and its matching pop operation in phase h. These sets also keep necessary information about changes in stacks t, $t \neq i$, between a push and a matching pop operation of stack-i. $N_{i,h}$ are defined using a careful induction on h.

We generalize the set of conditions $N_{i,h}$ from [2] to $N_{i,j,h}$. Each element of $N_{i,j,h}$ codes relevant information for an order-j pop operation of stack-i in phase h. This requires us to study evolution of a higher order stack under a sequence of higher order operations. Unlike order-1 case, order-i push and order-i pop operations need not match in a one to one fashion, for $i > 1$. We define a suitable matching which may associate an unbounded many order-i pop operations with a single order-i push operation. We define sets $N_{i,j,h}$ which may contain pop scenarios for all such matching order-i pop operations associated to a single order-i push operation. To take into account relevant differences in these pop operations, though they match the same push operation, some additional information is kept in each element of $N_{i,j,h}$. This requires induction on j, for each fixed h in the definition of $N_{i,j,h}$.

Equipped with sets $N_{i,j,h}$ we define a reduction from a k phase *hmpds* game to a finite state game as in [12,11,2]. In this game with each $push_{i,j}$ move o in phase

p player 0 guesses a $\theta \subseteq \cup_{x=p}^{k} N_{i,j,x}$. The guessed set θ needs to include conditions for each $pop_{i,j}$ operation that may match the $push_{i,j}$ move o. The player 1 now either continues the game after o till a matching $pop_{i,j}$ move occurs or it chooses a scenario in θ and sets the game to a configuration resulting after a pop in this scenario. Solving the finite state game leads to deciding which player in the original $hmpds$ game has a winning strategy. As a byproduct of this method as in [12,11,2], we also get that winning strategy in a bounded phase $hmpds$ game can be executed by a bounded phase higher order multi-stack pushdown automaton ($hmpda$).

The complexity of our algorithm is a tower of exponential of height $n \cdot k$ for solving a k phase n-$hmpds$ parity game. We do not know if this is optimal, however when specialized to the case of $hpds$ ($l = 1$) it reduces to the optimal bound and when specialized to the case of $mpds$ ($n = 1$), it reduces to the best known upper bound of [2]. Despite the high computational complexity, we think that from mathematical view point it is interesting to find classes of infinite graphs over which parity games can be solved effectively. Our technique gives a unified proof of solving $hpds$ and $mpds$ parity games from basic principles.

Finally to relate this work to reachability problems, the title of this conference, we note that solving reachability problem over $hmpds$ is a special case of solving parity games. In particular our results imply that there is an algorithm $\mathcal{A}$ which takes a $hmpds$ $\mathcal{H}$ a number k and two configurations u, v of $\mathcal{H}$ and answers yes if in the configuration graph of $\mathcal{H}$, v can be reached from u in at most k phases, otherwise $\mathcal{A}$ answers no. Further, if the answer is yes then a k phase path from u, v can also be produced by the algorithm. Similar result holds for two player reachability between configurations of bounded phase $hmpds$.

2   Preliminaries

Definition 1. *Let Γ be a finite stack alphabet and let $\bot$ be a symbol s.t. $\bot \notin \Gamma$. The Set of order-i stacks over Γ, S_i for $i \geq 0$ is defined inductively as follows.*

- $S_0 = \Gamma \cup \{\bot\}$ *(we consider an element of Γ as an order-0 stack).*
- $S_1 = \{[\bot, s_1, \ldots, s_v] \mid s_1 \ldots s_v \in \Gamma, \ v \geq 0\}$.
- $S_{i+1} = \{[s_1, \ldots, s_v] \mid s_1 \ldots s_v \in S_i, \ v \geq 1\}$, *for $i \geq 1$.*

We also define $\bot_0 = \bot$ and for $i \geq 0$, $\bot_{i+1} = [\bot_i]$. Note that $\bot_i \in S_i$.
Stack $\bot_i$, for $i > 0$, is called the empty stack of order-i.
We use $order(s)$ to denote the order of a stack s.

The symbol $\bot$ is used to mark bottom of an order-1 stack.

Definition 2. *Let s be an order-j stack for $j \geq 0$. The topmost order-i element ($i \leq j$) of s is defined as*
$$top_i(s) = \begin{cases} s & if \quad order(s) = i \\ top_i(s_u) & if \quad order(s) > i \text{ and } s = [s_1, \ldots, s_u] \end{cases}$$

Definition 3. *Operations $push_i, pop_i$ on stacks of order $\geq i$ are defined as follows. Let $s = [s_1, \ldots, s_u]$ and $b \in \Gamma$.*

$push_1^b(s) = [s_1, \ldots, s_u, b]$, *if $order(s) = 1$.*
$push_1^b(s) = [s_1, \ldots, s_{u-1}, push_1^b(s_u)]$, *if $order(s) > 1$.*
For $i > 1$,
$push_i(s) = [s_1, \ldots, s_u, s_u]$, *if $order(s) = i$*
$push_i(s) = [s_1, \ldots, s_{u-1}, push_i(s_u)]$, *if $order(s) > i$*
For $i \geq 1$,
$pop_i(s) = [s_1, \ldots, s_{u-1}]$, *if $order(s) = i$ and $u > 1$.*
$pop_i(s) = [s_1, \ldots, s_{u-1}, pop_i(s_u)]$, *if $order(s) > i$, $pop_i(s_u)$ is defined.*
In keeping with conventional notation, a $push_{i+1}$ operation pushes an element of order-i and pop_{i+1} operation pops an element of order-i. Note that $pop_i(s)$ is defined iff $top_i(s)$ has more than one element.

2.1 Order-n Multi-stack Pushdown Systems

A order-n multi-stack pushdown system is the same as a multi-stack pushdown system except that each stack is a nested stack of order-n. On each stack the *push* and *pop* operations of order-0 to order-$(n-1)$ can be performed. A formal definition is given below.

Definition 4. *An order-n multi-stack pushdown system (n-hmpds) is given as a tuple $(Q, \Gamma, \perp, l, \delta, q_0)$, where Q is a finite set of states, l is the number of stacks, Γ is the stack alphabet with $\perp$ as in definition 1 and q_0 is the initial state. The transition function δ is given as $\delta = (\bigcup_{j=1}^{n} \delta_{ins,j}) \cup (\bigcup_{j=1}^{n} \delta_{rem,j})$, where*

- $\delta_{ins,j} \subseteq Q \times (\Gamma \cup \perp) \times Q \times [1 \ldots l] \times \Gamma, 1 \leq j \leq n$.
- $\delta_{rem,j} \subseteq Q \times (\Gamma \cup \perp) \times Q \times [1 \ldots l], 1 \leq j \leq n$.

$(q, \gamma, q', i, \gamma') \in \delta_{ins,m}$ denotes *push* transition of order-$(m-1)$ in stack numbered i and $(q, \gamma, q', i) \in \delta_{rem,m}$ denotes *pop* transition of order-$(m-1)$ in stack numbered i. These moves are also referred to as $push_{i,m}$ and $pop_{i,m}$ respectively. Values q, γ are the state of *hmpds* and the topmost symbol of stack-i before the transition, q' is the state of *hmpds* after the transition. Symbol γ' is the symbol pushed when *push* transition is of order-0 ($m = 1$). For $m > 1$, in transition $(q, \gamma, q', i, \gamma') \in \delta_{ins,m}$ symbol γ' plays no role. We keep this extra symbol to avoid treating the case $m = 1$ separately. For notational convenience, we stipulate in the following that for $m > 1$, if $(q, \gamma, q', i, \gamma') \in \delta_{ins,m}$ then $\gamma' = \gamma$.

Formal definition of *hmpds* configurations and transitions on them is as follows.

Definition 5. *A configuration of n–hmpds $\mathcal{H} = (Q, \Gamma, l, \delta, q_0)$ is a tuple $(q, s_1, \ldots, s_l)$, where $q \in Q$ and $s_i \in S_n$ for $1 \leq i \leq l$. One step transition on configurations of $\mathcal{H}$ is defined as below.*

- $(q, s_1, \ldots, s_l) \xrightarrow{t} (q', s_1', \ldots, s_l')$ *if $top_0(s_i) = \gamma$, $t = (q, \gamma, q', i, \gamma') \in \delta_{ins,m}$,*
  $s_j' = s_j$ *for $j \neq i, 1 \leq j \leq l$ and if $m = 1$ then $s_i' = push_1^{\gamma'}(s_i)$ else*
  $s_i' = push_m(s_i)$.

$-\ (q, s_1, \ldots, s_l) \xrightarrow{t} (q', s'_1, \ldots, s'_l)$ if $top_0(s_i) = \gamma$, $t = (q, \gamma, q', i) \in \delta_{rem,m}$, $s'_i = pop_m(s_i)$ and $s'_j = s_j$ for $j \neq i$, $1 \leq j \leq l$.

The initial configuration of n-hmpds *is defined as* $(q_0, \perp_n, \ldots, \perp_n)$.

Following is the standard definition of the reflexive and transitive closure of relation $\xrightarrow{t}$.

Definition 6. *A multi-step transition between configurations of mpds, on say sequence* $t_1 t_2 \ldots t_n$ *of hmpds moves,* $c \xrightarrow{t_1 t_2 \ldots t_n} d$ *is defined as follows.* $c \xrightarrow{t_1 t_2 \ldots t_n} d$ *iff either* $n = 0$ *and* $c = d$ *or there is a* c' *s.t.* $c \xrightarrow{t_1} c'$ *and* $c' \xrightarrow{t_2 \ldots t_n} d$. *We write* $c \twoheadrightarrow d$ *for a multi-step transition from* c *to* d *when the sequence of* hmpds *moves is not relevant.*

Following is a straightforward adaptation of the notion of phases for *mpds* [5] to *hmpds*.

Definition 7. *A phase is a sequence of* hmpds *transitions where pop moves (of any order) are performed only on a single stack (though in a single phase push moves may be performed on any stack). A k-phase bounded run of a n-hmpds is one which can be partitioned into* k *contiguous segments such that each segment is a single phase.*

It is clear that phase change occurs when a *pop* operation is performed on a stack other than the stack on which it was performed last. We extend the notion of a configuration to $(q, h, r, s_1, \ldots, s_l)$ where $(q, s_1, \ldots, s_l)$ is a configuration as before, h is a natural number recording the phase and $r \in [1, l]$, is the stack number on which the last *pop* operation was performed. In the initial configuration $h = 1$ and $r = 0$.

Definition 8. *One step transition on extended configurations of* $\mathcal{H}$ *is defined as* $(q, h, r, s_1, \ldots, s_l) \xrightarrow{t} (q', h', r', s'_1, \ldots, s'_l)$ *where* $(q, s_1, \ldots, s_l) \xrightarrow{t} (q', s'_1, \ldots, s'_l)$ *and*

$-\ $ *if* $t \in \delta_{ins,m}$ *then* $h' = h$ *and* $r' = r$

$-\ $ *if* $t = (q, -, q', i) \in \delta_{rem,m}$ *then* $r' = i$ *and*
$$h' = np(h, r, i) = \begin{cases} h & \text{if } r = 0 \text{ or } r = i \\ h + 1 & \text{if } r \neq i \end{cases}.$$

The initial extended configuration of n-hmpds *is defined as* $(q_0, 0, 1, \perp_n, \ldots, \perp_n)$. *Function* $np(h, r, i)$ *gives the new phase after the transition.*

To consider only k-phase bounded runs of hmpds, we restrict ourselves to extended configuration graph where each vertex has phase less than or equal to k. In the sequel we use the word configuration for the extended configuration defined above. $top_{i,j}$ refers to top_j of stack-i and $push_{i,j}$, $pop_{i,j}$ refer to transitions of the form $(-, -, -, i, -) \in \delta_{ins,j}$, $(-, -, -, i) \in \delta_{rem,j}$ respectively.

2.2 Parity Games

We assume the reader to be familiar with standard notions of two player parity games, such as game graph, plays, a winning strategy and parity winning condition, see [14].

A 2-player k-phase *hmpds* parity game is given as $(\mathcal{H}, Q_0, Q_1, M, \Omega, k)$, where $\mathcal{H} = (Q, \Gamma, l, \delta, q_0)$ is an *hmpds*, $Q = Q_0 \oplus Q_1$ is a partition of states in player 0 and player 1, M is a finite set of priorities and $\Omega : Q \to M$ is a priority assignment to each state in Q.

Vertices of our game graph are configurations of the form $(q, h, r, s_1, \ldots, s_l)$, where $h \leq k$, of *hmpds*. Edge relation of this game graph is given by transition relation ' $\to$ ' in definition 8. A vertex $(q, -, \ldots, -)$ belongs to player i iff $q \in Q_i$. Priority of a vertex $(q, -, \ldots, -)$ is defined as $\Omega(q)$. A player can move from c to c' only if $c \to c'$. A play is a sequence of legal moves starting from the initial configuration. By our choice of the vertex set all plays in this game are $k - phase$ bounded. That is a player can not make a move that takes the play into $(k+1)^{th}$ phase.

Winning condition for a maximal play (play which can not be extended further) ρ is defined as follows. If ρ is finite then the player whose turn it is to move at the last vertex of ρ loses. If ρ is infinite then a priority $i \in M$ is said to be visited infinitely often iff there are infinitely many vertices with priority i in ρ. ρ is winning for player 0 iff the minimum, among the set of priorities visited infinitely often in ρ, is even.

Informally, having a winning strategy for player i, means that regardless of player $(1 - i)$'s moves, player i can always play a move such that he wins the resulting play. We will always consider games which start in a predefined initial configuration. A game is called winning for player i if player i has a winning strategy in it starting from the initial configuration.

Given a winning strategy τ for player 0, in game $\mathcal{G}$, by a τ-play we mean a play of $\mathcal{G}$ in which all moves of player 0 are according to τ. For vertices c, c' of $\mathcal{G}$, $c \xrightarrow{\tau} c'$ and $c \xrightarrow{\tau}{}^* c'$ mean that c' is reachable from c in a τ-play in one move or in an arbitrary number of moves respectively.

3 Main Ideas in Solving a HMPDS Game

In higher order stacks a sub-stack after being pushed can be copied implicitly several times before being popped. As an example consider an order-2 stack $r = [[\perp a]]$. Here a is a stack symbol which we consider as an order-0 stack. An order-1 push operation o_1 pushing a symbol b, leads to $r_1 = [[\perp ab]]$. Now consider an order-2 push o_2 which copies the topmost order-1 stack, this leads to $r_2 = [[\perp ab][\perp ab]]$. In r_2 the topmost order-1 stack $[\perp ab]$ is created by o_2 by a direct copy of the topmost order-1 stack below. The second occurrences of order-0 stacks a, b however are not created by any order-1 *push* directly. These occurrences get created automatically from the occurrences in the sub-stack being copied by o_2. We call such an occurrence an *auto copy* of the corresponding occurrence below. A motivation for considering the *auto copy* relation is to associate with each sub-stack of a higher order stack a *push* operation. This is needed for matching each *pop* operation with a *push* operation. If s is an *auto copy* of t then we associate the same *push* operation with s as with t. So in our example we associate with the second occurrence of b, operation o_1. A pop operation which pops a sub-stack z is said to match the push operation associated with z. This results in matching several pop operations with a single push operation, unlike in order-1 case. For example, if we do an order-1 pop o_3 on r_2, followed by an order-2 pop o_4, followed by an order-1 pop o_5 then both o_3 and o_5 match o_1.

As mentioned in the introduction, we generalize the set of conditions $N_{i,h}$ from [2] to $N_{i,j,h}$. Each element of $N_{i,j,h}$ codes relevant information for an order-j pop operation of stack-i in phase h. A set of such pop scenarios $\theta \subseteq \cup_{x=1}^{k} N_{i,j,x}$, where k is the number of phases allowed, is associated with an order-j push operation o. θ captures conditions for any pop operation o' matching o. If s is a stack created by o then a matching o' may pop either s or a stack s' which is obtained by several steps of *auto copying* of s. For an $order - (n-1)$ sub-stack no *auto copy* can be made. An order-j sub-stack s, $j < n-1$, however may be *auto copied* by pushing an order-x, $x > j$, stack t containing s. That is, s may be *auto copied* by a $push_{i,m}$, $m > j+1$, note that in our notation a $push_{i,m}$ pushes a stack of order-$(m-1)$ in stack-i. Therefore, if s' is an *auto copy* of s then s' may be contained in different sub-stacks of order $> j$ than s was. So popping conditions for an *auto copy* of s may depend on popping conditions of sub-stacks of order $> j$ containing this *auto copy* of s. This is handled by induction on j (from $j = n-1$ down to $j = 0$) while defining $N_{i,j,h}$ for a fixed h. Therefore sets $N_{i,j,h}$ are defined by a nested induction. The outer induction on h, the phase number, takes care of the contents of the other stacks in a popping scenario, as in [2]. The inner induction on j, for a fixed h, handles popping conditions of higher order sub-stacks containing an order-j stack.

4   Copying by Higher Order Push Operations

In this section we formally define some useful notions related to copying of higher order stacks as they evolve under a sequence of *hpds* transitions. We begin by defining labeled stacks and extend *hpds* transitions to them. Labeled stacks are used as a tool to define a copy relation on sub-stacks of a higher order stack and to define matching between $push_i$ and pop_i operations in a sequence of *hpds* transitions. A sub-stack of a higher order stack s is a stack of any order contained in s.

Definition 9. *A labeled order-n stack is an order-n stack s such that each sub-stack of s (including s itself) is labeled by a non-negative integer. We show this label as a superscript at the end of a stack. For example $[[\perp^0 a^1]^0 [\perp^0 a^1]^2]^0$ is a labeled order-2 stack.*

Definition 10. *We extend* hpds *transitions to labeled stacks as follows.*
$push_1^b([a_1^{l_1}, \ldots, a_u^{l_u}]^m) = [a_1^{l_1}, \ldots, a_u^{l_u}, b^l]^m$, *where l is a fresh label.*
$push_1^b([s_1^{l_1}, \ldots, s_u^{l_u}]^m) = [s_1^{l_1}, \ldots, s_{u-1}^{l_{u-1}}, push_1^b(s_u^{l_u})]^m$, *if $order(s_u) > 1$.*
For $i > 0$,

$push_{i+1}([s_1^{l_1}, \ldots, s_u^{l_u}]^m) = [s_1^{l_1}, \ldots, s_u^{l_u}, s_u^l]^m$, *if* $order(s_u) = i$ *and* l *is a fresh label.*

$push_{i+1}([s_1^{l_1}, \ldots, s_u^{l_u}]^m) = [s_1^{l_1}, \ldots, s_{u-1}^{l_{u-1}}, push_{i+1}(s_u^{l_u})]^m$, *if* $order(s_u) > i$.

Here $l_1, \ldots, l_u, l, m$ are labels. A fresh label is a label not occurring in input (including its sub-stacks) to the push operation.

The pop operations extend to a labeled stack in a straightforward way, where a stack along with its label is popped. There is no change in the labels of the remaining stacks.

Each $push_{i+1}$ move creates a stack of order-i with a unique label. It also copies labeled stacks of order $< i$ without changing their labels.

Example 1. Let $t = [[\perp^0]^0]^0$ be an order-2 stack. In our examples, stacks grow from left to right. After a $push_1^a$ operation on t we get the stack $t_1 = [[\perp^0 a^1]^0]^0$. A $push_2$ on t_1 gives $t_2 = [[\perp^0 a^1]^0[\perp^0 a^1]^2]^0$. A further $push_1^b$ operation on t_2 gives $t_3 = [[\perp^0 a^1]^0[\perp^0 a^1 b^3]^2]^0$.

We define below a simple ordering on sub-stacks s_1, s_2 of a given stack s based on if s_1 occurs below s_2 in s. This is used in a later definition.

Definition 11. *Given a stack s of order-j and its two sub-stacks t_1 and t_2 of order-i ($i < j$), we define "$t_1 \leq t_2$ in s" if either of the following holds.*

- $s = [s_1, \ldots, s_u]$, $t_1 = s_i$ and $t_2 = s_j$ for $1 \leq i \leq j \leq u$.
- $s = [s_1, \ldots, s_u]$, t_1 is a sub-stack of s_i, t_2 is a sub-stack of s_j, for $1 \leq i \leq j \leq u$.
- $s = [s_1, \ldots, s_u]$, t_1, t_2 are sub-stacks of s_i for $1 \leq i \leq u$ and $t_1 \leq t_2$ in s_i.

This definition is by induction on $j - i$.

Definition 12. *The initial order-n stack $[\ldots[\perp]\ldots]$ is labeled as $[\ldots[\perp^0]^0 \ldots]^0$, that is every sub-stack in it is labeled by 0. Let s be a labeled order-n stack obtained by starting the labeled initial stack and applying some hpds moves. Let s_1, s_2 be two order-i sub-stacks of s with same labels and let $s_1 \leq s_2$ in s. Then we say that s_2 is an auto copy of s_1 and denote it as $s_1 \preceq_i s_2$.*

*If a pop move pops a stack with label l then we say that it **matches** the push move associated with l.*

Example 2. If on order-3 stack $[[[\perp^0]^0]^0]^0$ the sequence of operations $push_1^a$, $push_1^a$, $push_2$, $push_3$, pop_1, $push_1^b$ is performed then we get $t = [[[\perp^0 a^1 a^2]^0[\perp^0 a^1 a^2]^3]^0[[\perp^0 a^1 a^2]^0[\perp^0 a^1 b^5]^3]^4]^0$. Here order-2 stack $[\perp^0 a^1 b^5]^3$ is a copy of the stack $[\perp^0 a^1 a^2]^3$. This shows that a stack can be an *auto copy* of another stack even if their contents are not the same. The *auto copy* relation that we have defined establishes a relation between 'containers' not their contents.

Example 3. In the example 1, in $t_3 = [[\perp^0 a^1]^0[\perp^0 a^1 b^3]^2]^0$ the 'a' (order-0 stack) above is a copy of the 'a' below. If on t_3 we perform the sequence pop_1, pop_1, pop_2, pop_1 then both second and the last pop_1 operations match the $push_1^a$ operation performed on t. Both these operations pop 'a' associated with this $push_1^a$ operation.

The *auto copy* relation that we have defined can be informally understood as follows. A $push_j$ for $j > 1$, on stack s creates a new instance of $top_{j-1}(s)$ and pushes it on s. Creating a new instance of $top_{j-1}(s)$ automatically creates new instances of all stacks (of lower order) contained in $top_{j-1}(s)$. We refer to these new instances of order $< j - 1$ as auto copies of their corresponding stacks in $top_{j-1}(s)$. However, we do not consider the new order $j - 1$ instance as an auto copy of $top_{j-1}(s)$. The full *auto copy* relation is obtained by taking reflexive, transitive closure of the basic *auto copy* step described above.

There can be many pop_j operations (for $j < n$) matching a single $push_j$ operation. This is a departure from order-1 case. This happens because a stack of order-$(j - 1)$ may be *auto copied* several times by $push_r$, $r > j$, operations. A pop_j operation popping any of these *auto copies* is associated with the same $push_j$.

5   Reducing *n-HMPDS* Game to Finite State Game

5.1   Popping Scenarios

Let $\mathcal{H} = (Q, \Gamma, l, \delta, q_0)$ be n-mpds and let $\mathcal{G} = (\mathcal{H}, Q_0, Q_1, M, \Omega : Q \to M)$ be a game structure on $\mathcal{H}$, where $Q = Q_0 \oplus Q_1$ and $M = \{0, \ldots, max\}$ is the set of priorities assigned to vertices of the game graph.

We now define sets $N_{i,j,p}$. These sets are defined simultaneously for all i by a nested induction on p, j. The outer induction is on p (starting with $p = k$ down to $p = 1$). For each fixed value of p, an induction on j (starting with $j = n - 1$ down to $j = 0$) is done. Intuitively $N_{i,j,p}$ is the set of scenarios or constraints to be met for doing a $pop_{i,j+1}$ when the configuration resulting after this *pop* is in phase p. This scenario also keeps information about possible scenarios of future *pop* operations. This leads to the induction definition.

Definition 13. *In this definition we assume that $q \in Q$, $\overline{\gamma} \in (\Gamma \cup \{\bot\})^n$ and $\overline{m} \in M^{l \times n}$. Also, $u \in [1, l]$, $r \in (j, n)$ and $h, p \in [1, k]$, where the intervals shown are integer intervals.*

$$N_{i,n-1,p} = \{(a_1, \ldots, a_{i-1}, (p, q, \overline{\gamma}, \overline{m}), a_{i+1}, \ldots, a_l) \mid a_u \in A_{u,p+1} \; for \; u \neq i\}$$

$$N_{i,j,p} = \{(a_1, \ldots, a_{i-1}, (p, q, \overline{\gamma}, \overline{m}, b_{i,j+1}, \ldots, b_{i,n-1}), a_{i+1}, \ldots, a_l) \mid$$
$$a_u \in A_{u,p+1} \; for \; u \neq i, \; b_{i,r} \subseteq \cup_{x=p}^{k} N_{i,r,x}\}$$

where $A_{i,h}$ are auxiliary sets given as
$$A_{i,k+1} = \{\emptyset\}, \quad A_{i,h} = \{(T_0, \ldots, T_{n-1}) \mid T_j \subseteq \cup_{x=h}^{k} N_{i,j,x} \}$$

The auxiliary sets $A_{i,h}$ in the above definition are used to keep the definition a bit compact, $A_{i,h}$ do not have any other role.

Each $e = (a_1, \ldots, a_{i-1}, (p, q, \overline{\gamma}, \overline{m}, b_{i,j+1}, \ldots, b_{i,n-1}), a_{i+1}, \ldots, a_l) \in N_{i,j,p}$ represents a scenario for a $pop_{i,j+1}$ operation. Entries $p, q, \overline{\gamma}, \overline{m}$ in e refer to a $pop_{i,j+1}$ transition from configuration c to $c' = (q, p, i, s_1, \ldots, s_l)$, where $\gamma_t = top_{t,0}(c)$, for $1 \leq t \leq l$, $t \neq i$ (γ_i is not needed in e, we keep it for compact notation), $m_{u,v}$ is the minimum priority visited since $push_{u,v+1}$ corresponding to $top_{u,v}(c)$. Data

a_u, $u \neq i$ in e refer to scenarios for $pop_{u,x}$, $0 \leq x < n$, and $b_{i,y}$ refer to scenarios for $pop_{i,y}$, $j + 1 < y \leq n$ *after* the $pop_{i,j+1}$ in scenario e.

The motivation for a_u, $u \neq i$ in e is similar to that in the definition of $N_{i,p}$ in [2], except that in stack u now not just $pop_{u,0}$ but $pop_{u,0}, \ldots, pop_{u,n-1}$ moves are possible. A tuple in set $A_{u,h}$ gives popping scenarios for $pop_{u,0}, \ldots, pop_{u,n-1}$ when such a pop move results in a configuration of phase $\geq h$. These are denoted by $T_0, \ldots, T_{n-1}$ respectively. Any pop operation in stack-u, $u \neq i$, subsequent to $pop_{i,j+1}$ move in phase p, can only be made in a phase $> p$. Therefore the entries a_u in e are taken from sets $A_{u,p+1}$.

For each $b_{i,r}$, $j + 1 \leq r < n$, note that $N_{i,j,p}$ contains scenarios about popping not just a single stack but also about all *auto copies* of it. This is because constraints for all matching $pop_{i,j+1}$ are to be guessed at the time of a $push_{i,j+1}$. As discussed in section 3, if a sub-stack t is popped by a $pop_{i,j+1}$ then order-$(j+1), \ldots,$order-$(n-1)$ stacks, $t_{j+1}, \ldots, t_{n-1}$ containing t need not be the same stacks in which push associated to t was done. Stack t may be an *auto copy* of stack s, where s is the stack actually pushed by the push associated to t. Values $q, \overline{\gamma}, \overline{m}$ allowed in a scenario which pops t also depend on popping scenarios for $t_{j+1}, \ldots, t_{n-1}$ containing t. Scenario e contains an allowable combination of $q, \overline{\gamma}, \overline{m}$ and popping scenarios for $t_{j+1}, \ldots, t_{n-1}$ explicitly. These Scenarios are $b_{i,j+1}, \ldots, b_{i,n-1}$ in the definition above. In general these are some subsets of $\cup_{x=p}^{k} N_{i,j+1,x}, \ldots, \cup_{x=p}^{k} N_{i,n-1,x}$ respectively.

For the base case, $e \in N_{i,x,k}$ and $a_u = \emptyset$ for $u \neq i$ as in this case k is the last phase so a pop in any stack other than i will not occur after this pop.

Below, we use notation like $\overline{B} = (B_{i,j} | 1 \leq i \leq l, 0 \leq j < n)$, for double indexed sets. We also use $\overline{B}[C/(i,j)]$ to mean the indexed set which is same as $\overline{B}$ except at index (i,j) where it is C. For single indexed set we use sequence like notation with $\overline{T}$ and $\overline{T}[C/i]$ as obvious counterparts of $\overline{B}$, $\overline{B}[C/(i,j)]$respectively. For double indexed sets $\overline{B}$ as above we use $\overline{B}_i$ for $B_{i,0}, \ldots, B_{i,n-1}$.

We follow [11] in presentation of our finite state game. Most important vertices of the finite state game (FSG) are of the form $Check(q, p, r, \overline{\gamma}, \overline{B}, \overline{m})$, where $q \in Q$, $p \in [1, k]$, $r \in [0, l]$, $\overline{\gamma} = \gamma_1 \ldots \gamma_l$ with each $\gamma_i \in \Gamma \cup \{\bot\}$. Finally $\overline{B} = (B_{i,j} | 1 \leq i \leq l, 0 \leq j < n)$ and $\overline{m} = (m_{i,j} | 1 \leq i \leq l, 0 \leq j < n)$. Intuitively vertex $Check(q, p, r, \overline{\gamma}, \overline{B}, \overline{m})$ asserts the following about a *hmpds* configuration c.

- q is the state of the configuration.
- p is the current phase.
- r is the number of stack on which last pop operation was done (initially it is set to 0).
- γ_i is the topmost symbol of stack i.
- $B_{i,j} \subseteq \cup_{x=1}^{k} N_{i,j,x}$ is a set of scenarios for $pop_{i,j}$ move on c.
- $m_{i,j}$ records the minimum priority visited in the play (till the current instant) since the $push_{i,j}$ move associated to $top_{i,j}(c)$ was played.

Apart from *Check* vertices there are also some auxiliary vertices.

A bit more notation. We use $E \uparrow p$ for $E \cap (\cup_{x=p}^{k} N_{i,j,x})$. For a sequence $\overline{T} = (T_i \mid 1 \leq i \leq m)$, $\overline{T} \uparrow p$ stands for $(T_i \uparrow p \mid 1 \leq i \leq m)$. For a double index $\overline{T}, \overline{T} \uparrow p$ is defined similarly.

5.2   The Finite State Game (FSG)

Each *hmpds* transition gives rise to some FSG transitions. We group transitions of FSG according to *hmpds* transitions (shown in bold).

1. $(\mathbf{q}, \boldsymbol{\gamma_i}, \mathbf{q'}, \mathbf{i}, \boldsymbol{\gamma'}) \in \boldsymbol{\delta_{\text{ins},j}}$ where $1 \leq i \leq l$ and $1 \leq j \leq n$.

   This gives rise to transitions:
   (a) $Check(q, p, r, \overline{\gamma}, \overline{B}, \overline{m}) \rightarrow Push_{i,j}(p, r, \overline{\gamma}, \overline{B}, \overline{m}, q', \gamma')$
   (b) $Push_{i,j}(p, r, \overline{\gamma}, \overline{B}, \overline{m}, q', \gamma') \rightarrow Claim_{i,j}(p, r, \overline{\gamma}, \overline{B}, \overline{m}, q', \gamma', C)$,
      for $C \subseteq \cup_{h=p}^{k} N_{i,j-1,h}$
   (c) $Claim_{i,j}(p, r, \overline{\gamma}, \overline{B}, \overline{m}, q', \gamma', C) \rightarrow Check(q', p, r, \overline{\gamma}[\gamma'/i], \overline{B}[C/(i, j - 1)], \overline{m'})$,

   $$\text{where} \quad m'_{x,y} = \begin{cases} \Omega(q') & \text{if } x = i \text{ and } y = j - 1 \\ min(m_{i,j}, \Omega(q')) & \text{otherwise} \end{cases}$$

   (d) To check the game after a matching $pop_{i,j}$ operation.

   $Claim_{i,j}(p, r, \overline{\gamma}, \overline{B}, \overline{m}, q', \gamma', C) \rightarrow Jump_{i,j}(q'', h, \overline{\gamma}, \overline{\gamma''}, \overline{m'}, \overline{B'}, \overline{m})$
   for any $(a_1, \ldots, a_{i-1}, z, a_{i+1}, \ldots, a_l) \in C$ where
   $a_r = (B'_{r,0}, \ldots, B'_{r,n-1})$ for $1 \leq r \leq l$, $r \neq i$
   $z = (h, q'', \overline{\gamma''}, \overline{m'}, B'_{i,j}, \ldots, B'_{i,n-1})$,
   $B'_{i,t} = B_{i,t}$ for $0 \leq t < j$.

   (e) $Jump_{i,j}(q'', h, \overline{\gamma}, \overline{\gamma''}, \overline{m'}, \overline{B'}, \overline{m}) \rightarrow Check(q'', h, i, \overline{\gamma''}[\gamma_i/i], \overline{B'}, \overline{m''})$,

   $$\text{where} \quad m''_{x,y} = \begin{cases} min(m'_{x,y}, \Omega(q'')) & \text{if } x \neq i \text{ or } y \geq j \\ min(m_{x,y}, m'_{i,j-1}, \Omega(q'')) & \text{if } x = i \text{ and } y < j \end{cases}$$

2. $(\mathbf{q}, \boldsymbol{\gamma_i}, \mathbf{q'}, \mathbf{i}) \in \boldsymbol{\delta_{\text{rem},j}}$, $1 \leq i \leq l$.

   This gives rise to transitions:
   (a) $Check(q, p, r, \overline{\gamma}, \overline{B}, \overline{m}) \rightarrow Win_0$ if $D \in B_{i,j-1}$ and $p' \leq k$
   (b) $Check(q, p, r, \overline{\gamma}, \overline{B}, \overline{m}) \rightarrow Win_1$ if $D \notin B_{i,j-1}$ and $p' \leq k$

   where $p' = np(p, r, i)$ and if $p' \leq k$ then
   - $D = (C_1, \ldots, C_{i-1}, z, C_{i+1}, \ldots, C_l)$
   - $z = (p', q', \overline{\gamma}, \overline{m}, B'_{i,j}, \ldots, B'_{i,n-1})$
   - $C_r = (B'_{r,0}, \ldots, B'_{r,n-1})$ for $1 \leq r \leq l$, $r \neq i$

   $$\text{and} \quad B'_{x,y} = \begin{cases} B_{x,y} \uparrow (p' + 1) & \text{if } x \neq i \\ B_{x,y} \uparrow p' & \text{if } x = i \text{ and } y \geq j \end{cases}$$

Priority of a vertex v in FSG, denoted by $\lambda(v)$, is defined as follows.

- $\lambda(Check(q, \ldots)) = \Omega(q)$, $\lambda(Push_{i,j}(\ldots)) = \lambda(Claim_{i,j}(\ldots)) = max$.
- $\lambda(Jump_{i,j}(q, h, \overline{\gamma}, \overline{\gamma'}, \overline{m'}, \overline{B}, \overline{m})) = m'_{i,j-1}$,
      where $\overline{m'} = (m'_{x,y} \mid x \in [1, l], y \in [0, n))$.

Vertices of the form $Check(q, \ldots)$ belong to player-z, $z \in \{0,1\}$, iff $q \in Q_z$. Vertices $Push_{i,j}(\ldots)$, belong to player-0 whereas vertices $Claim_{i,j}(\ldots)$, belong to player-1. Vertices $Jump_{i,j}(\ldots)$ belong to player 0. Vertices Win_0 and Win_1 belong to player 1 and player 0 respectively. As there are no transitions from Win_0 and Win_1, by our convention Win_0 and Win_1 are winning for player 0 and player 1 respectively.

We explain the transition rules of FSG in some detail below.

- Rule (1.a) transfers the game to player-0's vertex ($Push_{i,j}$ vertex) regardless of the player to whom $Check$ vertex in belongs. This is because *player 0* only can make a claim about popping scenarios.
- In rule (1.b), C is the set of scenarios for all $pop_{i,j}$ matching the $push_{i,j}$. The same set of popping scenarios is maintained for any *auto copy* of this stack.
- In rule (1.c), player 1 sets the game to the configuration after the $Push_{i,j}$ move. Sub-stacks $top_{x,y}$ in the new configuration are *auto copies* of sub-stacks $top_{x,y}$, in configuration before, for $x \neq i$ or $y \neq j$. Therefore $B_{x,y}$, $m_{x,y}$ for $x \neq i$ or $y \neq j$ remain unchanged.
- In rule (1.d-e), the game is verified after a matching $pop_{i,j}$ move with popping scenario in C. We show the transitions for an arbitrary scenario $(a_1, \ldots, a_{i-1}, z, a_{i+1}, \ldots, a_l) \in C$. Let c' be a configuration arising after a *pop* in this scenario. The phase of c' is h and $m'_{x,y}$ are priorities corresponding to $top_{x,y}(c')$, for $x \neq i$ or $y \geq j$.
  Note that if a $push_{i,j}$ is done in configuration c then for $0 \leq t < j$, $top_{i,t}(c')$ is an *auto copy* of $top_{i,t}(c)$. Therefore we have $B'_{i,t} = B_{i,t}$ as the set of popping scenarios remain same for the *auto copies*.
  The value $m'_{i,j-1}$ gives the minimum priority visited between $push_{i,j}$ and the matching $pop_{i,j}$ in the play. This explains the expression $m''_{i,t}$, for priorities corresponding to $top_{i,t}(c')$ for $0 \leq t < j$.
  The $Jump$ vertex is to capture the min priority between $push_{i,j}$ and the matching $pop_{i,j}$ (it is $m'_{i,j-1}$ in the present case) in FSG path (1.a-b-d-e).
- In rule (2), condition $p' \leq k$ refers to the fact that a *pop* move is possible only if the resulting phase is $\leq k$. Given $p' \leq k$, the transition (2.a) represents the case where *pop* move satisfies the popping condition, the transition (2.b) represents complement of this case.
  The main step is to define $D \in N_{i,j,p}$ based on various popping conditions in the current configuration of the play. In $B_{x,y}$ there may be some popping scenarios for phase $< p'$. We remove these scenarios by using operator $\uparrow$. More specifically, we keep scenarios of phase $\geq p'$ for popping in stack i and scenarios of phase $> p'$ for popping in stacks other than i as after the present *pop*, other *pops* can occur only in these phases. The $B'_{x,y}$ defined in this way form the desired components of D so that $D \in N_{i,j,p}$.

6   Relating Winning in n-*HMPDS* Game and the FSG

Our main theorem is the following.

Theorem 1. *A hmpds game is winning for player 0 (from initial configuration $(q_0, \perp_n, \ldots, \perp_n)$) iff FSG is winning for player 0 (from initial configuration*

$Check(q_0, 1, 0, \overline{\mathbb{I}}, \overline{\emptyset}, \overline{0}))$. *Further, if* hmpds *game is winning for player* 0 *then player* 0 *has a winning strategy in* hmpds *game that is computable by an order-n multi-stack automaton.*

Proof. (FSG to *hmpds* game) Assuming that there is a winning strategy for player-0 in FSG from $Check(q_0, 1, 0, \overline{\mathbb{I}}, \overline{\emptyset}, \overline{0})$, we design an l stack deterministic $n-$hmpda $\mathcal{S}$ which executes a winning strategy τ of player 0 in *hmpds* game starting from *hmpds* configuration $(q_0, \perp_n, \ldots, \perp_n)$. The automaton $\mathcal{S}$ is an l stack deterministic hmpda with an input and an output tape. It reads moves of player 1 from the input tape and outputs moves of player 0 on the output tape. Detailed construction of $\mathcal{S}$ and the correctness proof of $\mathcal{S}$ is given in full version of this paper [1].

(*Hmpds* game to FSG) Proof of this direction is given in [1].

Idea of the proofs in both directions is similar to that in [11,2], but we need to deal with operations on higher order stacks. □

6.1   Complexity of Solving the Game

By the reduction in section 5, to solve a *hmpds* game it suffices to solve an associated FSG. In this section we estimate size of the FSG and the complexity of solving it. Let us define a class of functions $exp_n(m)$ iteratively as follows. $exp_1(m) = 2^m$ and for $n \geq 1$, $exp_{n+1}(m) = 2^{exp_n(m)}$. Roughly, $exp_n(m)$ is a tower of exponentials of height n. Let $\mathcal{H}$ be an *hmpds* and $\mathcal{G}$ be an *hmpds* game on $\mathcal{H}$ as in section 5.1. For a set A, we let $|A|$ denote its cardinality.

By a simple complexity analysis, whose details are omitted due to lack of space, we get $|N_{i,0,k-1}| \leq exp_{n-1}(z)$ and the number of vertices in FSG is $exp_{n \cdot k}(O(z))$, where $z = |Q| \cdot |M|^{l \times n} \cdot |\Gamma|$ and $|M| > 1$. It follows by [13] that our FSG can be solved and the winning strategy can be constructed in time bounded by $exp_{n \cdot k}(O(z))$, with z as above.

We can code $e \in N_{i,j,k}$ more economically by noting that we need to keep only $m_{i,j}, \ldots, m_{i,n-1}$ in it, in particular $m_{u,x}$, $u \neq i$ need not be stored as there is no *pop* in stack u after a *pop* corresponding to e. This leads to size of FSG and the time to solve it as $exp_{n \cdot k}(O(|Q| \cdot |M| \cdot l \cdot |\Gamma|))$.

7   Conclusion

In this paper we have defined higher order multi-stack pushdown systems (hmpds). We have shown that parity games on bounded phase *hmpds* are effectively solvable and a winning strategy executable by higher order multi-stack automata can be synthesized effectively. It remains open if the complexity bound given in the paper to solve these games can be improved. Recently we have also shown that winning regions in parity games on bounded phase hmpds are regular.

Acknowledgments. Financial support for this work is provided by Research I Foundation.

References

1. Seth, A.: Games on Higher Order Multi-Stack Pushdown Systems, full version, http://www.cse.iitk.ac.in/users/seth/RP09/fullversion
2. Seth, A.: Games on Multi-Stack Pushdown Systems. In: Artemov, S., Nerode, A. (eds.) LFCS 2009. LNCS, vol. 5407, pp. 395–408. Springer, Heidelberg (2008)
3. Carayol, A., Hague, M., Meyer, A., Ong, C.-H.L., Serre, O.: Winning Regions of Higher-Order Pushdown Games. In: Proc: LICS 2008, pp. 193–204. IEEE Computer Society, Los Alamitos (2008)
4. Hague, M., Murawski, A.S., Luke Ong, C.-H., Serre, O.: Collapsible Pushdown Automata and Recursion Schemes. In: Proc: LICS 2008, pp. 452–461. IEEE Computer Society, Los Alamitos (2008)
5. Madhusudan, P., Parlato, G., La Torre, S.: A Robust Class of Context-Sensitive Languages. In: Proc: LICS 2007, pp. 161–170. IEEE Computer Society, Los Alamitos (2007)
6. Madhusudan, P., Parlato, G., La Torre, S.: Context-Bounded Analysis of Concurrent Queue Systems. In: Ramakrishnan, C.R., Rehof, J. (eds.) TACAS 2008. LNCS, vol. 4963, pp. 299–314. Springer, Heidelberg (2008)
7. Qadeer, S., Rehof, J.: Context-bounded model checking of concurrent software. In: Halbwachs, N., Zuck, L.D. (eds.) TACAS 2005. LNCS, vol. 3440, pp. 93–107. Springer, Heidelberg (2005)
8. Knapik, T., Niwinski, D., Urzyczyn, P., Walukiewicz, I.: Unsafe Grammars and Panic Automata. In: Caires, L., Italiano, G.F., Monteiro, L., Palamidessi, C., Yung, M. (eds.) ICALP 2005. LNCS, vol. 3580, pp. 1450–1461. Springer, Heidelberg (2005)
9. Bouajjani, A., Meyer, A.: Symbolic reachability analysis of higher-order context-free processes. In: Lodaya, K., Mahajan, M. (eds.) FSTTCS 2004. LNCS, vol. 3328, pp. 135–147. Springer, Heidelberg (2004)
10. Cachat, T.: Higher order pushdown automata, the caucal hierarchy of graphs and parity games. In: Baeten, J.C.M., Lenstra, J.K., Parrow, J., Woeginger, G.J. (eds.) ICALP 2003. LNCS, vol. 2719, pp. 556–569. Springer, Heidelberg (2003)
11. Cachat, T.: Uniform solution of parity games on prefix-recognizable graphs. In: Proc. Infinity. ENTCS, vol. 68(6) (2002)
12. Walukiewicz, I.: Pushdown processes: games and model checking. Information and computation 164, 234–263 (2001)
13. Jurdziński, M.: Small Progress Measures for Solving Parity Games. In: Reichel, H., Tison, S. (eds.) STACS 2000. LNCS, vol. 1770, pp. 290–301. Springer, Heidelberg (2000)
14. Thomas, W.: Languages, automata and logic. In: Rozenberg, G., Salomaa, A. (eds.) Handbook of Formal Languages, vol. III, pp. 389–455. Springer, New York (1997)
15. Engelfriet, J.: Iterated pushdown automata and complexity classes. In: STOC 1983: Proceedings of the fifteenth annual ACM symposium on Theory of computing, pp. 365–373. ACM Press, New York (1983)
16. Damm, W., Goerdt, A.: An automata-theoretical characterization of the OI-hierarchy. Information and Control 71(1-2), 1–32 (1986)
17. Maslov, A.N.: Multilevel stack automata. Problems of Information Transmission 15, 1170–1174 (1976)

Limit Set Reachability in Asynchronous Graph Dynamical Systems

V.S. Anil Kumar[1], Matt Macauley[2], and Henning S. Mortveit[3]

[1] Department of Computer Science and Virginia Bioinformatics Institute,
Virginia Tech
[2] Department of Mathematical Sciences, Clemson University
[3] Department of Mathematics and Virginia Bioinformatics Institute, Virginia Tech

Abstract. Using the framework of sequential dynamical systems (SDSs), a class of asynchronous graph dynamical systems, we show how the notions of reachability and stability can be negatively correlated. Specifically, we show that certain threshold SDSs exhibit update sequence instability: Over certain graph classes, there exist initial configurations from where exponentially many fixed points are reachable under different update sequences, i.e., the ω-limit set has size $\Theta(2^{|V|})$. We establish this first for treewidth bounded graphs and then for random graphs in the $G(n, p)$ model of Erdős-Rényi for a large range of p. We also show that this update sequence instability is not present in dense graphs, suggesting that sparsity and tree-like structure plays an important role in the stability of the system. These dynamical systems arise in applications such as functional gene annotation, where threshold SDSs are employed to predict gene functions through a fixed point computation, based on an initial state (prediction) and a nongeneric choice of update sequence. The results in this paper should be viewed as cautionary advice in the construction and application of such algorithms. This paper also provides a starting point for a study of update sequence stochastic SDSs.

1 Introduction

Complex unstructured networks and dynamics on such networks arise in a number of applications involving natural and man-made physical and infrastructure systems, such as the Internet, the power grid and biological networks [1]. Graph dynamical systems (GDSs), which generalize the concept of finite cellular automata (CAs) (see e.g. [2]), are useful in understanding such systems. Graph dynamical systems are defined with respect to a finite set of vertices (also called cells or nodes) v[Y] = $\{1, \ldots, n\}$ where each vertex i has a state y_i taken from a finite set K. Moreover, each vertex has a *vertex function* f_i that governs the transition from $y_i(t)$ to $y_i(t + 1)$, taking as arguments the states associated to the neighborhood of vertex i. The application of f_i is called the *update* of vertex i. For CAs, the vertex functions are applied synchronously to generate the dynamical system map. In this paper, we focus on an important sub-class of GDSs,

O. Bournez and I. Potapov (Eds.): RP 2009, LNCS 5797, pp. 217–232, 2009.

called sequential dynamical systems (SDSs) whose dynamical system map is constructed as the *sequential* application of the vertex functions according to some fixed sequence w. These finite dynamical systems were introduced and studied in [5,4,3]; see [6] and the references therein for more details. Whereas finite CAs typically are defined over a regular graphs (often called a lattice) and are restricted to have a single common vertex function, GDSs and SDSs are defined over arbitrary finite graphs with no restrictions on the vertex functions.

The class of SDSs can be used to model and characterize a broad class of applications, algorithms and processes over graphs. Indeed, SDSs and their constituents (graph, vertex functions, update sequence) were originally designed as a dynamical system model of distributed systems where the notion of evaluation/update sequence is important. Examples of such applications include concurrent processes, distributed protocols, approximate discrete event simulations [7], image reconstruction [8], and functional gene annotation [9]. This last application directly motivates our paper, and is briefly described here.

Informally, a functional linkage network is a graph Y in which the set v[Y] of vertices represents a set of genes, with an edge $\{i, j\} \in$ e[Y] if there is evidence in experiments and/or databases that the corresponding genes co-express biological functions. The level of co-expression is measured by an edge weight w_{ij}. Given a new biological function and partial information about which genes express it, the goal is to estimate which remaining genes also express this function. Karaoz et al. [9] developed an algorithm for this problem based on an SDS framework with threshold functions (formally described in Section 2), which starts with an initial state vector $\mathbf{y} \in \{1, -1, 0\}^{|V|}$ (with the state y_i encoding "express"/"do not express"/"unknown" for gene i), and iteratively updates the vertex states by applying the vertex functions in an order specified by a randomly chosen permutation. It is shown that this process always ends at a fixed point, and the gene annotations are inferred from the vertex states at this fixed point. One important issue for the robustness of this method is how the choice of update sequence affects the fixed point (gene prediction) reached from a given initial state. This is related to the fundamental notion of dynamical stability of the system. In this paper, we explore the sensitivity of SDS dynamics to changes in the update sequence, and find that many classes of threshold SDSs are not stable in this sense. Even though we only consider the case where all edge weights are 1, similar instability results should hold for functional gene prediction and related algorithms. The results of this paper may thus be viewed as cautionary advice for the use and validity of sequential algorithms over graphs. Our work provides a nice example of how one can relate the structure of the GDS graph and the resulting dynamics. This line of work relating local structure and dynamics is valuable both from the point of view of theory and applications. We remark that our study of the dynamics as a function of the update sequence forms a natural path to the analysis of *update sequence stochastic* SDSs, a class of systems underlying many applications as well as computational paradigms such as the Gillespie algorithm [10]. We remark that other reachability problems for SDSs have been studied in [3].

Our Results. In this paper, we show that the class of *threshold* sequential dynamical systems exhibit *update sequence instability* for large classes of graphs relevant to many applications. Here, update sequence instability of SDSs over a given graph Y means (i) the number of fixed points of such SDSs is exponential or sub-exponential in n, the number of vertices of Y, and (ii) there exists states $\mathbf{y} \in K^n$ such that the collection of ω-limit states of $\mathbf{y}$ under the possible update sequences, denoted $\cup_{\pi \in S_Y} \omega_\pi(\mathbf{y})$, has exponential size in n. We show that update sequence instability is present in any 2-threshold SDS on trees, and more generally in graphs with bounded treewidth (a parameter that quantifies how "tree-like" the graph is, defined formally in Section 2). We also show that the $G(n, p)$ random graph class of Erdős-Rényi [11] exhibits sub-exponential instability for a range of p values; for $p = o(n^\epsilon/n)$, there are states for which $\cup_{\pi \in S_Y} \omega_\pi(\mathbf{y}) = \Omega(2^{n^{1-\epsilon}})$, where $\epsilon \in (0, 1)$. When p is constant, $G(n, p)$ does not exhibit exponential instability, and we show that this is generally true for dense graphs in which all vertices have degree $\Omega(n)$. In summary, sparse and tree-like graphs seem to be more likely to exhibit such instability, and this has important implications because a number of threshold systems, including biological networks [12] and power grids [13], involve sparse graphs, often with low treewidth.

Paper Outline. We describe the relevant definitions and notation in Section 2. In Section 4, we discuss instability in dense graphs, trees, and graphs of bounded treewidth, and in Sections 3 we extend this to the $G(n, p)$ family of random graphs. Due to space limitations, we have omitted several proofs.

2 Background and Definitions

Let Y be a finite, simple, undirected graph (henceforth just a graph) with vertex set $\mathrm{v}[Y] = \{1, 2, \ldots, n\}$ and edge set $\mathrm{e}[Y]$. For $v \in \mathrm{v}[Y]$, let $n[v]$ denote the sequence of vertices in increasing order contained in the 1-neighborhood of v, and denote the *degree* of vertex v by $d(v)$.

Let K be a finite set of *states*. Each vertex $v \in \mathrm{v}[Y]$ has a *vertex state* $y_i \in K$. An n-tuple

$$\mathbf{y} = (y_1, \ldots, y_n) \in K^n$$

is a *system state*, and the restriction of $\mathbf{y}$ to the vertices in $n[v]$ is written $\mathbf{y}[v]$. Each vertex v has a *vertex function* $f_{v,Y} \colon K^{d(v)+1} \longrightarrow K$, which alternatively can be encoded as a Y-*local function* $F_{v,Y} \colon K^n \longrightarrow K^n$, by

$$F_{v,Y}(y_1, \ldots, y_n) = (y_1, \ldots, y_{v-1}, f_{v,Y}(\mathbf{y}[v]), y_{v+1}, \ldots, y_n) . \tag{1}$$

We frequently omit the subscript Y if the underlying graph is clear from the context. We write a sequence of vertex functions as Y by $\mathfrak{f}_Y$, and the corresponding Y-local functions as $\mathfrak{F}_Y$. Finally, let W_Y be the set of *words* over $\mathrm{v}[Y]$ (the Kleene closure of $\mathrm{v}[Y]$) and $S_Y \subset W_Y$ the set of *permutations* of $\mathrm{v}[Y]$.

Definition 1 (Sequential Dynamical System). *A sequential dynamical system is a triple* $(Y, \mathfrak{F}_Y, w)$ *where* Y *is a graph,* $\mathfrak{f}_Y$ *is a sequence of* Y-*local functions and* $w = (w(1), \dots, w(m)) \in W_Y$ *is the update sequence. The associated* SDS *map* $[\mathfrak{F}_Y, w] \colon K^n \longrightarrow K^n$ *is defined by*

$$[\mathfrak{F}_Y, w] = F_{w(m)} \circ F_{w(m-1)} \circ \cdots \circ F_{w(1)} . \tag{2}$$

In contrast, if the Y-local maps are applied synchronously, we obtain a *generalized cellular automaton*. An SDS for which $w \in S_Y$ is called a *permutation* SDS. The *phase space* of a finite dynamical system $\phi \colon K^n \longrightarrow K^n$ is the directed graph $\Gamma(\phi)$ with vertex set K^n and edge set

$$e[\Gamma(\phi)] = \{(\mathbf{y}, \phi(\mathbf{y})) \mid \mathbf{y} \in K^n\} .$$

A key goal of graph dynamical system research is to unravel the properties of $\Gamma(\phi)$ from the defining constituents of the system, rather than from exhaustive computation. An *invariant set* of a map $\phi \colon K^n \longrightarrow K^n$ is a set M such that $\phi(M) \subset M$. The ω-*limit set* of $\mathbf{y} \in K^n$ is the set of periodic points $\mathbf{z} \in K^n$ such that $\phi^m(\mathbf{y}) = \mathbf{z}$ for some $m \geq 0$. The *basin of attraction* of an invariant set $M \subset K^n$, written $\mathcal{B}(M)$, is the collection of points in K^n whose ω-limit set is contained in M. Let Y be a graph, $\mathfrak{F}_Y$ be a sequence of Y-local functions, and let $\mathcal{P} \subset S_Y$ be a collection of permutation update sequences. For $\pi \in \mathcal{P}$, we write $\omega_\pi(\mathbf{y})$ for the ω-limit set of $\mathbf{y}$ under the SDS map $[\mathfrak{F}_Y, \pi]$. The ω-*limit set of* $\mathbf{y}$ *with respect to* $\mathcal{P}$ is

$$\omega_{\mathcal{P}}(\mathbf{y}) = \bigcup_{\pi \in \mathcal{P}} \omega_\pi(\mathbf{y}) .$$

The goal of this paper is to explore "update sequence instability," i.e., for a given state vector $\mathbf{y}$, how does changing the update sequence π effect the resulting limit set of $\mathbf{y}$? We restrict our attention to threshold functions (defined in Section 3), in which case the limit sets are fixed points. Our main observation is that the structure of the underlying graph has a significant impact on this form of stability. Central to our approach is the graph Star_n which has vertex set $\{0, 1, \dots, n\}$ and edges $\{0, k\}$ for $1 \leq k \leq n$. The Boolean majority function $\mathsf{maj}_k \colon \mathbb{F}_2^k \longrightarrow \mathbb{F}_2$ is a threshold function that returns 1 if the input vector has at least as many 1s as 0s, and returns 0 otherwise. Let Maj_i denote the Y-local majority function at vertex i with the corresponding vertex function $\mathsf{maj}_{d(i)+1}$, and let $\mathsf{Maj}_Y = (\mathsf{Maj}_i)_{i=0}^n$ denote the sequence of Y-local functions. The following example illustrates the concept of limit set instability of an SDS.

Example 1. Consider a permutation SDS over $Y = \mathsf{Star}_4$ with vertex functions induced by the Boolean majority function. Using update sequence $\pi_1 = (1, 0, 2, 3, 4)$, the state $\mathbf{y} = y_0 y_1 y_2 y_3 y_4 = 10000$ is mapped to 01000 which is a fixed point. On the other hand, for $\pi_0 = (0, 1, 2, 3, 4)$, the state 10000 is mapped to 00000. It is straightforward to show that by suitably choosing the permutation update sequence, the initial state 10000 can reach six different fixed points, and $\omega_{\mathcal{P}}(10000) = \{00000, 01000, 00100, 00010, 00001, 11111\}$. This is illustrated in Figure 1.

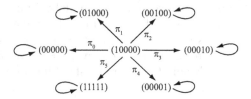

Fig. 1. The six fixed points reachable from the state 10000 in Example 1

A large part of the discussion in this paper is related to trees and graphs which are "tree-like" – this is captured formally by a measure called *treewidth* [14]. A tree-decomposition of a graph $G(V, E)$ is a pair $(\{X_i | i \in I\}, T = (I, F))$ with $\{X_i : i \in I\}$ a family of subsets of V, one for each node of T, and a tree T such that

- $\cup_{i \in I} X_i = V$;
- For each edge $(v, w) \in E$, there exists an $i \in I$ with $v, w \in X_i$;
- For all i, j, k, if j is on the path from i to k in T, then $X_i \cap X_k \subseteq X_j$.

The treewidth of a tree-decomposition $(\{X_i : i \in I\}, T = (I, F))$ is defined as $\max_{i \in I} |X_i| - 1$. The treewidth of a graph G is the minimum treewidth over all possible tree-decompositions of G. The smaller the treewidth, the closer the graph is to a tree – indeed a graph is a tree if and only if its treewidth is 1. Computing the exact treewidth is NP-complete in general [14], but many graphs that arise in practice have low treewidth [14].

3   Update Sequence Instability of Threshold SDSs

Threshold SDSs constitute a natural starting point when considering update sequence instability since (*i*) these SDSs only have fixed points as periodic points (see [4]), (*ii*) fixed points are independent of update sequence, and (*iii*) threshold systems are frequently used in modeling and applications. We remark that even though the set of fixed points is independent of the update sequence for a threshold SDS, the basins of attraction are generally not.

Let $\sigma : \mathbb{F}_2^m \longrightarrow \mathbb{N}$ be defined by $\sigma(\mathbf{y}) = \sum_{i=1}^m y_i$, using the convention of $0 \in \mathbb{N}$. A function $f : \mathbb{F}_2^m \longrightarrow \mathbb{F}_2$ is a *k-threshold function* if there exists $k \in \mathbb{N}$ such that $f(\mathbf{y}) = 1$ if $\sigma(\mathbf{y}) \geq k$, and $f(\mathbf{y}) = 0$ otherwise. An SDS is a *threshold* SDS if all vertex functions are threshold functions. Let $\mathsf{t}_i^k : \mathbb{F}_2^{d(i)+1} \longrightarrow \mathbb{F}_2$ denote the k-threshold vertex function at vertex i, and let T_i^k denote the associated Y-local function. For a graph Y, denote the sequence of Y-local k-threshold functions by

$$\mathbf{T}_Y^k = \left(\mathsf{T}_i^k \right)_{1 \leq i \leq n} .$$

Since all limit sets $\omega_\pi(\mathbf{y})$ of threshold SDSs consist of a single fixed point, the size of the limit set $\omega_\mathcal{P}(\mathbf{y})$ is a natural measure for *update sequence stability* in this case.

The following results are concerned with update sequence instability for threshold SDSs over star graphs, and are central to the remainder of this paper. The fact that many graphs have a large collection of star subgraphs allows us to take advantage of these results later. We only state the result for 2-threshold SDSs, but analogous bounds hold for other threshold functions, such as the majority function.

Proposition 1. *Let* $Y = \mathsf{Star}_n$, *let* $\pi \in S_Y$ *and let* $\psi \colon \mathbb{F}_2^{n+1} \longrightarrow \mathbb{F}_2^{n+1}$ *be the threshold SDS map* $[\mathbf{T}_Y^2, \pi]$. *Then* $|\mathsf{Fix}(\psi)| = 2^n$, *and there exists* $\mathbf{y} \in \mathbb{F}_2^{n+1}$ *with* $|\omega_{S_Y}(\mathbf{y})| = 2^n - n$. *The set* A_1 *of points* $\mathbf{y} \in \mathbb{F}_2^{n+1}$ *with* $|\omega_{S_Y}(\mathbf{y})| = 1$ *satisfies* $|A_1| = 2^n + n + 1$. *Let* $A_{>1}$ *be the complement of* A_1. *Then* $\lim_{n \to \infty} |A_{>1}| / 2^{n+1} = 1/2$ *and there exists states* $\mathbf{y}$ *for which* $\lim_{n \to \infty} |\omega_{S_Y}(\mathbf{y})| / |\mathsf{Fix}(\psi))| = 1$.

We note that Star_n is a sparse graph. For threshold SDSs it turns out that the update sequence instability observed in Proposition 1 does not arise in dense graph, as shown by the following results. Throughout, $\mathbf{0}$ and $\mathbf{1}$ denote the constant states $(0, 0, \ldots, 0)$ and $(1, 1, \ldots, 1)$, respectively.

Proposition 2. *A permutation threshold* SDS *over* K_n *has at most* $n + 1$ *fixed points. This bound is sharp.*

Proposition 3. *Let* Y *be a connected graph on* n *vertices with minimal degree* $d > (1 - \frac{1}{k})n$ *for* $k > 0$. *A threshold* SDS *over* Y *induced by* $\mathbf{T}_Y^k$ *has fixed points contained in* $T = \{\mathbf{0}, \mathbf{1}\}$.

As a special case, a threshold SDS induced by $\mathbf{T}_Y^2$ only has $\mathbf{0}$ and $\mathbf{1}$ as fixed points if the minimum degree d satisfies $d > \frac{n}{2}$.

We remark that bounds for the maximal transient length in threshold SDSs have been studied in [15].

4 Instability of 2-Threshold SDSs over Trees and Treewidth Bounded Graphs

Many graphs that occur in applications have a tree-like structure, or low treewidth. In this section, we show that 2-threshold SDSs over trees or treewidth bounded graphs exhibit exponential update sequence instability. The idea of the proof is to construct a collection $\mathcal{S}$ of "well-separated" Star_k subgraphs (henceforth referred to as a k-*star*) from the entire tree. The construction will be carried out in such a manner that Proposition 1 can be applied to each k-star in $\mathcal{S}$ independently. The result will follow once we demonstrate that $|\mathcal{S}|$ is sufficiently large (linear in n). The instability for the individual star subgraphs leads to exponential instability for the tree Y, and shows how dynamics over subgraphs can influence the global dynamics. We first describe the details for the case of a tree, and then discuss how the proof can be extended to the case of treewidth bounded graphs.

In the following, Y denotes a tree with n vertices rooted at some arbitrary but fixed vertex r. Having a root allows us to partially order $v[Y]$ (with r as the maximum element). For simplicity, we assume that n is larger than some constant and Y is a connected graph. Also, we assume that Y is not a line – it can be verified easily that a line graph can be decomposed into $\Theta(n)$ well separated stars, and has exponential instability. Each vertex $v \neq r$ has a unique covering element $p_Y(v)$ called the *parent* of v. We refer to the set $C_Y(v) = \{v' \in \mathrm{v}[Y] \mid p_Y(v') = v\}$ as the *children* of v. Moreover, the elements of the set $L(Y) = \{v \in Y \mid C_Y(v) = \varnothing\}$ are the *leaves* of Y, and vertices contained in $I(Y) := \mathrm{v}[Y] \setminus L(Y)$ are *internal* vertices. Writing $\ell(Y)$ and $i(Y)$ for the number of leaf and internal vertices, respectively, we clearly have $n = i(Y) + \ell(Y)$. A k-*chain* is a sequence of vertices $(v_1, \dots, v_k)$ such that $C(v_i) = \{v_{i+1}\}$ for $1 \leq i < k$ and $|C(v_k)| = 1$. A k-chain is *maximal* if it is not contained in a $(k+1)$-chain. The graph subscript will be omitted if it is clear from the context.

We first outline our construction of the set $\mathcal{S}$, making the convention that each element $S_i \in \mathcal{S}$ has center vertex v_{i_0} and edges $\{v_{i_0}, v_{i_j}\}$ for $1 \leq j \leq k$ where $k \geq 2$. When establishing the set $\mathcal{S}$ we will construct a set $\Pi(S_i)$ of permutations for each $S_i \in \mathcal{S}$. These sets of permutations, along with a suitably defined initial state vector $X \in \mathbb{F}_2^n$, will be used to construct a set $\mathcal{P} \subset S_Y$ such that $\omega_{\mathcal{P}}(X)$ has size that is exponential in n. The set $\mathcal{S}$ will be a collection of "sufficiently" disjoint k-stars with $|\mathcal{S}| = \Theta(n)$.

The construction proceeds through iteration of two steps. The first step consists of removing from Y each leaf vertex whose parent has at least one non-leaf child and at most two leaf children. We refer to the vertices removed in this process as "lonely hanging vertices." In the second step, all vertices contained in maximal chains of length ≥ 2, with the exception of the maximal vertex of the chain, are removed from Y. Clearly, these chains are pairwise disjoint. For these chains, we extract disjoint 2-star subgraphs and add them to the set $\mathcal{S}$. The removal of chains may re-introduce lonely hanging vertices, so the two steps are repeated until no lonely hanging vertices remain. At this point, the graph that results from Y is a collection of connected components, and for every leaf vertex in this graph there is at least one other leaf vertex with the same parent. For every parent of a leaf vertex, we add to $\mathcal{S}$ the star consisting of that parent and all of its children that are leaf vertices. This completes the construction of $\mathcal{S}$. Vertices not in $\mathcal{S}$ are assigned to the set B.

The next step is to construct a set $\mathcal{P}$ of permutations and initial state $\mathbf{y} \in \mathbb{F}_2^n$. For every $S_i \in \mathcal{S}$, assign state 1 to leaf vertices, and state 0 to the center vertex. For every 2-star $S_i \in \mathcal{S}$, let $P(S_i) = \{(s_{i_0}, s_{i_1}, s_{i_2}), (s_{i_1}, s_{i_2}, s_{i_0})\}$. For every k-star $S_i \in \mathcal{S}$ with $k > 2$, let $P(S_i)$ be the set of permutations of $\mathrm{v}[S_i]$ where s_{i_0} appears in the middle position, i.e., with precisely $\lfloor k/2 \rfloor$ vertices preceding it. These steps are described in detail below.

Part I: Vertex removal. The following two steps are repeated as long as possible, i.e., until lonely hanging vertices or maximal chains (defined below) no longer exist.

1. *(Removal of "lonely hanging vertices")* For each vertex v such that $C(v) \cap I(Y) \neq \varnothing$ and $1 \leq |C(v) \cap L(Y)| \leq 2$, remove all the vertices in $C(v) \cap L(Y)$. Denote the (cumulative) set of vertices deleted in Step 1 by B_1.

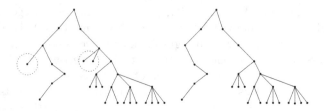

Fig. 2. Step 1: Removal of "lonely hanging vertices"

2. *(Removal of maximal chains).* For any maximal chain $C = (v_1, \ldots, v_k)$ in Y with $k \geq 2$, delete the vertices from the sub-chain $C' = (v_2, \ldots, v_k)$.

Fig. 3. Step 2: Removal of maximal chains

As mentioned above, when there are no more lonely hanging vertices or maximal chains, Part I is complete. At this point, let $\mathcal{C} = \{C_1', C_2', \ldots, C_m'\}$ be the set of deleted chains, and let $\mathcal{Y} = \{Y_1, Y_2, \ldots, Y_k\}$ be the set of connected components of Y that results from the deletion of the chains in $\mathcal{C}$ and the lonely hanging vertices.

Part II: Construction of k-stars, of sub-configurations and of sub-permutations

3. *(Extraction of 2-stars from chains).* For each chain $C_i' \in \mathcal{C}$ where $k_i := |C_i'| < 7$, add the vertices of C_i' to the set B_3^S. For the remaining chains ($k \geq 7$), perform the following steps to assign subsets of vertices to the set B_3^L:
   (a) Renumber the vertices in each chain, so $C_i' = \{v_{i_1}, \ldots, v_{i_k}\}$ while preserving the relation $p_Y(v_{i_j}) = v_{i_j-1}$.
   (b) For each $j \leq i_k - 3$ such that $j \equiv 4, 9 \pmod{10}$, form the 2-star $S = \{v_{j-1}, v_j, v_{j+1}\}$ and add it to $\mathcal{S}$.
   (c) For each such 2-star $S = \{v_{j-1}, v_j, v_{j+1}\}$, let $P(S) = \{(v_j, v_{j-1}, v_{j+1}), (v_{j-1}, v_{j+1}, v_j)\}$, and set $y_{v_{j-1}} = \mathbf{y}_{v_{j+1}} = 1$ and $y_{v_j} = 0$.
   (d) Add all vertices in C_i' that are not included in any 2-star to B_3^L, and set $B_3 = B_3^S \cup B_3^L$.
4. *(Extraction of k-stars from remaining components).* For each component $Y_i \in \mathcal{Y}$, perform the following steps:
   (a) Let $D = \{v \in Y_i : C(v) \cap L(Y_i) \neq \varnothing\}$. Note that by Steps 1 and 2, $|C(v) \cap L(Y_i)| \neq 1$ for all $v \in Y_i$.
   (b) For each vertex $v \in D$ add the star $S = \{v\} \cup (C(v) \cap L(Y_i))$ to $\mathcal{S}$. Set $y_v = 0$ and set $y_u = 1$ for all other vertices u in S.

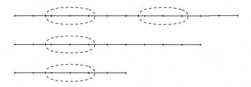

Fig. 4. Step 3: Extraction of 2-stars from chains

(c) Number the leaves in S by $\{w_1, \ldots, w_k\}$. If $k = 2$, define $P(S) = \{(v, w_1, w_2), (w_1, w_2, v)\}$. If $k > 2$, define $P(S)$ to be the set of permutations of the form

$$(w_{\pi(1)}, \ldots, w_{\pi(\lfloor k/2 \rfloor)}, v, w_{\pi(\lfloor k/2 \rfloor + 1)}, \ldots, w_{\pi(k)}),$$

where π is a permutation of the vertices contained in $C(v) \cap L(Y_i)$.
(d) Add all vertices u not contained in any star in S to the set B_4.

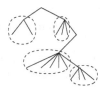

Fig. 5. Step 4: Extraction of k-stars from remaining components

Part III: Construction of the set $\mathcal{P}$ of permutations

5. *(Construction of permutations of Y).* Let $B = B_1 \cup B_3 \cup B_4$. Define $P(\mathcal{S})$ to be the set of all permutations π such that (i) $\pi[S] \in P(S)$, for all $S \in \mathcal{S}$, where $\pi[S]$ is the restriction of π to S, and (ii) all vertices not in B appear before those in B, i.e., for all vertices $u, v \in \mathrm{v}[Y]$, if $u \notin B$ and $v \in B$, then $\pi(u) < \pi(v)$.

We first discuss several useful lemmas below before proving our main result in Theorem 1.

Lemma 1. *Let $C' \in \mathcal{C}$ and let S be a star constructed from vertices in C. Then for any vertex $v \in S$ the inequality $d(v, B_4) \geq 3$ holds.*

Proof. Let $C' = \{v_1, \ldots, v_k\}$. By construction, the 2-stars formed from vertices in C' only involve vertices v_3 through v_{k-2}, and the distance of both v_3 or v_{k-2} to the closest vertex in B_4 is at least 3.

Lemma 2. *Let $Y_i \in \mathcal{Y}$ and let S be a star constructed from vertices in Y_i. Then $d(v, B_4 \setminus Y_i) \geq 2$ for any leaf vertex $v \in S$.*

Proof. Since v is a leaf in S it only has one other neighbor in Y_i, namely its parent, which is not in B_4 because it is also in S.

Lemma 3. *Let T be a tree such that for each vertex $v \in T$, either $|C(v)| = 0$ or $|C(v)| \geq 2$. Then $i(T) \leq \ell(T) - 1$.*

Proof. The proof is by induction on $i(T)$. Let v denote the root of T. The lemma is clearly true when $i(T) = 1$, since in this case $\ell(T) = |C(v)| \geq 2$.

For the inductive step, assume that $C(v) = \{w_1, \ldots, w_k\}$. We may assume that $C(v) \cap L(T) = \varnothing$, because if the result holds in this case, it clearly holds in the case where there are additional edges of the form $\{v, w\}$ where $w \in L(T)$. Let T_j denote the subtree rooted at vertex w_j. By induction, $i(T_j) \leq \ell(T_j) - 1$. We have, $i(T) = \sum_j i(T_j) + 1 \leq \sum_j \ell(T_j) - k + 1 \leq \ell(T) - 1$, since $k \geq 2$, by assumption.

Lemma 4. *For each component $Y_i \in \mathcal{Y}$, the total number of vertices from Y_i added to stars in $\mathcal{S}$ is at least $|Y_i|/4$.*

Proof. By construction, there are no k-chains in Y_i of length 3 or more. We first contract all k-chains of length $k \leq 2$:

1. Let v_1 be a maximal 1-chain, with $p(v_1) = v_0$ and $C(v_1) = \{v_2\}$. Identify the vertices v_1 and v_2, and replace them by a new vertex v_1', with $p(v_1') = v_0$ and $C(v_1') = C(v_2)$.
2. Let v_1, v_2 be a maximal 2-chain, with $v_0 = p(v_1)$ and $C(v_2) = \{v_3\}$. Identify the vertices v_1, v_2 and v_3, and replace them by a new vertex v_1', with $p(v_1') = v_0$ and $C(v_1') = C(v_3)$.

Let Y_i' be the resulting tree after all possible contractions above. Because of the above contractions, for any vertex $v \in Y_i'$, either $|C(v)| = 0$ or $|C(v)| \geq 2$. By Lemma 3, $i(Y_i') \leq \ell(Y_i') = \ell(Y_i)$. Also, the contraction step identifies at most three vertices into one. Therefore, $i(Y_i) \leq 3i(Y_i') \leq 3\ell(Y_i)$. Since all leaves in Y_i are part of stars, the lemma follows.

Lemma 5. $\sum_{S \in \mathcal{S}} |S| = \Theta(n)$.

Proof. We will show that the vertices in $B = B_1 \cup B_3 \cup B_4$ can be charged to the vertices in $\mathcal{S}$, with a constant charge per vertex. We begin with the vertices in B_4. By Lemma 4, $|Y_j \cap \mathcal{S}| \geq |Y_j|/4$. Therefore, each vertex in $Y_j \cap \mathcal{S}$ gets a charge of at most 3 from the vertices in B_4.

Next, consider the vertices in B_3^S, and let $C_i' \in \mathcal{C}$ of length $k < 7$. Since C_i (the chain C_i' plus one parent vertex v_0) was a maximal chain in Y, either $p(v_0)$ or $C(v_k)$ is a vertex with degree at least 3 and therefore, at least one of these vertices lies in some $Y_j \in \mathcal{Y}$. We will charge all vertices in C_i' to the vertices in one such Y_j. A given vertex in $Y_j \cap \mathcal{S}$ now can have a charge of at most 3 from vertices in B_4, and at most 6 from vertices in B_3^S, for a total charge of at most 9.

Moving on to the vertices in B_3^L, consider a k-chain $C_i' \in \mathcal{C}$ where $k \geq 7$, which we know contains $\lfloor \frac{k-6}{5} \rfloor$ 2-stars from $\mathcal{S}$. The quantity $|C_i' \cap B_3^L|/|C_i' \cap \mathcal{S}|$ converges to $2/5$ as $k \to \infty$, but is never larger than $8/3$ (which occurs when $k = 11$). Therefore, the vertices in B_3^L can be accounted for by a charging at most 3 such vertices to a distinct vertex in $C_i \cap S$, and these vertices are disjoint from each Y_j.

Thus far, every vertex in S has a charge of at most 9. Finally, we have to account for B_1 – the lonely hanging vertices. The parents of these vertices can be in either $B_3 \cup B_4$ or some $S_i \in S$, and each such parent can only have two children from B_1. Therefore, vertices in B_1 can be assigned to vertices in $B_3 \cup B_4 \cup S$ with a constant charge of 20/9 per vertex, which as we've argued can be charged to the vertices in S with a constant charge of 9. Together, we conclude the vertices in B_1 can be charged to vertices in S with a constant charge of 20. Since the vertices in $B_3 \cup B_4$ can be charged to vertices S with a constant charge of 9, we now charge the vertices in B to those in S with a constant charge of 29. Therefore, $\sum_{S \in S} |S| \geq n/29$.

Lemma 6. $|P(S)| = 2^{\Omega(n)}$.

Proof. By construction, for each star $S_i \in S$, we have $|P(S)| \geq 2$. Also by construction, for any transversal $\{\pi_i[S_i] \mid S_i \in S\}$, there is an update sequence $\pi \in P(S)$ with that $\pi[S_i] = \pi_i[S_i]$. Together, we conclude that

$$|P(S)| \geq \prod_{S \in S} |P(S)| \geq \prod_{S \in S} 2 \geq \prod_{i=1}^{n/29} 2 = 2^{\Theta(n)},$$

and hence $|P(S)| = 2^{\Omega(n)}$.

Lemma 7. *Starting at the configuration X, the states of all vertices in $B_1 \cup B_3$ remain forever 0, for all update sequences $\pi \in P(S)$.*

Proof. Any vertex added to B_1 or B_3 initially has state 0, and it can have at most one neighbor with initial state 1. The remaining neighbors have state 0, and are updated later. Therefore, such a vertex remains in state 0.

Theorem 1. *If Y is a tree, the 2-threshold SDS on Y exhibits exponential update sequence instability, i.e.,*

$$\left| \bigcup_{\pi \in S_Y} \omega_\pi(X) \right| = \Theta(2^n).$$

Proof. We will show that starting with the initial configuration $\mathbf{y}$, a distinct fixed point is reached for each permutation $\pi \in P(S)$. The theorem then follows from Lemma 6. The specific structure of the clusters in S and the construction of the set $P(S)$ is crucial in the proof.

Fix any permutation $\pi \in P(S)$. Let $\mathbf{y}' = \omega_\pi(\mathbf{y})$, the fixed point reachable from $\mathbf{y}$. By Lemma 7, all vertices in B_1 or B_3 have state 0 in $\mathbf{y}'$. Pick any star $S \in S$, and .set $S = \{v, w_1, \ldots, w_k\}$, with vertex v being the center of S. There are three cases to consider.

1. S is a 2-star in some $C'_i \in C$. In this case, vertices w_1, w_2 each have one neighbor that is not in S, and $(y_v, y_{w_1}, y_{w_2}) = (0, 1, 1)$. If $\pi[S] = (v, w_1, w_2)$,

then all vertices in S turn to state 1 and remain there. If $\pi[S] = (w_1 w_2, v)$, these vertices turn to state 0. Since w_1 and w_2 each have at most one other neighbor, which might be in state 1, these vertices remain in state 0 for all subsequent updates. Therefore, both choices of $\pi[S]$ lead to fixed points that differ on the states of vertices in S.

2. S is a 2-star in some $Y_i \in \mathcal{Y}$. Because of Step 1 in Part I of the algorithm, $C(v) \not\subset L(Y)$. In this case, all neighbors of w_1 and w_2 not in S are in $B_1 \cup B_3$ by construction, which are guaranteed to remain in state 0 by Lemma 7, and thus $(y_v, y_{w_1}, y_{w_2}) = (0, 1, 1)$. If $\pi[S] = (v, w_1, w_2)$, the vertices in S all turn to state 1, and remain in that state in $\mathbf{y}'$. If $\pi[S] = (w_1, v, w_2)$, these vertices all turn to state 0. Since all neighbors of w_1, and w_2 not in S are in $B_1 \cup B_3$, which are always in state 0, and there is at most one neighbor of v that is not in state 0 (possibly $p(v)$), the states of vertices in S do not change. Therefore, the two choices $\pi[S]$ lead to fixed points that differ on the states of vertices in S.

3. S is a k-star in some $Y_i \in \mathcal{Y}$. Note that in this case, S can only be part of some Y_i. We have $y_v = 0$ and $y_{w_j} = 1$ for $j = 1, \ldots, k$. By construction, $\pi[S]$ must be of the form

$$(w_{i_1}, \ldots, w_{i_{\lfloor k/2 \rfloor}}, v, w_{i_{\lfloor k/2 \rfloor + 1}}, \ldots, w_{i_k}),$$

Also by construction, all neighbors of the w_js not in S are in $B_1 \cup B_3$, which remain in state 0, by Lemma 7. Therefore, after one full update per π, the vertices $w_{i_1}, \ldots, w_{i_{\lfloor k/2 \rfloor}}$ turn to state 0, and remain there forever. Vertex v and the vertices $w_{i_{\lfloor k/2 \rfloor + 1}}, \ldots, w_{i_k}$ turn to state 1, and remain in this state. Note that it is possible that v has neighbor not in s that could be in state 1, but this does not affect its state. Therefore, each choice of $\pi[S]$ leads to a different fixed point.

Next, consider two permutations $\pi, \pi' \in P(\mathcal{S})$. It is easy to choose $S \in \mathcal{S}$ such that $\pi[S] \neq \pi'[S]$. By the above discussion, the vertices in S end up in different states under these two update sequences. Therefore, each permutation in $P(\mathcal{S})$ leads to a different fixed point, and the theorem follows from Lemma 6.

Extension to Treewidth bounded graphs. The exponential instability for trees (Theorem 1) also holds for treewidth bounded graphs, and we sketch the main differences below. Our proof relies not on the basic definition of treewidth from Section 2, but an equivalent characterization in terms of graph separators [14]: A graph with constant treewidth has recursive separators of constant size. In other words, there is a constant sized subset S of nodes so that the graph $Y \setminus S$ gets partitioned into graphs Y_1 and Y_2, each with $\Theta(n)$ nodes, and both satisfying this property recursively. Thus, the recursive separators form a tree in which each node corresponds to subset of k nodes in the original graph Y, where k (the treewidth) is a constant. Our construction for such a graph is also similar – we partition it into a large number of "well-separated stars", each of which exhibits exponential instability.

5   Instability of 2-Threshold SDSs over $G(n, p)$

5.1   A Motivating Example

We now consider threshold SDSs induced by $\mathbf{T}_Y^2$ on $G(n, p)$, that is, when the graph Y is chosen from the Erdős-Rényi random graph model $G(n, p)$. We show that this class of SDSs also exhibits the same update sequence instability as shown for Star_n. As a motivating example, let Y be any graph containing disjoint sets $S_1, \ldots, S_\ell \subset \mathrm{v}[Y]$ such that each induced subgraph $Y[S_i]$ is isomorphic to Star_3. Let $\mathrm{v}[S_i] = \{w_i, v_{i,1}, v_{i,2}, v_{i,3}\}$, where w_i is the center vertex, and let $\mathcal{P}$ be a set of update sequences over Y satisfying the following conditions.

- For all $\pi \in \mathcal{P}$, the vertices in S_i appear before those in S_j for $i < j$.
- π restricted to S_i is either $(w_i, v_{i,1}, v_{i,2}, v_{i,3})$ or $(v_{i,1}, v_{i,2}, v_{i,2}, w_i)$.
- The elements of $V_0 = \mathrm{v}[Y] \setminus \cup_i S_i$ have an arbitrary but fixed sequence for all $\pi \in \mathcal{P}$.

Clearly, we can construct $\mathcal{P}$ to have size 2^ℓ. Let $V_S = \cup_i S_i$, and let $\mathbf{y} \in \mathbb{F}_2^n$ be the state where each leaf vertex $v_{i,j} \in S_i$ has state 1, and all other vertices have state 0. Let $\mathfrak{F}_Y$ be the sequence of local functions with 2-threshold functions at all vertices in V_S, and constant 0 functions (i.e., n-threshold functions) at all vertices in V_0. For any star S_i, if the center vertex is updated before any leaves of S_i, all vertices in S_i end up at state 0. However, if the center vertex is updated after all three leaves, all vertices in S_i end up at state 1. Moreover, these vertex-states can never change again, so we can make the following conclusion.

Proposition 4. *If $\mathfrak{F}_Y$ is the sequence described above, then*

$$\omega_{\mathcal{S}_y}(\mathbf{y}) \geq \omega_{\mathcal{P}}(\mathbf{y}) = 2^\ell , \tag{3}$$

i.e., any SDS over $\mathfrak{F}_Y$ exhibits exponential update sequence instability.

We next will extend this to the $G(n, p)$ random graph model where $p > c/n$ for some constant $c > 0$, by showing that there always exists $\ell = O(n)$ disjoint Star_3-subgraphs (albeit slightly different requirements then above).

5.2   Update Sequence Instability in $G(n, p)$

Following the earlier discussion, we show that there are many disjoint stars in $G(n, p)$. We do this in two steps. First, we show that with high probability, a Star_3 exists, along with the extra properties we need, and second, we iteratively ($O(n)$ times) choose such a subgraph, and remove it and its neighbors.

First, we show that a Star_3 with extra properties exists with high probability. Let $Y < K_n$ be a random graph, and set $V = \mathrm{v}[Y]$. For $V' \subset V$, let

$$\deg_Y(V') = \left|\{v \in V \setminus V' \mid \exists u \in V', \{u, v\} \in \mathrm{e}[Y]\}\right| ,$$

and let $\deg_Y(v, V') = |N(v) \cap V'|$, the number of neighbors of $v \in V$ contained in V'. We will drop the subscript Y whenever it is clear from the context. Let p

be a fixed probability, and c any constant. For any subgraph A of Y, define the function

$$\delta_Y(A) = \begin{cases} 1 & \forall v \in \mathrm{v}[A], \ \deg_Y(v, Y \setminus A) \le cnp, \\ 0 & \text{otherwise.} \end{cases}$$

This function is 1 if each vertex in A is adjacent to no more than cnp vertices not in A. Define the indicator functions

$$x_A \colon G(n,p) \longrightarrow \{0,1\}, \qquad x_A(Y) = \begin{cases} 1 & Y[\mathrm{v}[A]] = A \\ 0 & \text{otherwise,} \end{cases}$$

$$X_A \colon G(n,p) \longrightarrow \{0,1\}, \qquad X_A(Y) = \begin{cases} 1 & Y[\mathrm{v}[A]] = A \text{ and } \delta_Y(A) = 1 \\ 0 & \text{otherwise.} \end{cases}$$

We are now ready to define Graph Property 1.

Definition 2 (Graph Property 1). *A graph Y has* Property 1 *if $X_A(Y) = 1$ for some* Star$_3$ *subgraph A.*

We will show that with high probability, a random graph Y satisfies Property 1. Let $\mathcal{A}$ be the set of all Star$_3$ subgraphs of K_n. Define the function

$$X \colon G(n,p) \longrightarrow \mathbb{N} \cup \{0\}, \qquad X(Y) = \sum_{A \in \mathcal{A}} X_A(Y).$$

Observe that Y satisfies Property 1 if $X(Y) > 0$. From the definition of X_A and δ_Y, if A is a Star$_3$ subgraph, then

$$E[X_A] = \Pr[X_A(Y) = 1] = p^3(1-p)^3 \Pr[\delta_Y(A) = 1]. \tag{4}$$

The following theorem shows that a random graph $Y \in G(n,p)$ satisfies Property 1 with high probability by showing that $\Pr[X(Y) = 0] = O(n^{-1})$, using the Second Moment method [11].

Theorem 2. $\Pr[X = 0] = O(n^{-1})$.

Theorem 2 implies that for $Y \in G(n,p)$, there exists with high probability a Star$_3$ graph with the extra properties we require. We now extend this to show that there exists many Star$_3$ subgraphs in Y with these properties.

Lemma 8. *If $p = o(\frac{n^\epsilon}{n})$ for some constant $\epsilon \in (0,1)$, then a random graph from $G(n,p)$ contains $\ell = n^{1-\epsilon}$ disjoint sets $S_1, \ldots, S_\ell$ such that*

(i) *$G[S_i]$ is a* Star$_3$, *for each i;*
(ii) *$N(S_i) \cap N(S_j) = \varnothing$, for $i \ne j$, with probability at least $1 - o\left(\frac{1}{n^\epsilon}\right)$.*

Proof. Repeatedly apply Theorem 2, i.e., for $i = 1, \ldots, \ell$,

1. Choose a set S_i such that $G[S_i]$ is a Star$_4$, and $\deg(S_i) = O(np)$;
2. Remove $S_i \cup N(S_i)$ from the current graph.

By Theorem 2, after $S_1, \ldots, S_i$ have been chosen, $n - \Theta(npi)$ vertices remain. Since $p = o(\frac{n^\epsilon}{n})$ and $i \leq \ell = n^{1-\epsilon}$, we have $npi = o(n)$ for each i. Therefore, the probability that the above iterative procedure fails (i.e., set S_i does not exist) is at most

$$O\left(\sum_i \frac{1}{(n - npi)^2 p}\right) = O\left(\frac{\ell}{n^2 p}\right) = O\left(\frac{n^{1-\epsilon}}{n}\right) = o\left(\frac{1}{n^\epsilon}\right).$$

Therefore, the iterative procedure succeeds with probability at least $1 - o\left(\frac{1}{n^\epsilon}\right)$.

Corollary 1. *Threshold systems in $G(n,p)$ contain initial configurations from which $\Omega(2^{n^{1-\epsilon}})$ different fixed points can be reached by changing the update sequence, with high probability, for $p = o(\frac{n^\epsilon}{n})$.*

6   Summary

In this paper, we have demonstrated how update sequence instability is present in threshold SDSs for broad classes of graphs, and have outlined the potential consequences this may have for robustness and validity of algorithms based on such systems. Of course, the presence of instability implies that there are initial configurations for which the reachability question has an affirmative answer for many final configurations.

Our study of dynamics and long-term behavior under multiple update sequences also provides insight into *update sequence stochastic* sequential dynamical systems, a theory of stochastic SDSs that is currently being developed. In this setting, one studies the probability space of SDS maps induced by a set $\Omega = \{(\pi_1, p_1), \ldots, (\pi_k, p_k)\}$ of update sequences with matching probabilities such that $\sum_i p_i = 1$. This is naturally captured through Markov chains, and current work provides insight into their structure.

From a mathematical point of view, a key step in our work is connecting the existence of sufficiently many subgraphs of given types to the global dynamics of the system. Results and theory relating subgraph structure and global dynamics for graph dynamical systems in a rigorous manner are highly desirable. We hope this paper can motivate further research along these lines.

Acknowledgements. The research of the first and third authors was partially supported by the following grants: NSF Nets Grant CNS-0626964, NSF HSD Grant SES-0729441, CDC Center of Excellence in Public Health Informatics Grant 2506055-01, NIH-NIGMS MIDAS project 5 U01 GM070694-05, DTRA CNIMS Grant HDTRA1-07-C-0113, NSF NETS CNS-0831633, and CNS-0845700.

References

1. Newman, M.: The structure and function of complex networks. SIAM Review 4 (2003)
2. Ilachinski, A.: Cellular Automata: A Discrete Universe. World Scientific Publishing Company, Cambridge (2001)
3. Barrett, C.L., Hunt III, H.B., Marathe, M.V., Ravi, S.S., Rosenkrantz, D.J., Stearns, R.E.: Reachability problems for sequential dynamical systems with threshold functions. Theor. Comput. Sci. 295, 41–64 (2003)
4. Barrett, C.L., Mortveit, H.S., Reidys, C.M.: Elements of a theory of simulation III, equivalence of sds. Applied Mathematics and Computation 122, 325–340 (2001)
5. Mortveit, H.S.: Sequential Dynamical Systems. PhD thesis, Norwegian University of Science and Technology (2000)
6. Mortveit, H.S., Reidys, C.M.: An Introduction to Sequential Dynamical Systems. Universitext. Springer, Heidelberg (2007)
7. Korniss, G., Novotny, M.A., Guclu, H., Toroczkai, Z., Rikvold, P.A.: Suppressing roughness of virtual times in parallel discrete-event simulations. Science 299, 677–679 (2003)
8. Besag, J.: On the statistical analysis of dirty pictures. Journal of the Royal Statistical Society – Series B 48, 259–302 (1986)
9. Karaoz, U., Murali, T., Letovsky, S., Zheng, Y., Ding, C., Cantor, C.R., Kasif, S.: Whole-genome annotation by using evidence integration in functional-linkage networks. In: Proceedings of the National Academy of Sciences, vol. 101, pp. 2888–2893 (2004)
10. Gillespie, D.T.: Exact stochastic simulation of coupled chemical reacitions. Journal of Physical Chemistry 81, 2340–2361 (1977)
11. Alon, N., Spencer, J.: The Probabilistic Method. Wiley, Chichester (2000)
12. Yamaguchi, A., Aoki, K., Mamitsuka, H.: Graph complexity of chemical compounds in biological pathways. Genome Informatics 14, 376–377 (2003)
13. Atkins, K., Chen, J., Kumar, V.S.A., Marathe, A.: Structural properties of electrical networks. International Journal of Critical Infrastructure (2007)
14. Bodlaender, H.L.: A tourist guide through treewidth. Acta Cybernetica 11, 1–22 (1993)
15. Barrett, C.L., Hunt III, H.B., Marathe, M.V., Ravi, S.S., Rosenkrantz, D.J., Stearns, R.E.: On some special classes of sequential dynamical systems. Annals of Combinatorics 7, 381–408 (2003)

Author Index

Abdulla, Parosh Aziz   36
Atig, Mohamed Faouzi   1, 51

Barbuti, Roberto   64
Boichut, Yohan   79
Bouajjani, Ahmed   1

Chaouch-Saad, Mouna   93
Charron-Bost, Bernadette   93
Collins, Pieter   107

Delzanno, Giorgio   36

Héam, Pierre-Cyrille   79
Habermehl, Peter   51
Henzinger, Thomas   3

Jobstmann, Barbara   3

Kouchnarenko, Olga   79
Kumar, V.S. Anil   217

Lehtonen, Eero   120
Levi, Francesca   64
Lugiez, Denis   127

Macauley, Matt   217
Majster-Cederbaum, Mila   189
Maler, Oded   24
Manuel, Amaldev   141
Margenstern, Maurice   154
Maubert, Bastien   166
McMillan, Kenneth L.   176
Merz, Stephan   93
Milazzo, Paolo   64
Minnameier, Christoph   189
Mortveit, Henning S.   217

Pinchinat, Sophie   166

Ramanujam, R.   141
Rezine, Ahmed   36

Scatena, Guido   64
Seth, Anil   203
Shen, Alexander   26

Vardi, Moshe Y.   35

Wolf, Verena   3

Zapreev, Ivan S.   107
Zuck, Lenore D.   176